Natural Bioactive Compounds from Fruits and Vegetables as Health Promoters
Part I

Edited by

Luís Rodrigues da Silva

CICS – UBI – Health Sciences Research Centre
University of Beira Interior
Covilhã
Portugal

&

Branca Silva

CICS – UBI – Health Sciences Research Centre
University of Beira Interior
Covilhã
Portugal

Natural Bioactive Compounds From Fruits and Vegetables as Health Promoters

Authors: Luís R. Silva and Branca Silva

ISBN (eBook): 978-1-68108-239-4

ISBN (Print): 978-1-68108-240-0 © 2016, Bentham eBooks imprint.

Published by Bentham Science Publishers – Sharjah, UAE.

First published in 2016.

advertisements or ideas contained in the Work.

Limitation of Liability:

In no event will Bentham Science Publishers, its staff, editors and/or authors, be liable for any damages, including, without limitation, special, incidental and/or consequential damages and/or damages for lost data and/or profits arising out of (whether directly or indirectly) the use or inability to use the Work. The entire liability of Bentham Science Publishers shall be limited to the amount actually paid by you for the Work.

General:

1. Any dispute or claim arising out of or in connection with this License Agreement or the Work (including non-contractual disputes or claims) will be governed by and construed in accordance with the laws of the U.A.E. as applied in the Emirate of Dubai. Each party agrees that the courts of the Emirate of Dubai shall have exclusive jurisdiction to settle any dispute or claim arising out of or in connection with this License Agreement or the Work (including non-contractual disputes or claims).

2. Your rights under this License Agreement will automatically terminate without notice and without the need for a court order if at any point you breach any terms of this License Agreement. In no event will any delay or failure by Bentham Science Publishers in enforcing your compliance with this License Agreement constitute a waiver of any of its rights.

3. You acknowledge that you have read this License Agreement, and agree to be bound by its terms and conditions. To the extent that any other terms and conditions presented on any website of Bentham Science Publishers conflict with, or are inconsistent with, the terms and conditions set out in this License Agreement, you acknowledge that the terms and conditions set out in this License Agreement shall prevail.

Bentham Science Publishers Ltd.
Executive Suite Y - 2
PO Box 7917, Saif Zone
Sharjah, U.A.E.
Email: subscriptions@benthamscience.org

BENTHAM SCIENCE

CONTENTS

FOREWORD

For centuries, humans have considered food only as an "energy" source for survival. Clarification of nutritional relevant components, as protein, fat, carbohydrates, minerals and vitamins, was determinant to understand metabolic needs, and to adjust consumption patterns. However, this oversimplified definition of food resulted in processed foods composed by mixtures of ingredients rich in these components, while diet is increasingly claimed as being responsible for the most common diseases of modern society: cardiovascular diseases, obesity, and cancer.

When we look upon food from this simplified perspective, it is as if we are regarding food without its "soul". Indeed, although being difficult to demonstrate causality between food and health, there is now appreciable epidemiologic evidence for the protective role of diets rich in fruits and vegetables, being the Mediterranean diet an interesting example. These foods have thousands of components without nutritional essentiality that have been neglected. The interest in these components has increased tremendously in the last two decades, seeking to identify the dietary bioactive components (*i.e.*, those that have a measurable impact on human health), their amounts, and availability. Simultaneously, it is also becoming clear that each one of these components has different effects and potencies when ingested alone or when taking its part in the complex network of molecules present in whole foods. These are amazing days for food scientists because we are closer to understand these bioactive compounds, while the consumer is following closely scientific advances, being increasingly interested in the health properties of foods.

The editors took an enormous and successful effort to assemble a huge variety of knowledge on different natural bioactive components in foods, bringing together experts working in different fields of food composition and health. This first issue was written to provide readers a comprehensive review of bioactive constituents in fruits from different parts of the world. This assembled knowledge allows the reader to attribute a "health-value" to these foods in a more clear way, understand the care needed to preserve their bioactivity, while also adding value to fruits residues (peels, pulp, seeds, and stones) that are frequently neglected by industry. Therefore, this book is designed for food scientists, nutritionists, pharmaceuticals, physicians, food industrials, as well as for health-conscious consumers. More similar comprehensive reviews on other natural food products will be certainly welcomed by readers.

José Alberto Pereira
Mountain Research Centre (CIMO)
School of Agriculture

Polytechnic Institute of Bragança
Portugal
&
Susana Casal
REQUIMTE / Bromatology Service
Faculty of Pharmacy
University of Porto
Portugal

PREFACE

Plants have been widely used as food and medicines, since they provide, not only essential nutrients required for human life, but also other bioactive compounds which play important roles in health promotion and disease prevention, commonly known as phytochemicals. Moreover, in the recent years, the impact of lifestyle and dietary choices for human health has increased the interest in fruits and vegetables, as well as in foods enriched with bioactive compounds and nutraceuticals. In fact, epidemiological studies have consistently shown that the Mediterranean diet, characterized by the daily consumption of fruits and vegetables, is strongly associated with reduced risk of developing a wide range of chronic diseases, such as cancer, diabetes, neurodegenerative and cardiovascular diseases.

Phytochemicals are secondary metabolites present in fruits and vegetables in low concentrations that have been hypothesized to reduce the risk of several pathological conditions. There are thousands of dietary phytochemicals, namely flavonoids, phenolic acids, glucosinolates, terpenes, alkaloids, between many other classes of compounds, which present different bioactivities, such as antioxidant, antimutagenic, anticarcinogenic, antimicrobial, anti-inflammatory, hypocholesterolemic, hypoglycemic and other clinically relevant activities. The evidence suggests that the health benefits of consuming fruits and vegetables are attributed to the additive and synergistic interactions between these phytocomponents. Therefore, nutrients and bioactive compounds present in fruits and vegetables should be preferred instead of unnatural and expensive dietary supplements.

In this ebook, we provide an overview about the different classes of phytochemicals commonly found in fruits and vegetables, highlighting their chemical structures, occurrence in fruits and vegetables, biological importance and mechanisms of action. Part (I) is particularly focused on Mediterranean and Tropical fruits.

Luís Rodrigues da Silva & Branca Silva
CICS – UBI – Health Sciences Research Centre
University of Beira Interior
Portugal

List of Contributors

Ana R. Nunes
CICS – UBI – Health Sciences Research Centre, University of Beira Interior, Covilhã, Portugal

Marco G. Alves
CICS – UBI – Health Sciences Research Centre, University of Beira Interior, Covilhã, Portugal

Pedro F. Oliveira
CICS – UBI – Health Sciences Research Centre, University of Beira Interior, Covilhã, Portugal
ICBAS – UMIB – Department of Microscopy, Laboratory of Cell Biology, Institute of Biomedical Sciences Abel Salazar and Unit for Multidisciplinary Research in Biomedicine, University of Porto, Porto, Portugal

Luís R. Silva
CICS – UBI – Health Sciences Research Centre, University of Beira Interior, Covilhã, Portugal
IPCB – ESALD – Polytechnic Institute of Castelo Branco, School of Health Dr. Lopes Dias, Castelo Branco, Portugal
LEPABE – Department of Chemical Engineering, Faculty of Engineering, University of Porto, Porto, Portugal

Branca M. Silva
CICS – UBI – Health Sciences Research Centre, University of Beira Interior, Covilhã, Portugal

Amílcar Duarte
Center for Mediterranean Bioresources and Food (MeditBio), Faculty of Sciences and Technology, University of Algarve, Faro, Portugal

Catarina Carvalho
Tecnoparque Colombia Nodo Rionegro, SENA, Colombia

Graça Miguel
Center for Mediterranean Bioresources and Food (MeditBio), Faculty of Sciences and Technology, University of Algarve, Faro, Portugal

Andrea Catalina Galvis-Sánchez
REQUIMTE, Department of Chemical Engineering, Faculty of Engineering, University of Porto, Rua Dr. Roberto Frias, 4200-465, Porto, Portugal

Ada Rocha
Faculty of Nutrition and Food Sciences, University of Porto, Rua Dr. Roberto Frias, 4200-465, Porto, Portugal
LAQV@REQUIMTE, Porto, Portugal

Juliana Vinholes
Embrapa Temperate Agriculture, Pelotas, Brazil

Daniel Pens Gelain
Center of Oxidative Stress Research, Department of Biochemistry, Federal University of Rio Grande do Sul, Porto Alegre, Brazil

Márcia Vizzotto
Embrapa Temperate Agriculture, Pelotas, Brazil

Ana Paula Duarte
CICS-UBI – Health Sciences Research Centre, Faculty of Health Sciences, University of Beira Interior, Covilhã, Portugal

Ângelo Luís	CICS-UBI – Health Sciences Research Centre, Faculty of Health Sciences, University of Beira Interior, Covilhã, Portugal
Fernanda C. Domingues	CICS-UBI – Health Sciences Research Centre, Faculty of Health Sciences, University of Beira Interior, Covilhã, Portugal
Amadeo Gironés-Vilaplana1	Research Group on Quality, Safety and Bioactivity of Plant Foods, Department of Food Science and Technology, Campus University Espinardo, CEBAS (CSIC), Murcia, Spain
Cristina García-Viguera	Research Group on Quality, Safety and Bioactivity of Plant Foods, Department of Food Science and Technology, Campus University Espinardo, CEBAS (CSIC), Murcia, Spain
Diego A. Moreno	Research Group on Quality, Safety and Bioactivity of Plant Foods, Department of Food Science and Technology, Campus University Espinardo, CEBAS (CSIC), Murcia, Spain
Raúl Domínguez-Perles	Centre for the Research and Technology for Agro-Environment and Biological Sciences, Universidade de Trás-os-Montes e Alto Douro (CITAB-UTAD), Quinta de Prados, Vila Real, Portugal
Iris Feria Romero	Unidad de Investigación Médica en Enfermedades Neurológicas. Centro Médico Nacional "Siglo XXI", Instituto Mexicano del Seguro Social, México, D.F., México
Christian Guerra-Araiza	Unidad de Investigación Médica en Farmacología. Centro Médico Nacional "Siglo XXI". , Instituto Mexicano del Seguro Social, México, D.F., México
Hermelinda Salgado Ceballos	Unidad de Investigación Médica en Enfermedades Neurológicas. Centro Médico Nacional "Siglo XXI", Instituto Mexicano del Seguro Social, México, D.F., México
Juan Gallardo	Unidad de Investigación en Enfermedades Nefrológicas. Hospital de Especialidades. Centro Médico Nacional "Siglo XXI", Instituto Mexicano del Seguro Social, México, D.F., México
Julia J. Segura-Uribe	Unidad de Investigación Médica en Enfermedades Neurológicas. Centro Médico Nacional "Siglo XXI", Instituto Mexicano del Seguro Social, México, D.F., México
Sandra Orozco-Suárez	Unidad de Investigación Médica en Enfermedades Neurológicas. Centro Médico Nacional "Siglo XXI", Instituto Mexicano del Seguro Social, México, D.F., México
Renan C. Chisté	UCIBIO, REQUIMTE, Department of Chemical Sciences, Faculty of Pharmacy, University of Porto, Porto, Portugal
Eduarda Fernandes	UCIBIO, REQUIMTE, Department of Chemical Sciences, Faculty of Pharmacy, University of Porto, Porto, Portugal

Aline Pereira   Natural Products Core, Plant Morphogenesis and Biochemistry Laboratory,
                           Federal University of Santa Catarina, Florianopolis, Brazil

Rodolfo Moresco   Natural Products Core, Plant Morphogenesis and Biochemistry Laboratory,
                           Federal University of Santa Catarina, Florianopolis, Brazil

Marcelo Maraschin   Natural Products Core, Plant Morphogenesis and Biochemistry Laboratory,
                           Federal University of Santa Catarina, Florianopolis, Brazil

Natural Bioactive Compounds from Fruits and Vegetables as Health Promoters

2

CHAPTER 1

Bioactive Compounds and Health-Promoting Properties of *Ficus carica* (L.): A Review

Ana R. Nunes[1], Marco G. Alves[1], Pedro F. Oliveira[1,2], Luís R. Silva[1,3,4], Branca M. Silva[1,*]

[1] *CICS – UBI – Health Sciences Research Centre, University of Beira Interior, 6201-506 Covilhã, Portugal*

[2] *ICBAS – UMIB – Department of Microscopy, Laboratory of Cell Biology, Institute of Biomedical Sciences Abel Salazar and Unit for Multidisciplinary Research in Biomedicine, University of Porto, 4050-313 Porto, Portugal*

[3] *IPCB – ESALD – Polytechnic Institute of Castelo Branco, School of Health Dr. Lopes Dias, 6000-767, Castelo Branco, Portugal*

[4] *LEPABE – Department of Chemical Engineering, Faculty of Engineering, University of Porto, 4200-465 Porto, Portugal*

Abstract: *Ficus carica* (L.), also known as fig, is a deciduous tree belonging to the Moraceae family, and was one of the first plants cultivated. Figs, *F. carica* fruits, are an important component of the Mediterranean diet and can be consumed either fresh or dried, or used for jam production. Notably, this fruit has a great economic importance in many countries, including Portugal, due to its nutritional value and medicinal properties. This botanical species is a good and low cost natural source of bioactive compounds, such as organic acids, phenolic compounds, minerals, amino acids, fibers and others. Several biological activities of *F. carica* have been reported illustrating a high beneficial health potential for this species. In this chapter, we will discuss the phytochemical composition, nutritional value and biological activities of F. carica, particularly the leaf, fruit and latex. The potential use of *F. carica* to prevent and treat a wide range of diseases will also be discussed. This natural product may be a promising candidate for the development of new nutraceuticals and food supplements. However, despite the advances in phytomedicine area, the molecular mechanisms by which fig

* **Corresponding author Branca M. Silva:** CICS – UBI – Health Sciences Research Centre, University of Beira Interior, 6201-506 Covilhã, Portugal; Email: bmcms@ubi.pt.

derivatives contribute to health improvement remain largely unknown.

Keywords: *Ficus carica*, Fig, Organic acids, Phenolic compounds, Volatile compounds.

INTRODUCTION

Plants have been used since ancient times as foods and medicines to prevent and treat diseases, due to their nutritional role and chemical properties [1]. In the last years, modern societies have recognized the interest of phytotherapy as an attractive alternative to conventional medicine [2, 3].

The bioactive compounds of plants have served as basis for the development of new drugs and biological products. On the other hand, they are an attractive alternative option to conventional drugs [2, 3]. Noteworthy, a large number of natural products have been introduced over the last years in the pharmaceutical industry [4]. According to the World Health Organization (WHO), treatments with herbal medicine or vegetable extracts are practiced by approximately 80% of the world's population [5]. Several studies of our team have been performed to explore the medicinal properties of natural products and their bioactive compounds, particularly of tea (*Camellia sinensis*), quince (*Cydonia oblonga*), walnut (*Juglans regia*), olive (*Olea europaea*), and fig (*Ficus carica*) and their main phytocomponents, against oxidative stress (OS) and related diseases, including diabetes mellitus (DM), cardiovascular diseases (CVD), male infertility, neurodegenerative disorders, and cancer [6 - 33]. These studies have provided compelling evidence that the antioxidant compounds present in plant infusions, fruits, and vegetables are effective against these illnesses, protecting the organism from oxidative damage.

F. carica is a deciduous tree belonging to the Moraceae family, and one of the earliest cultivated fruit trees. The common fig is a tree native to southwest Asia and the eastern Mediterranean. The *F. carica* fruit is an important harvest worldwide, and it is part of the Mediterranean diet [34] and may be consumed fresh, dried or used to prepare jams. Besides fruits, other parts of the plant like leaves, seeds, bark, tender shoots, and latex also have phytomedicinal potential. In the last decade, several studies have been performed to determine the chemical composition of this species, namely in leaves, fruits, and latex. It was shown that this plant is an excellent source of phenolic compounds, organic acids, volatile compounds, phytostherols, dietary fiber, sugars, among others [34 - 37]. Of note, some authors reported that figs have even higher phenolic content than tea or red wine [38] which are well-known for their great content in polyphenols and consequently strong phytomedicinal properties. In traditional medicine, *F. carica* leaves, fruits, and roots, have been used to treat gastrointestinal, respiratory, cardiovascular, and inflammatory disorders [39]. Nowadays, several biological properties of *F. carica* components have been evaluated and confirmed [40, 41]. For example, it has been reported that the consumption of figs helps and prevents vein blockage [42], and that the decoction of its leaves has an hypoglycemic action in type 1 diabetic individuals [36]. These pharmacological properties are at least in part due to fig's high content in antioxidants. The pharmaceutical industry is giving more attention to medicinal plants and its chemical properties that may translate into biological activities relevant for human health. In fact, fruits and vegetables may represent a precious resource for the development of new drugs/therapies against several health problems. Throughout this chapter we will discuss the recent findings concerning the phytochemical composition, nutritional value, and biological properties of *F. carica*, focusing on the fruits, leaves and latex. The relationship between *F. carica* phytochemical composition and the health effects already reported will be highlighted.

FICUS CARICA ORIGIN AND PRODUCTION

F. carica belongs to the Moraceae family. Moraceae is an angiosperm plant family with more than 800 species of trees, shrubs, hemiepiphytes, climbers, and creepers, that exists in the tropics and subtropics worldwide [43]. It is a tree native to southwest Asia and the eastern Mediterranean countries, growing in mediterranean and dryer warm-temperate climates [44]. It is one of the oldest fruit trees cultivated, being an important crop worldwide [45]. Countries like Turkey, Egypt, Morocco, Spain, Portugal, Greece, California, Italy, Brazil and others with typically mild winters and hot dry summers are the major producers of edible figs [39]. *F. carica* is a deciduous tree with numerous spreading branches, and its root system is typically shallow and spreading. Its foliage is single, alternate and large, deeply lobed with three or seven lobes, rough and hairy on the upper surface and soft, hairy on the underside [46] (Fig. **1**). The fruit has a tough peel (pure, green, green suffused with brown, brown or purple), often cracking upon ripeness, and laticiferous tissues that contain the usual organelles of plant cells, such as nucleus, mitochondria, vacuoles and ribosomes, among others [47], and has a protein hydrolytic enzyme, ficin [48]. All these components are rich in phytochemicals with several bioactive properties.

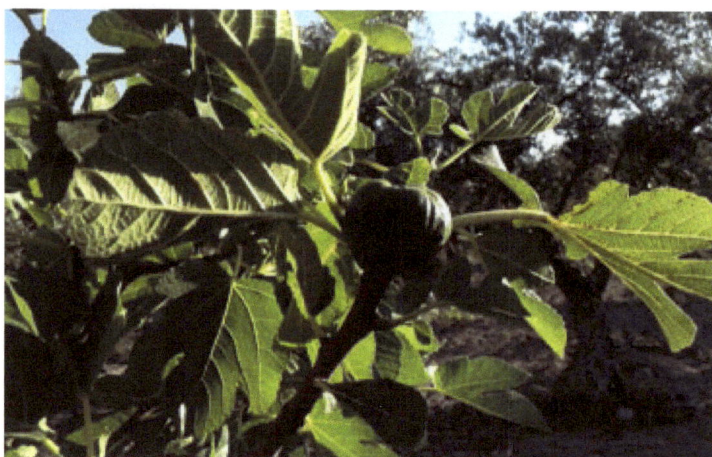

Fig. (1). Leaves and fruit of *F. carica*. This plant is very branched, and its leaves are large with three or seven lobules, rough and hairy on the upper surface and soft, hairy on the underside. The fig flowers are not visible because they are within the fig. Fig fruit is has a tough peel, which can be pure, green, green suffused with brown, brown or purple, and it is very rich in nutrients and phytochemicals.

NUTRITIONAL PROPERTIES AND CHEMICAL COMPOSITION OF *FICUS CARICA*

F. carica fruit has been a typical component of the health-promoting Mediterranean diet [34], which can be consumed fresh or dried. In fact, fresh fig is a good source of water and nutrients, such as minerals, vitamins, carbohydrates, fibers, proteins, and fats [34, 46]. Moreover, it is very rich in phytochemicals with high antioxidant potential [19, 20, 33], such as phenolic compounds.

Major Components

Fresh fig presents a relatively low total caloric content (70 Kcal per 100 g of edible portion) [49]. As expected, dried fig has a considerably higher total caloric content (234 Kcal per 100 g of edible portion) due to its lower water content and consequently greater proportion of carbohydrates [49]. It is important to underline that dried figs are very rich in sugars, whereby the quantity ingested should be moderate. Other relevant groups are proteins, fats, fibers, and minerals [49] (Fig. **2**).

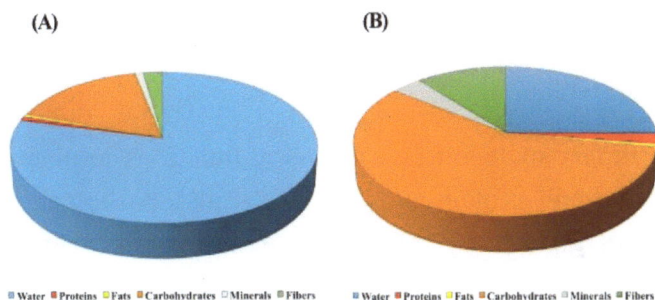

(A)   **(B)**

■ Water ■ Proteins ▣ Fats ■ Carbohydrates ▢ Minerals ■ Fibers   ■ Water ■ Proteins ▣ Fats ■ Carbohydrates ▢ Minerals ■ Fibers

Fig. (2). Nutritional composition of fresh (A) and dried (B) fig (adapted from [49]). Fresh fig contains higher moisture content and lower carbohydrates, proteins, fibers and minerals contents than dried fig. In both, the amount of fats is very low.

Macronutrients

Figs are nutrient dense fruits, especially the dried ones. According to the Food Composition Table published by Ricardo Jorge Institute, the national reference laboratory and health observatory in Portugal, the macronutrients present in fresh and dried fig are carbohydrates, proteins, fibers, and fats [49]. As would be expected, fresh fig contains more water than dried fig; so, dried fig presents higher concentrations of proteins, carbohydrates, and fibers [49]. These fruits have a very low lipid content [49]. *F. carica* contains fourteen fatty acids (myristic, pentadecylic, palmitic, margaric, *cis*-10-heptadecenoic, stearic, oleic, elaidic, linoleic, arachidic, heneicosylic, behenic, tricosylic, and lignoceric) identified by GC-ITMS [13]. The fatty acids total content is very similar in fresh and dried figs (approximately 0.6 g per 100 g) [49]. Linoleic acid is a polyunsaturated fatty acid (PUFA), and both fig types contain approximately 0.3 g per 100 g of this essential PUFA [49]. Moreover, linolenic acid was the predominant fatty acid (53.1%) in dried fig fruit, followed by linoleic acid (21.1%), palmitic acid (13.8%), and oleic acid (9.8%) [50]. In fact, monounsaturated fatty acids (MUFA) and saturated fatty acids (SFA) are less abundant in fig fruit [49].

F. carica leaves and latex also contain several fatty acids. Linolenic acid methyl ester, stearic acid ethyl ester, and linoleic acid ethyl ester were the most abundant ones found in leaves [51]. Palmitic, arachidic, and behenic acids were the major fatty acids present in the latex. In fact, latex is essentially constituted by SFA (161.77 mg per kg) [13].

Amino acids are primary metabolites with many functions in plant metabolism. Besides being involved in protein synthesis, they also participate in metabolic networks that control growth and adaptation to the environment. In the plant secondary metabolism, amino acids are involved in the biosynthesis of phenolic compounds, glucosinolates, cyanogenic heterosides, and alkaloids [52]. The amino acid profile of *F. carica* latex was determined by Oliveira *et al.*, [13] using an HPLC/UV method. This study showed that the latex of *F. carica* contains five essential amino acids (leucine, tryptophan, phenylalanine, lysine, and histidine), and eight non-essential amino acids (asparagine, alanine, glutamine, serine, glycine, ornithine, tyrosine, and cysteine), in a total of thirteen amino acids. Tryptophan, cysteine, and tyrosine were the main amino acids [13]. Some of these metabolites are crucial to health. For example, glutamine and glycine have an important role in the treatment of brain metabolism imbalance and as neurotransmitters [53]. Cysteine, one of the major amino acids identified, is necessary for the detoxification of harmful toxins from the body, protecting the brain and liver from alcohol and drug damage [54].

Organic acids are primary metabolites, which can be found in great amounts in all plants, especially in fruits. These compounds also have antioxidant properties, playing an important role against several diseases [17, 18]. For example, the oxalic acid is an organic acid found in higher amounts in green leaves, while citric, malic and tartaric acids are commonly present in fruits and berries [21]. The organoleptic characteristics of fruits and vegetables are greatly influenced by organic acids total content and proportion. Fresh fig contains reduced amounts of organic acids (approximately 0.17 g per 100 g) [49]. Oxalic, citric, malic, shikimic, and fumaric acids were found in pulp and peels of fig fruits. However, the content of these compounds vary according to the variety [16].

Oliveira *et al.*, [16] determined the organic acid profile of *F. carica* leaves, and the pulps and peels of fig. The authors found six organic acids: oxalic, citric, malic, quinic, shikimic, and fumaric acids (Fig. **3**). Quinic and shikimic acids in *F. carica* were identified in the leaves for the first time in this study [16].

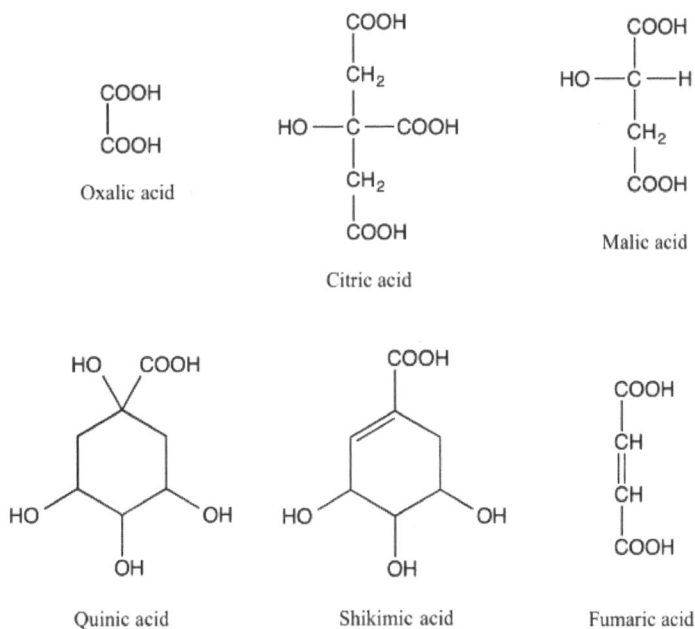

Fig. (3). Chemical structures of the main organic acids of *F. carica*. Oxalic, citric, malic and fumaric acids are dicarboxylic acids; while quinic and shikimic acids monocarboxylic acids.

Leaves are characterized by higher contents of organic acids than the other materials. It can be explained by the photosynthetic capacity of the leaves. In the photosynthetic process, light energy is converted in glucose and ATP. Then, glucose can be used in the synthesis of different metabolites, including organic acids. It should be emphasized that quinic acid was only detected in leaves [16], making it a possible chemical marker of the leaves.

Regarding the *F. carica* latex, the organic acids profile was composed by the same six compounds of the leaves [13]. The most abundant are malic and shikimic acids (26% of total organic acids content), followed by quinic acid (24%) [13].

Minor Components

Micronutrients

F. carica fruit contains several vitamins, namely vitamins A, C, B$_6$, α-tocopherol, riboflavin, carotene, and folates. Carotene, the vitamin A precursor, is present in high quantities in fresh and dried fig [49]. Noteworthy, fig fruit contains good levels of B-complex group of vitamins [49]. These compounds have important physiological and biochemical functions, namely as co-factors for metabolism of carbohydrates, proteins, and fats, and with high antioxidant potential.

Potassium is the major mineral present in fresh and dried figs [49]. This element is a crucial component of cell and body fluids, essential in the control heart rate and blood pressure, preventing hypertension and stroke [55]. Calcium is another major mineral [49], and it is responsible to construction and maintenance of bones and teeth. Moreover, fresh fig is also rich in phosphorus, and dried fig in magnesium [49]. All these minerals are relevant to health, preventing, for example, cardiovascular disorders and bone demineralization [56].

Phytochemicals

The phytochemical composition of *F. carica* includes phytosterols, phenolic compounds, volatile compounds, and other secondary metabolites. These compounds have been found in the leaves, fruits, and latex.

Phytosterols

Phytosterols are formed in ornamental plants, medicinal herbs, edible plants, shrubs, and trees. Vegetable oils contain high concentrations of these secondary metabolites [48]. In the last years, phytosterols have aroused much interest due to its potential effects in reducing serum cholesterol levels [57]. These compounds are poorly absorbed, but inhibit the absorption of intestinal cholesterol, including recirculating endogenous biliary cholesterol, a key step in cholesterol elimination

[58]. Moreover, these metabolites are involved in the regulation of membrane fluidity, in adaptation of membranes to temperature [59] and participation in cellular differentiation and proliferation [60].

The dried fig contains considerable amounts of small seeds rich in oil [61], which may explain its level of these metabolites. Campesterol, stigmasterol, sitosterol, and fucosterol were the main phytosterols found in fruits of *F. carica* [50].

Five sterols were identified in the *F. carica* leaves by Oliveira *et al.*, [32]: betulol, lanosterol, lupeol acetate, *β*-amyrin, and *β*-sitosterol (Fig. **4**). The major component identified was lanosterol. In a general way, the phytosterols profile is similar among and within cultivars. However, Oliveira *et al.*, [32] found that betulol was the only compound displaying some variation between cultivars, and *β*-sitosterol was the less variable compound. Jeong *et al.*, reported that fig fruit has 433 mg of phytosterols per 100 g dry weight of fruit [50].

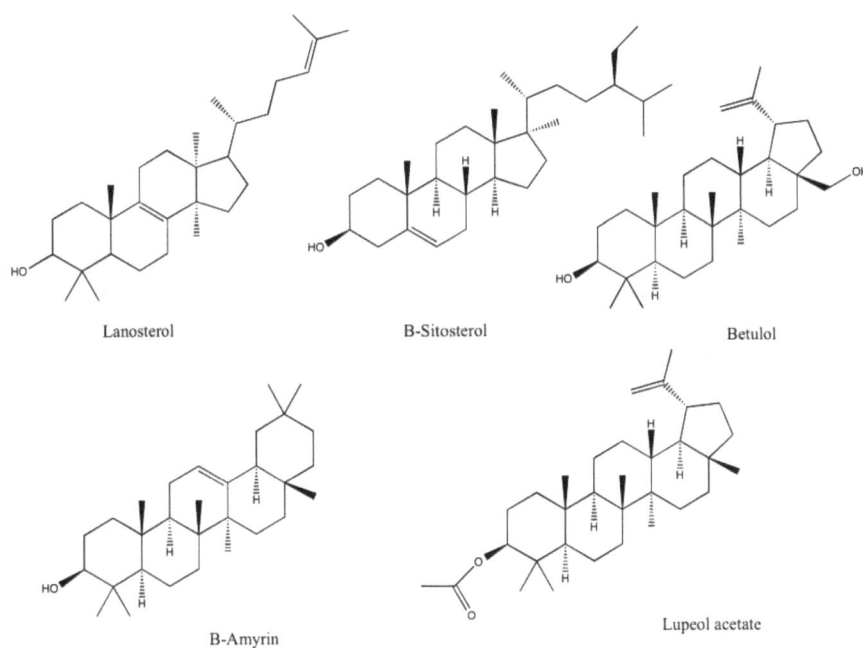

Fig. (4). Chemical structures of the major phytosterols present in *F. carica*. Lanosterol is a tetracyclic triterpene, a precursor of sterols; β-Sitosterol has a similar structure to cholesterol; β-amyrin and lupeol acetate are pentacyclic triterpenes.

Recently, Oliveira *et al.*, [13] identified seven phytosterols (β-sitosterol, betulol, lupeol, lanosterol, lupeol acetate, β-amyrin, and α-amyrin) in *F. carica* latex. Of note, β-sitosterol was the main phytosterol, while α-amyrin was the minor one.

Phenolic Compounds

Phenolic compounds are widely distributed in fruits and vegetables. They are involved in the prevention of damage caused by free radicals in proteins, carbohydrates, lipids and DNA [62]. In human health, they are able to act as reducing agents, hydrogen donors, free radicals scavengers, singlet oxygen quenchers, metals chelators, enzyme inhibitors. They also interact with cell signaling, and can promote cell survival [14, 15]. In the last few decades, these compounds have attracted much attention from scientific community due to their health-promoting properties, mainly for their antioxidant properties. Recently, *in vivo* studies performed by our team showed that natural products rich in phenolic compounds, have antioxidant, antidiabetic, neuroprotective and cardioprotective properties [6, 7]. These compounds have been shown to have the capacity to delay or prevent diseases associated to OS, such as DM, CVD, neurodegenerative diseases, and cancer [6, 7, 9, 15]. Several studies reported that fig can promote important health benefits due to its high content in phenolic compounds [45, 63]. Oliveira *et al.*, [16] reported the presence of several phenolic acids, such as hydroxycinnamic acids (3-*O*- and 5-*O*-caffeoylquinic acids, and ferulic acid), flavonoid glycosides (quercetin-3-*O*-glucoside and quercetin-3-*O*-rutinoside), and furanocoumarins (psoralen and bergapten) in the aqueous extract of the leaves of *F. carica* (Fig. **5**). These authors found 3-*O*-caffeoylquinic acid in this species for the first time. Moreover, leaves of *F. carica* contain higher content in polyphenols than fruit pulps [16]. Another study, showed that a methanolic extract of *F. carica* leaves contains high amounts of psoralen and bergapten [32].

Phenolic compounds are present in fresh and dried figs in a considerable amount [33, 63]. The dried fig skin presents the highest amount of phenolic compounds. In fact, according to Vallejo *et al.*, [63], dried figs contain higher amounts of these compounds than the pulp and fresh fruits. On the other hand, the major phenolic component found in dried fruits was the quercetin 3-*O*-rutinoside [63]. Furthermore, Oliveira *et al.*, [16] demonstrated that different varieties of *F. carica* present distinct phenolic composition (of pulp and peel). According to the same authors, the quercetin 3-*O*-rutinoside content in peels was significantly higher than that found in pulps and leaves. In addition, Faleh *et al.*, [33] revealed that phenolic composition does not depend on the phenotype or the variety. *F. carica* latex is still poorly studied. As far as we know, the presence of phenolic compounds in the latex has not been reported yet.

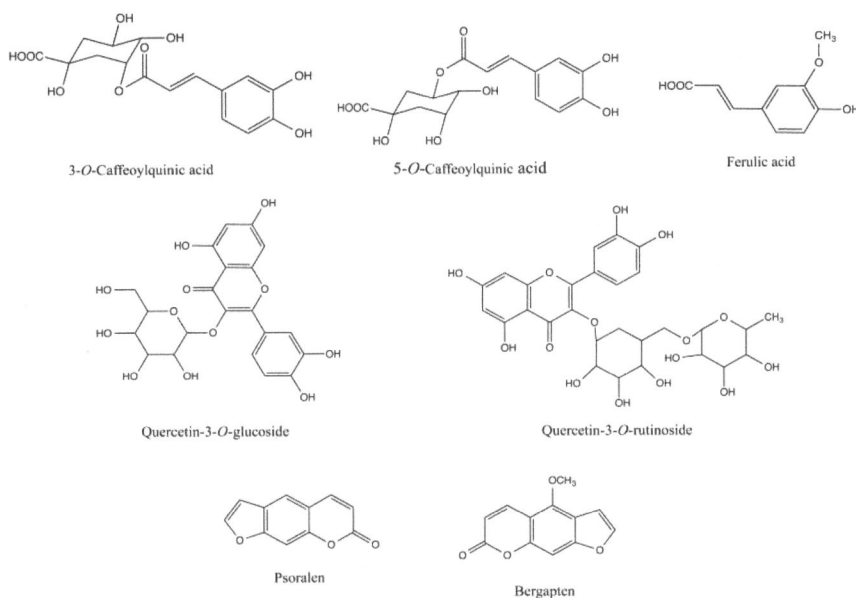

3-*O*-Caffeoylquinic acid   5-*O*-Caffeoylquinic acid   Ferulic acid

Quercetin-3-*O*-glucoside   Quercetin-3-*O*-rutinoside

Psoralen

Bergapten

Fig. (5). Chemical structures of the main phenolic compounds found in *F. carica*. The 3-O-caffeoylquinic acid, 5-O-caffeoylquinic acid, and ferulic acid are hydroxycinnamic acids derivatives (C6-C3 structure). Quercetin-3-O-glucoside and quercetin-3-O-rutinoside are flavanol (C6-C3-C6 structure) glycosides. Psoralen and bergapten are furanocoumarins (C2-C6-C3 structure).

Anthocyanins are members of the flavonoid group, with a strong antioxidant capacity [64]. The studies about the detailed pigment composition in fig are scarce. However, these pigments are present in dark fig skins and constitute the main coloring compounds [34]. Studies conducted by Solomon *et al.*, [34] and Dueñas *et al.*, [45] characterized the qualitative and quantitative pigment composition of skin and pulp from commercial varieties of fig with different color. Both groups identified fifteen anthocyanin pigments, most of them containing cyanidin as aglycone. Rutinose and glucose were present as substituting sugars, as well as acylation with malonic acid. Minor levels of peonidin 3-O-rutinoside in pulp were also found. In addition, they reported the detection of 5-carboxypyranocyanidin-3-rutinoside, a cyanidin 3-O-rutinoside dimer and five condensed pigments containing C–C linked anthocyanins and flavanol (catechin and epicatechin) residues [45]. More recently, Solomon *et al.*, [40] investigated the antioxidant properties of cyanidin 3-O-rutinoside, and concluded that this compound inhibits lipid peroxidation. Moreover, cyanidin 3-O-rutinoside showed a strong chelating activity toward Fe^{2+} [40].

Volatile Compounds

The volatile compounds are responsible by the flavour and aroma quality in fresh and processed fruits [19]. Furthermore, plant volatiles are involved in species-specific ecological interactions and are associated with defensive and attractive roles [65]. *F. carica* fruits (peels and pulps) have higher contents of aldehydes and monoterpenes, whereas sesquiterpenes are present in lower amounts, comparatively to leaves [19]. The higher content of monoterpenes in the fig fruit may be related with their important role in the attraction of specific pollinators, for example fig/wasp relationship [66]. In general, Oliveira *et al.*, [19] identified several compounds in fig, such as 3-methyl-butanal, 2-methyl-butanal, (*E*)-2 - pentanal, (*E*)-2-hexenal, heptanal, (Z)-2-hexanal octanal, nonanal, 1-penten-3-ol, 3-methylbutanol, benzyl alcohol, (*E*)-2-nonenol, phenylethyl alcohol, 6-methyl-5-hepten-2-one, methyl hexanoate, methyl salicylate, ethyl salicylate, limonene, menthol, α-pinene, β-pinene, linalool, eucalyptol, α-cubenene, copaene, β-caryophyllene, τ-muurolene, τ cadinene, germacrene D, β-cyclocitral, and eugenol [19]. Methyl salicylate was found in leaves and fruits of *F. carica*, being present in higher amounts in leaves. This compound acts as an antimicrobial and insecticide agent [67]. Another study conducted by Oliveira *et al.*, [30] showed that volatile compounds are present in larger amounts in leaves, following peels and pulps [30].

Oliveira *et al.*, [19] identified fifty nine volatile and semi-volatile compounds using GC-ITMS, in the leaves and fruits (peels and pulps) of five Portuguese varieties of *F. carica*. They have found aldehydes, alcohols, ketones, esters, monoterpenes, sesquiterpenes, norisoprenoid, and other compounds. Leaves contained: 3-methyl-butanal, 2-methyl-butanal, (*E*)-2-pentanal, hexanal, (*E*)- 2-hexenal, 1-penten-3-ol, 3-methyl-1-butanol, 2-methyl-1-butanol, 1-heptanol, benzyl alcohol, (*E*)-2-nonen-1-ol, phenylethyl alcohol, 3-pentanone, methyl butanoate, methyl hexanoate, hexyl acetate, ethyl benzoate, methyl salicylate, limonene, menthol, *α*-cubenene, *α*-guaiene, *α*-ylangene, copaene, *β*-bourbonene, *β*-elemene, *α*-gurjunene, *β*-caryophyllene, *β*-cubebene, aromadendrene, *α*-caryophyllene, *τ*-muurolene, *τ*-cadinene, *α*-muurolene, germacrene D, (+)-ledene, *β*-cyclocitral, and psoralen [19]. All classes of volatile compounds are present in high proportions, with the exception of aldehydes and monoterpenes. In general, sesquiterpenes class constituted the main class of volatile compounds in *F. carica* leaves [19].

Volatile compounds have also been found in *F. carica* latex. Oliveira *et al.*, [20] performed, for the first time, a study about the chemical composition of *F. carica* latex. The authors identified thirty-four compounds distributed by chemical classes: five aldehydes (pentanal, hexanal, heptanal, benzaldehyde, and octanal), seven alcohols (1-butanol-3-methyl, 1-butanol-2-methyl, 1-pentanol, 1-hexanol, 1-heptanol, phenylethyl alcohol, and phenylpropyl alcohol), one ketone (6-methyl-5-hepten-2-one), nine monoterpenes (*α*-thujene, *α*-pinene, *β*-pinene, limonene, eucalyptol, terpinolene, *cis*-linalool oxide, linalool, and epoxylinalol), nine sesquiterpenes (*α*-guaiene, *α*-bourbonene, *β*-caryophyllene, *trans*-*α*-bergamotene, *α*-caryophyllene, *τ*- muurolene, germacrene D, cadinene, and *α*-calacorene), and three other compounds (methyl salicylate, quinolone, and psoralene) [20]. The most abundant class in latex is sesquiterpenes (91% of total identified compounds).

BIOLOGICAL ACTIVITIES OF *FICUS CARICA*

The use of medicinal plants to treat diseases is nearly as old as humanity, and still is a current alternative to synthetic drugs in modern society. In the last years, there has been a growing interest to understand how natural products and its phytochemicals are able to improve human health. *F. carica* is no exception and its leaves, fruits, and latex are rich in potential health-promoting phytochemicals.

Antioxidant Activity

In recent years, natural antioxidants have under the spotlight due to their ability to scavenge free radicals, thereby inhibiting oxidation [68, 69]. Oxidation is a natural process occurring in most living organisms, but they possess enzymatic and non-enzymatic defenses systems against reactive oxygen species (ROS). However, aging, smoke, drugs, alcohol, and high-caloric diets contribute to a decrease in endogenous antioxidant defenses, resulting in oxidative disturbances [70]. In fact, OS has been linked to the development of several diseases, namely, DM, neurodegeneration, and CVD [71]. Noteworthy, it has been shown that the antioxidant potential of plants, such as *Camellia sinensis* (tea plant), is effective against any of these diseases [6, 7, 15].

F. carica is rich in phenolic compounds, which may act as excellent antioxidants. Several studies have been performed to evaluate the antioxidant activity of *F. carica* leaves, fruits, and latex. A study conducted by Oliveira *et al.*, [16] showed that pulps, peels and leaves extracts of *F. carica* exhibited 2,2'-diphenyl-1-picrylhydrazyl (DPPH) scavenging capacity, in a concentration-dependent way. Moreover, the leaves showed the strongest antiradical capacity against superoxide anion radical [16]. In fact, pulps contain a lower content of phenolic compounds than leaves, which may explain the fact that leaves have greater antioxidant activity.

Fig fruit color varies from the dark purple to green. Solomon *et al.*, [34] showed that dark purple fruits contain the highest levels of phenolic compounds, namely of anthocyanins, and exhibit the highest antioxidant capacity. Moreover, fruit skins possess higher content of phenolic compounds and higher antioxidant activity compared to the fruit pulp [34]. In the same study, it was also reported that cyanidin 3-*O*-rutinoside contributed to 92% of the total antioxidant capacity of the anthocyanin fraction.

According to Vision *et al.*, [62], the consumption of 40 g of dried figs increased the plasma antioxidant capacity. Similarly, Faleh *et al.*, [33] described that *F. carica* dried figs are able to scavenge free radicals successfully. Moreover, it is know that superoxide radical is implicated in cell damages and, on the other hand, nitric oxide reacts rapidly with this free radical. The antioxidant potential of *F. carica* latex was evaluated by using the DPPH assay [20]. The latex demonstrated strong antioxidant capacity, exhibiting protection against superoxide radical and nitric oxide. This antioxidant capacity may also be attributed to the presence of bioactive compounds such as organic acids and volatile compounds, known for their antioxidant properties [20].

Cardioprotective Activity

CVD are the major cause of death in developed countries [73]. The main risk factors for CVD are smoking, hypertension, dyslipidemia, DM, and sedentary lifestyle [74]. Atherosclerosis and inflammation also play key roles in heart diseases [75]. Phenolic compounds are natural antioxidants, which may inhibit lipid oxidation and attenuate the progress of atherosclerosis and thrombosis [76]. Moreover, epidemiological studies show that flavonoid consumption is beneficial for cardiovascular health, increasing plasma antioxidant capacity and reducing platelet reactivity [77].

The studies about the cardioprotective activity of *F. carica* are scarce. However, it was recently demonstrated that leaves of this species decrease ischemia/reperfusion-induced injuries [78] in isolated heart of Wistar rats. In fact, the methanolic extract of *F. carica* leaves also presented a significant effect on the reduction of infarct size [78]. The possible protection against myocardial infarction may be due to antioxidant capacity conferred by the flavonoid or other phenolic compounds in the extract but the mechanisms remain unknown.

Antidiabetic Activity

DM is a metabolic disorder and a major public health problem, reaching pandemic proportions. In recent years, interest has increased in using natural products for pharmacological purposes, as a form of complementary or replacement therapy, including to avoid the deleterious effects of DM. There are several reports showing that numerous plant extracts are efficient in reducing hyperglycaemia, causing fewer adverse effects and with lower costs than the conventional oral antidiabetic drugs [79, 80]. DM and CVD are strongly interlinked. In fact, diabetic individuals will probably develop CVD or die due to heart failure. *F. carica* demonstrated to be effective in the prevention/treatment of both diseases. The *F. carica* leaves extract induced a hypoglycaemic effect (oral or intraperitoneal administration) in streptozotocin-induced diabetic rats. However, the study does not explain the mechanism involved in lowering blood glucose levels [81]. A subsequent study conducted by the same authors, showed a clearly hypoglycaemic effect of *F. carica* leaves in diabetic rats. The extract decreased plasma glucose in diabetic rats from 27.9 to 19.6 mmol/L but did not present any effect in normal rats [82]. Moreover, plasma insulin levels were decreased by treatment in non-diabetic rats, and the lactate released was lower in untreated diabetic *vs.* untreated non-diabetic rats. In sum, *F. carica* extract tends to decrease the high glucose levels, showing a clear hypoglycaemic effect in diabetic individuals [82].

Hypocholesterolemic Activity

Unfortunately, individuals with DM are more prone to have high cholesterol levels, which contributes to the high prevalence of CVD in these individuals. By taking actions to manage cholesterol, individuals can reduce their chance of developing CVD and have a premature death. The aqueous decoction of *F. carica* leaves, after treatment with hydrochloric acid and sodium hydroxide, and extraction with chloroform, was found to improve the cholesterol levels of streptozotocin-induced diabetic rats [83]. This extract led to a decline in the levels of total cholesterol and a decrease in the total cholesterol/HDL cholesterol ratio. Another study showed that *F. carica* leaf extract in hyperlipidemic rats decreased the serum and liver cholesterol levels in a dose-dependent manner [84]. The effect of total extract on liver cholesterol was more pronounced than its effect on the serum cholesterol level [84]. The presence of flavonoids and other phytochemicals in *F. carica* leaves may be responsible for significantly lower cholesterol levels in both the plasma and liver [85]. Nevertheless, the type of flavonoids and the mechanism(s) of action in different leaf extracts on cholesterol levels remain a matter of debate. In fact, one or more constituents of the extracts may play a role in cholesterol synthesis or metabolic pathways.

Anticancer Activity

Cancer is also a pandemic disease, affecting millions of people in the world, and results from several genetic and metabolic alterations [86]. Moreover, cancer cells metabolism presents some unique properties related with a high glycolytic flux, a process known as the Warburg effect [87]. Genetic mutations, smoking, and unbalanced diet, are some factors that individual or collectively contribute to cancer development [88]. This disease can be largely preventable by modification in dietary habits [89]. Hashemi *et al.*, [90] investigated the effect of *F. carica* latex, in different concentrations, on stomach cancer line (NCBI: C-131). The authors reported that a concentration of 5 mg per mL inhibits stomach cancer cell line growth [90]. On the other hand, this concentration did not indicate any cytotoxic activity against normal cells *in vitro*. Similarly, Wang *et al.*, [91] studied the effect of fresh fig fruit latex on human glioma (U251) and showed that the fresh latex acted as an anticancer agent. A mixture of many phytochemicals present in *F. carica*, including the 6-*O*-acyl-β-d-glucosyl-β-sitosterol, have been isolated as an effective cytotoxic agent obtained from fig latex showing *in vitro* inhibitory effects on proliferation of various cancer cell lines [92].

Antimicrobial Activity

Antimicrobial agents are crucial against infectious diseases. Nowadays, the increase of bacteria antibiotic resistant is one of the most problems in the treatment against these pathogens. Thus, new antibacterial approaches are needed, and the use of medicinal plants can be a beneficial alternative. Jeong *et al.*, [93] used a methanolic extract of *F. carica* and demonstrated that this extract have a strong antibacterial activity against oral bacteria. The combination effects of the extract with ampicillin or gentamicin had a synergistic effect against oral bacteria that showed that figs could act as a natural antibacterial agent. Another study showed a total inhibition of activity of *Candida albicans*, although it has not had the same effect on *Cryptococcus neoforman* [94].

Other Bioactivities

The major biological activities of *F. carica* are cited above. However, *F. carica* can also be used for antipyretic, antispasmodic, and antimutagenic potentials [39]. Nevertheless, more studies are needed to clarify these activities.

CONCLUDING REMARKS

In last years, there has been a growing interest in natural products with antioxidant power. The possible biological activities of edible and medicinal plants are an emerging field of research to prevent and treat OS-linked diseases. *F. carica* (leaves, fruits, and latex) may be considered a promising natural source of phytochemicals, namely of phenolic compounds, organic acids, and volatile compounds. These compounds have beneficial properties to health, mainly due to its antioxidant power. Nevertheless, much research remains to be done to explain the molecular mechanisms involved in the health benefits produced by *F. carica* and its phytocomponents. Several biological activities have been reported, showing the phytomedicinal potential of leaves, fruits, and latex of *F. carica* in the prevention/therapy of a wide range of diseases. Thus, there is compelling evidence that these natural products are promising candidates for the development of new nutraceuticals and food supplements. However, more studies are needed to further confirm the preventive effects of *F. carica* phytocomponents and also the mechanisms by which the beneficial effects are achieved. Their beneficial potential to human health should deserve much consideration in the forthcoming years.

CONFLICT OF INTEREST

The author confirms that author has no conflict of interest to declare for this publication.

ACKNOWLEDGEMENTS

This work was supported by the Portuguese "Fundação para a Ciência e a Tecnologia" - FCT: M.G. Alves (SFRH/BPD/80451/2011); L.R. Silva (SFRH/BPD/105263/2014; CICS-UBI); UMIB (Pest-OE/SAU/UI0215/2014) and co-funded by FEDER *via* Programa Operacional Fatores de Competitividade - COMPETE/QREN & FSE and POPH funds.

REFERENCES

[1]   Marques V, Farah A. Chlorogenic acids and related compounds in medical plants and infusions. Food Chem 2009; 113: 1370-6.
       [http://dx.doi.org/10.1016/j.foodchem.2008.08.086]

[2]   Corns CM. Herbal remedies and clinical biochemistry. Ann Clin Biochem 2003; 40: 489-507.
        [http://dx.doi.org/10.1258/000456303322326407] [PMID: 14503986]

[3]   Barnes J. Quality, efficacy and safety of complementary medicines: fashions, facts and the future. Part
        II: Efficacy and safety. Br J Clin Pharmacol 2003; 55(4): 331-40.
        [http://dx.doi.org/10.1046/j.1365-2125.2003.01811.x] [PMID: 12680880]

[4]   Newman DJ, Cragg GM. Natural products as sources of new drugs over the 30 years from 1981 to
        2010. J Nat Prod 2012; 75(3): 311-35.
        [http://dx.doi.org/10.1021/np200906s] [PMID: 22316239]

[5]   Sen T, Samanta SK. Medicinal plants, human health and biodiversity: a broad review. Adv Biochem
        Eng Biotechnol 2015; 147: 59-110.
        [http://dx.doi.org/10.1007/10_2014_273] [PMID: 25001990]

[6]   Nunes AR, Alves MG, Tomás GD, *et al.* Daily consumption of white tea (Camellia sinensis (L.))
        improves the cerebral cortex metabolic and oxidative profile in prediabetic Wistar rats. Br J Nutr 2015;
        113(5): 832-42.
        [http://dx.doi.org/10.1017/S0007114514004395] [PMID: 25716141]

[7]   Alves MG, Martins AD, Teixeira NF, *et al.* White tea consumption improves cardiac glycolytic and
        oxidative profile of prediabetic rats. J Funct Foods 2015; 14: 102-10.
        [http://dx.doi.org/10.1016/j.jff.2015.01.019]

[8]   Dias TR, Alves MG, Bernardino RL, *et al.* Dose-dependent effects of caffeine in human Sertoli cells
        metabolism and oxidative profile: relevance for male fertility. Toxicology 2015; 328: 12-20.
        [http://dx.doi.org/10.1016/j.tox.2014.12.003] [PMID: 25486098]

[9]   Conde VR, Alves MG, Oliveira PF, Silva BM. Tea (Camellia Sinensis (L.)): a Putative Anticancer
        Agent in Bladder Carcinoma? Anticancer Agents Med Chem 2014.
        [http://dx.doi.org/10.2174/1566524014666141203143143] [PMID: 25495463]

[10]   Carvalho M, Silva BM, Silva R, Valentão P, Andrade PB, Bastos ML. First report on Cydonia oblonga
        Miller anticancer potential: differential antiproliferative effect against human kidney and colon cancer
        cells. J Agric Food Chem 2010; 58(6): 3366-70.
        [http://dx.doi.org/10.1021/jf903836k] [PMID: 20192210]

[11]   Martins AD, Alves MG, Bernardino RL, Dias TR, Silva BM, Oliveira PF. Effect of white tea
        (Camellia sinensis (L.)) extract in the glycolytic profile of Sertoli cell. Eur J Nutr 2014; 53(6): 1383-
        91.
        [http://dx.doi.org/10.1007/s00394-013-0640-5] [PMID: 24363139]

[12]   Dias TR, Alves MG, Tomás GD, Socorro S, Silva BM, Oliveira PF. White tea as a promising
        antioxidant medium additive for sperm storage at room temperature: a comparative study with green
        tea. J Agric Food Chem 2014; 62(3): 608-17.
        [http://dx.doi.org/10.1021/jf4049462] [PMID: 24372402]

[13]   Oliveira AP, Silva LR, Andrade PB, *et al.* Further insight into the latex metabolite profile of Ficus
        carica. J Agric Food Chem 2010; 58(20): 10855-63.
        [http://dx.doi.org/10.1021/jf1031185] [PMID: 20923221]

[14]   Costa RM, Magalhães AS, Pereira JA, *et al.* Evaluation of free radical-scavenging and antihemolytic

activities of quince (Cydonia oblonga) leaf: a comparative study with green tea *(Camellia sinensis)*. Food Chem Toxicol 2009; 47(4): 860-5.
[http://dx.doi.org/10.1016/j.fct.2009.01.019] [PMID: 19271320]

[15]   Nunes AR, Alves MG, Moreira PI, Oliveira PF, Silva BM. Can tea consumption be a safe and effective therapy against diabetes mellitus-induced neurodegeneration? Curr Neuropharmacol 2014; 12(6): 475-89.
[http://dx.doi.org/10.2174/1570159X13666141204220539] [PMID: 25977676]

[16]   Oliveira AP, Valentão P, Pereira JA, Silva BM, Tavares F, Andrade PB. Ficus carica L.: Metabolic and biological screening. Food Chem Toxicol 2009; 47(11): 2841-6.
[http://dx.doi.org/10.1016/j.fct.2009.09.004] [PMID: 19747518]

[17]   Silva BM, Andrade PB, Valentão P, Ferreres F, Seabra RM, Ferreira MA. Quince (Cydonia oblonga Miller) fruit (pulp, peel, and seed) and Jam: antioxidant activity. J Agric Food Chem 2004; 52(15): 4705-12.
[http://dx.doi.org/10.1021/jf040057v] [PMID: 15264903]

[18]   Oliveira AP, Costa RM, Magalhães AS, *et al.* Targeted metabolites and biological activities of Cydonia oblonga Miller leaves. Food Res Int 2012; 46: 496-504.
[http://dx.doi.org/10.1016/j.foodres.2010.10.021]

[19]   Oliveira AP, Silva LR. Volatile profiling of *Ficus carica* varieties by HS-SPME and GC–IT-MS. Food Chem 2010; 123: 548-57.
[http://dx.doi.org/10.1016/j.foodchem.2010.04.064]

[20]   Oliveira AP, Silva LR, Ferreres F, *et al.* Chemical assessment and *in vitro* antioxidant capacity of Ficus carica latex. J Agric Food Chem 2010; 58(6): 3393-8.
[http://dx.doi.org/10.1021/jf9039759] [PMID: 20158255]

[21]   Oliveira AP, Pereira JA, Andrade PB, Valentão P, Seabra RM, Silva BM. Organic acids composition of Cydonia oblonga Miller leaf. Food Chem 2008; 111(2): 393-9.
[http://dx.doi.org/10.1016/j.foodchem.2008.04.004] [PMID: 26047441]

[22]   Silva BM, Andrade PB, Martins RC, *et al.* Quince (Cydonia oblonga miller) fruit characterization using principal component analysis. J Agric Food Chem 2005; 53(1): 111-22.
[http://dx.doi.org/10.1021/jf040321k] [PMID: 15631517]

[23]   Silva BM, Andrade PB, Ferreres F, Seabra RM, Oliveira MB, Ferreira MA. Composition of quince (Cydonia oblonga Miller) seeds: phenolics, organic acids and free amino acids. Nat Prod Res 2005; 19(3): 275-81.
[http://dx.doi.org/10.1080/14786410410001714678] [PMID: 15702641]

[24]   Vinha AF, Ferreres F, Silva BM, *et al.* Phenolic profiles of Portuguese olive fruits *(Olea europaea L.)*: influence of cultivar and geographical origin. Food Chem 2005; 89: 561-8.
[http://dx.doi.org/10.1016/j.foodchem.2004.03.012]

[25]   Meirinhos J, Silva BM, Valentão P, *et al.* Analysis and quantification of flavonoid compounds from Portuguese olive *(Olea europaea L.)* leaf cultivars. Nat Prod Res 2005; 19(2): 189-95.
[http://dx.doi.org/10.1080/14786410410001704886] [PMID: 15715265]

[26]   Oliveira AP, Pereira JA, Andrade PB, Valentão P, Seabra RM, Silva BM. Phenolic profile of Cydonia

oblonga Miller leaves. J Agric Food Chem 2007; 55(19): 7926-30.
[http://dx.doi.org/10.1021/jf0711237] [PMID: 17711340]

[27]   Magalhães AS, Silva BM, Pereira JA, Andrade PB, Valentão P, Carvalho M. Protective effect of quince (Cydonia oblonga Miller) fruit against oxidative hemolysis of human erythrocytes. Food Chem Toxicol 2009; 47(6): 1372-7.
[http://dx.doi.org/10.1016/j.fct.2009.03.017] [PMID: 19306906]

[28]   Moderno PM, Carvalho M, Silva BM. Recent patents on Camellia sinensis: source of health promoting compounds. Recent Pat Food Nutr Agric 2009; 1(3): 182-92.
[http://dx.doi.org/10.2174/2212798410901030182] [PMID: 20653539]

[29]   Carvalho M, Ferreira PJ, Mendes VS, *et al.* Human cancer cell antiproliferative and antioxidant activities of *Juglans regia L.* Food Chem Toxicol 2010; 48(1): 441-7.
[http://dx.doi.org/10.1016/j.fct.2009.10.043] [PMID: 19883717]

[30]   Oliveira AP, Silva LR, Andrade PB, *et al.* Determination of low molecular weight volatiles in Ficus carica using HS-SPME and GC/FID. Food Chem 2010; 121: 1289-95.
[http://dx.doi.org/10.1016/j.foodchem.2010.01.054]

[31]   Carvalho M, Jerónimo C, Valentão P, *et al.* Green tea: a promising anticancer agent for renal cell carcinoma. Food Chem 2010; 122: 49-54.
[http://dx.doi.org/10.1016/j.foodchem.2010.02.014]

[32]   Oliveira A, Baptista P, Andrade PB, *et al.* Characterization of *Ficus carica L.* cultivars by DNA and secondary metabolites analysis: Is genetic diversity reflected in the chemical composition? Food Res Int 2012; 49: 710-9.

[33]   Faleh E, Oliveira A, Valentão P, *et al.* Influence of Tunisian *Ficus carica* fruit variability in phenolic profiles and *in vitro* radical scavenging potential. Braz J Pharmacog 2012; 22: 1282-9.

[34]   Solomon A, Golubowicz S, Yablowicz Z, *et al.* Antioxidant activities and anthocyanin content of fresh fruits of common fig *(Ficus carica L.).* J Agric Food Chem 2006; 54(20): 7717-23.
[http://dx.doi.org/10.1021/jf060497h] [PMID: 17002444]

[35]   Veberic R, Colaric M, Stampar F. Phenolic acids and flavonoids of fig fruit *(Ficus carica L.)* in the northern Mediterranean region. Food Chem 2008; 106: 153-7.
[http://dx.doi.org/10.1016/j.foodchem.2007.05.061]

[36]   Teixeira DM, Patão RF, Coelho AV, da Costa CT. Comparison between sample disruption methods and solid-liquid extraction (SLE) to extract phenolic compounds from *Ficus carica* leaves. J Chromatogr A 2006; 1103(1): 22-8.
[http://dx.doi.org/10.1016/j.chroma.2005.11.047] [PMID: 16343519]

[37]   Vaya J, Mahmood S. Flavonoid content in leaf extracts of the fig *(Ficus carica L.)*, carob (Ceratonia siliqua L.) and pistachio (Pistacia lentiscus L.). Biofactors 2006; 28(3-4): 169-75.
[http://dx.doi.org/10.1002/biof.5520280303] [PMID: 17473377]

[38]   Vinson JA, Hao Y, Su X, *et al.* Phenol antioxidant quantity and quality in foods: vegetables. J Agric Food Chem 1998; 46: 3630-4.
[http://dx.doi.org/10.1021/jf980295o]

[39]   Mawa S, Husain K, Jantan I. (Moraceae): Phytochemistry, traditional uses and biological activities.

Evid Based Complement Alternat Med 2013; 2013: 8.
[http://dx.doi.org/10.1155/2013/974256]

[40] Solomon A, Golubowicz S, Yablowicz Z, *et al.* Protection of fibroblasts (NIH-3T3) against oxidative damage by cyanidin-3-rhamnoglucoside isolated from fig fruits *(Ficus carica L.)*. J Agric Food Chem 2010; 58(11): 6660-5.
[http://dx.doi.org/10.1021/jf100122a] [PMID: 20443626]

[41] Lazreg-Aref H, Mars M, Fekih A, Aouni M, Said K. Chemical composition and antibacterial activity of a hexane extract of Tunisian caprifig latex from the unripe fruit of *Ficus carica*. Pharm Biol 2012; 50(4): 407-12.
[http://dx.doi.org/10.3109/13880209.2011.608192] [PMID: 22136172]

[42] Wang L, Jiang W, Ma K, *et al.* The production and research of fig *(Ficus carica L.)* in China. Acta Hortic 2003; 605: 191-6.

[43] Frodin DG. History and concepts of big plant genera. Taxon 2004; 2004: 3.
[http://dx.doi.org/10.2307/4135449]

[44] Ghada B, Ahmed BA, Khaled C, *et al.* Molecular evolution of chloroplast DNA in fig *(Ficus carica L.)*: Footprints of sweep selection and recent expansion. Biochem Syst Ecol 2010; 38: 563-75.
[http://dx.doi.org/10.1016/j.bse.2010.06.011]

[45] Dueñas M, Pérez-Alonso JJ, Santos-Buelga C, *et al.* Anthocyanin composition in fig *(Ficus carica L.)*. J Food Compos Anal 2008; 21: 107-15.
[http://dx.doi.org/10.1016/j.jfca.2007.09.002]

[46] Barolo MI, Ruiz Mostacero N, López SN. *Ficus carica L.* (Moraceae): an ancient source of food and health. Food Chem 2014; 164: 119-27.
[http://dx.doi.org/10.1016/j.foodchem.2014.04.112] [PMID: 24996314]

[47] Kim JS, Kim YO, Ryu HJ, Kwak YS, Lee JY, Kang H. Isolation of stress-related genes of rubber particles and latex in fig tree *(Ficus carica)* and their expressions by abiotic stress or plant hormone treatments. Plant Cell Physiol 2003; 44(4): 412-4.
[http://dx.doi.org/10.1093/pcp/pcg058] [PMID: 12721382]

[48] Badgujar SB, Patel VV, Bandivdekar AH, Mahajan RT. Traditional uses, phytochemistry and pharmacology of *Ficus carica*: a review. Pharm Biol 2014; 52(11): 1487-503.
[http://dx.doi.org/10.3109/13880209.2014.892515] [PMID: 25017517]

[49] Instituto Nacional de Saúde Dr. Ricardo Jorge. Tabela dos alimentos: Figo. Available from: http://www.insa.pt/sites/INSA/Portugues/AreasCientificas/AlimentNutricao/AplicacoesOnline/Tabela Alimentos/PesquisaOnline/Paginas/DetalheAlimento.aspx?ID=IS650.

[50] Jeong W-S, Lachance PA. Phytosterols and Fatty Acids in Fig *(Ficus carica, var. Mission)* Fruit and Tree Components. J Food Sci 2001; 66: 278-81.
[http://dx.doi.org/10.1111/j.1365-2621.2001.tb11332.x]

[51] Marrelli M, Statti GA, Tundis R, Menichini F, Conforti F. Fatty acids, coumarins and polyphenolic compounds of *Ficus carica L.* cv. Dottato: variation of bioactive compounds and biological activity of aerial parts. Nat Prod Res 2014; 28(4): 271-4.
[http://dx.doi.org/10.1080/14786419.2013.841689] [PMID: 24087937]

[52]   Hounsome N, Hounsome B, Tomos D, Edwards-Jones G. Plant metabolites and nutritional quality of vegetables. J Food Sci 2008; 73(4): R48-65.
[http://dx.doi.org/10.1111/j.1750-3841.2008.00716.x] [PMID: 18460139]

[53]   Wu G. Amino acids: metabolism, functions, and nutrition. Amino Acids 2009; 37(1): 1-17.
[http://dx.doi.org/10.1007/s00726-009-0269-0] [PMID: 19301095]

[54]   Meletis CD, Barker JE. Therapeutic uses of aminoacids. Altern Complement Ther 2005; 11: 24-8.
[http://dx.doi.org/10.1089/act.2005.11.24]

[55]   McCune LM, Kubota C, Stendell-Hollis NR, Thomson CA. Cherries and health: a review. Crit Rev Food Sci Nutr 2011; 51(1): 1-12.
[http://dx.doi.org/10.1080/10408390903001719] [PMID: 21229414]

[56]   Soni MG, Thurmond TS, Miller ER III, Spriggs T, Bendich A, Omaye ST. Safety of vitamins and minerals: controversies and perspective. Toxicol Sci 2010; 118(2): 348-55.
[http://dx.doi.org/10.1093/toxsci/kfq293] [PMID: 20861067]

[57]   Ostlund RE Jr. Phytosterols and cholesterol metabolism. Curr Opin Lipidol 2004; 15(1): 37-41.
[http://dx.doi.org/10.1097/00041433-200402000-00008] [PMID: 15166807]

[58]   Ostlund RE Jr. Phytosterols in human nutrition. Annu Rev Nutr 2002; 22: 533-49.
[http://dx.doi.org/10.1146/annurev.nutr.22.020702.075220] [PMID: 12055357]

[59]   Bouvier F, Rahier A, Camara B. Biogenesis, molecular regulation and function of plant isoprenoids. Prog Lipid Res 2005; 44(6): 357-429.
[http://dx.doi.org/10.1016/j.plipres.2005.09.003] [PMID: 16289312]

[60]   Piironen V, Lindsay DG, Miettinen TA, *et al.* Plant sterols: biosynthesis, biological function and their importance to human nutrition. J Sci Food Agric 2000; 80: 939-66.
[http://dx.doi.org/10.1002/(SICI)1097-0010(20000515)80:7<939::AID-JSFA644>3.0.CO;2-C]

[61]   Kolesnik AA, Kakhniashvili TA, Zherebin YL, *et al.* Lipids of the fruit of *Ficus carica*. Chem Nat Compd 1986; 22: 394-7.
[http://dx.doi.org/10.1007/BF00579808]

[62]   Vinson JA, Zubik L, Bose P, Samman N, Proch J. Dried fruits: excellent *in vitro* and *in vivo* antioxidants. J Am Coll Nutr 2005; 24(1): 44-50.
[http://dx.doi.org/10.1080/07315724.2005.10719442] [PMID: 15670984]

[63]   Vallejo F, Marín JG, Tomás-Barberán FA. Phenolic compound content of fresh and dried figs *(Ficus carica L.)*. Food Chem 2012; 130: 485-92.
[http://dx.doi.org/10.1016/j.foodchem.2011.07.032]

[64]   Lila MA. Anthocyanins and human health: an *in vitro* investigative approach. J Biomed Biotechnol 2004; 2004: 306-13.
[http://dx.doi.org/10.1155/S111072430440401X]

[65]   Schwab W, Davidovich-Rikanati R, Lewinsohn E. Biosynthesis of plant-derived flavor compounds. Plant J 2008; 54(4): 712-32.
[http://dx.doi.org/10.1111/j.1365-313X.2008.03446.x] [PMID: 18476874]

[66]   Grison-Pigé L, Hossaert-McKey M, Greeff JM, Bessière JM. Fig volatile compounds--a first

comparative study. Phytochemistry 2002; 61(1): 61-71.
[http://dx.doi.org/10.1016/S0031-9422(02)00213-3] [PMID: 12165303]

[67] James DG, Price TS. Field-testing of methyl salicylate for recruitment and retention of beneficial insects in grapes and hops. J Chem Ecol 2004; 30(8): 1613-28.
[http://dx.doi.org/10.1023/B:JOEC.0000042072.18151.6f] [PMID: 15537163]

[68] Jin D, Hakamata H, Takahashi K, Kotani A, Kusu F. Determination of quercetin in human plasma after ingestion of commercial canned green tea by semi-micro HPLC with electrochemical detection. Biomed Chromatogr 2004; 18(9): 662-6.
[http://dx.doi.org/10.1002/bmc.370] [PMID: 15386501]

[69] Wongkham S, Laupattarakasaem P, Pienthaweechai K, Areejitranusorn P, Wongkham C, Techanitiswad T. Antimicrobial activity of Streblus asper leaf extract. Phytother Res 2001; 15(2): 119-21.
[http://dx.doi.org/10.1002/ptr.705] [PMID: 11268109]

[70] Dias TR, Tomás GD, Teixeira NF, *et al.* White Tea *(Camellia Sinensis (L.))*: Antioxidant Properties and Beneficial Health Effects. Int J Food Sci Nutr Diet 2013; 2: 1-15.

[71] Valko M, Leibfritz D, Moncol J, Cronin MT, Mazur M, Telser J. Free radicals and antioxidants in normal physiological functions and human disease. Int J Biochem Cell Biol 2007; 39(1): 44-84.
[http://dx.doi.org/10.1016/j.biocel.2006.07.001] [PMID: 16978905]

[72] Beckman JS. Oxidative damage and tyrosine nitration from peroxynitrite. Chem Res Toxicol 1996; 9(5): 836-44.
[http://dx.doi.org/10.1021/tx9501445] [PMID: 8828918]

[73] Go AS, Mozaffarian D, Roger VL, *et al.* Heart disease and stroke statistics--2013 update: a report from the American Heart Association. Circulation 2013; 127(1): e6-e245.
[http://dx.doi.org/10.1161/CIR.0b013e31828124ad] [PMID: 23239837]

[74] Okrainec K, Banerjee DK, Eisenberg MJ. Coronary artery disease in the developing world. Am Heart J 2004; 148(1): 7-15.
[http://dx.doi.org/10.1016/j.ahj.2003.11.027] [PMID: 15215786]

[75] Hansson GK. Inflammation, atherosclerosis, and coronary artery disease. N Engl J Med 2005; 352(16): 1685-95.
[http://dx.doi.org/10.1056/NEJMra043430] [PMID: 15843671]

[76] Kinsella JE, Frankel E, German B, *et al.* Possible mechanisms for the protective role of antioxidants in wine and plant foods: physiological mechanisms by which flavonoids, phenolics, and other phytochemicals in wine and plant foods prevent or ameliorate some common chronic diseases are discussed. Food Technol 1993; 47: 85-9.

[77] Kris-Etherton PM, Keen CL. Evidence that the antioxidant flavonoids in tea and cocoa are beneficial for cardiovascular health. Curr Opin Lipidol 2002; 13(1): 41-9.
[http://dx.doi.org/10.1097/00041433-200202000-00007] [PMID: 11790962]

[78] Allahyari S, Delazar A, Najafi M. Evaluation of general toxicity, anti-oxidant activity and effects of ficus carica leaves extract on ischemia/reperfusion injuries in isolated heart of rat. Adv Pharm Bull 2014; 4 (Suppl. 2): 577-82.

[PMID: 25671192]

[79] Gupta RK, Kesari AN, Murthy PS, Chandra R, Tandon V, Watal G. Hypoglycemic and antidiabetic effect of ethanolic extract of leaves of Annona squamosa L. in experimental animals. J Ethnopharmacol 2005; 99(1): 75-81.
[http://dx.doi.org/10.1016/j.jep.2005.01.048] [PMID: 15848023]

[80] Abolfathi AA, Mohajeri D, Rezaie A, *et al.* Protective effects of green tea extract against hepatic tissue injury in streptozotocin-induced diabetic rats 2012.
[http://dx.doi.org/10.1155/2012/740671]

[81] Pérez C, Domínguez E, Ramiro JM, *et al.* A study on the glycaemic balance in streptozotocin-diabetic rats treated with an aqueous extract of *Ficus carica* (fig tree) leaves. Phytother Res 1996; 10: 82-3.
[http://dx.doi.org/10.1002/(SICI)1099-1573(199602)10:1<82::AID-PTR776>3.0.CO;2-R]

[82] Pérez C, Domínguez E, Canal JR, Campillo JE, Torres MD. Hypoglycaemic activity of an aqueous extract from ficus carica (fig tree) leaves in streptozotocin diabetic rats. Pharm Biol 2000; 38(3): 181-6.
[http://dx.doi.org/10.1076/1388-0209(200007)3831-SFT181] [PMID: 21214459]

[83] Canal JR, Torres MD, Romero A, Pérez C. A chloroform extract obtained from a decoction of Ficus carica leaves improves the cholesterolaemic status of rats with streptozotocin-induced diabetes. Acta Physiol Hung 2000; 87(1): 71-6.
[http://dx.doi.org/10.1556/APhysiol.87.2000.1.8] [PMID: 11032050]

[84] Rassouli A. Effects of Fig tree () leaf extracts on serum and liver cholesterol levels in hyperlipidemic rats Ficus carica. Int J Vet Res 2010; 2: 77-80.

[85] Borradaile NM, de Dreu LE, Barrett PH, Behrsin CD, Huff MW. Hepatocyte apoB-containing lipoprotein secretion is decreased by the grapefruit flavonoid, naringenin, *via* inhibition of MTP-mediated microsomal triglyceride accumulation. Biochemistry 2003; 42(5): 1283-91.
[http://dx.doi.org/10.1021/bi026731o] [PMID: 12564931]

[86] López-Lázaro M. A new view of carcinogenesis and an alternative approach to cancer therapy. Mol Med 2010; 16(3-4): 144-53.
[http://dx.doi.org/10.2119/molmed.2009.00162] [PMID: 20062820]

[87] Oliveira PF, Martins AD, Moreira AC, Cheng CY, Alves MG. The Warburg effect revisited--lesson from the Sertoli cell. Med Res Rev 2015; 35(1): 126-51.
[http://dx.doi.org/10.1002/med.21325] [PMID: 25043918]

[88] Noonan DM, Benelli R, Albini A. Angiogenesis and cancer prevention: a vision. Recent Results Cancer Res 2007; 174: 219-24.
[http://dx.doi.org/10.1007/978-3-540-37696-5_19] [PMID: 17302199]

[89] Fresco P, Borges F, Diniz C, Marques MP. New insights on the anticancer properties of dietary polyphenols. Med Res Rev 2006; 26(6): 747-66.
[http://dx.doi.org/10.1002/med.20060] [PMID: 16710860]

[90] Hashemi SA, Abediankenari S, Ghasemi M, Azadbakht M, Yousefzadeh Y, Dehpour AA. The Effect of Fig Tree Latex *(Ficus carica)* on Stomach Cancer Line. Iran Red Crescent Med J 2011; 13(4): 272-5.

[PMID: 22737478]

[91]   Wang J, Wang X, Jiang S, *et al.* Cytotoxicity of fig fruit latex against human cancer cells. Food Chem Toxicol 2008; 46(3): 1025-33.
[http://dx.doi.org/10.1016/j.fct.2007.10.042] [PMID: 18078703]

[92]   Rubnov S, Kashman Y, Rabinowitz R, Schlesinger M, Mechoulam R. Suppressors of cancer cell proliferation from fig *(Ficus carica)* resin: isolation and structure elucidation. J Nat Prod 2001; 64(7): 993-6.
[http://dx.doi.org/10.1021/np000592z] [PMID: 11473446]

[93]   Jeong M-R, Kim H-Y, Cha J-D. Antimicrobial activity of methanol extract from *Ficus carica* leaves against oral bacteria. J Bacteriol Virol 2009; 39: 97-102.
[http://dx.doi.org/10.4167/jbv.2009.39.2.97]

[94]   Aref HL, Salah KB, Chaumont JP, Fekih A, Aouni M, Said K. *In vitro* antimicrobial activity of four *Ficus carica* latex fractions against resistant human pathogens (antimicrobial activity of *Ficus carica* latex). Pak J Pharm Sci 2010; 23(1): 53-8.
[PMID: 20067867]

CHAPTER 2

Bioactive Compounds of Citrus as Health Promoters

Amílcar Duarte[1,*], Catarina Carvalho[2], Graça Miguel[1]

[1] *Center for Mediterranean Bioresources and Food (MeditBio), Faculty of Sciences and Technology, University of Algarve, 8005-139 Faro, Portugal*

[2] *Tecnoparque Colombia Nodo Rionegro, SENA, Colombia*

Abstract: Citrus are a group of fruit species, quite heterogeneous in many aspects, including chemical composition of the fruit. Since ancient times, some citrus fruits were used to prevent and cure human diseases. In the recent decades, it has been demonstrated that fruits can actually help prevent and cure some diseases and above all, they are essential in a balanced diet. Citrus fruits, as one of the groups of fruit species, with greater importance in the world, have been studied for their effects on human health. Some species of citrus were referred as potential antioxidant based therapy for heart disease, cancer and inflammation. Fruit peels and seeds have also high antioxidant activity. The health benefits of citrus fruit have mainly been attributed to the high level of bioactive compounds, such as phenols (*e.g.*, flavanone glycosides, hydroxycinnamic acids), carotenoids and vitamin C. These compounds are present in the fruit pulp and hence in the juice. But some bioactive compounds can be found in parts of the fruit which usually are not used for human food. The content of bioactive compounds depends on the species and cultivar, but also depends on the production system followed in the orchard. Citrus fruits, their derivatives and their by-products (peel, pulp and oil) are reach in different bioactive compounds and its maturity, postharvest and agroindustry processes influence their composition and concentration. The aim of this chapter was to review the main bioactive compounds of the different components of citrus and their relationship to health.

Keywords: Ascorbic acid, Clementine, Coumarin, Grapefruit, Hesperidin, Lemon, Lime, Mandarin, Narirutin, Orange.

[*] **Corresponding author Amílcar Duarte:** Faculty of Sciences and Technology, University of Algarve, 8005-139 Faro, Portugal; Email: aduarte@ualg.pt.

INTRODUCTION

Since ancient times the citrus fruit have been considered important to human health. In some older books, they were even considered as a medicine.

In the Virgil's Georgics, likely published in 29 BC, the author mentions the beneficial effects of citron (*Citrus medica* L.) on health. He wrote that *"Persia yields the acidic juices and lasting flavor of the fruitful citrus tree, none other more efficacious – if ever wicked stepmothers have poisoned the drinking cups by mixing herbs and harmful spells – none better to bring immediate help and drive the lethal toxins from the body. The tree itself grows large and much resembles the bay laurel (and, did it not release its own distinctive scent, would be a bay). No wind can cause its leaves to fall, its flower clings fast indeed, and the Persians use it to sweeten bad breath and treat congestion in the old."* [1]. About a thousand years later, Ibn Sina (Avicenna) in the *Canon of Medicine* completed in 1025, mentioned the citron leaves as one of the drugs to stroke [2]. Although some of the effects have never been proven scientifically, other effects mentioned by the ancient authors have been proven by modern medicine [3].The oldest references mentioned only the citron because it was the only citrus they knew. The orange, lemon, and tangerine trees we know today were then unknown to the authors, because they were only grown in limited areas of Asia or because they only appeared later, as a result of mutations or hybridization between other citrus fruits.

The maritime discoveries were important to the spread of citrus species on all continents. But in addition, during long sea travel, it became clear that inclusion of citrus in the diet of seamen was the solution to prevent or even cure scurvy, a historical killer of seafarers for centuries [4].

The discovery of vitamin C and its high content in the citrus explains part of the beneficial effects of these fruits on health. Moreover, in the recent decades, numerous works on other bioactive compounds present in citrus fruits have been published.

Citrus fruits are characterized by the distinct aroma and delicious taste, and have been recognized as an important food. They are part of our daily diet, playing key roles in supplying nutrients and energy and in human health promotion [5].

According to Kitts [6], bioactive compounds are "extranutritional" constituents naturally present in small amounts in the food matrix, produced upon either *in vivo* or industrial enzymatic digestion. Ascorbic acid, carotenoids (lycopene and β-carotene), limonoids, flavonoids, essential oils, vitamin-B complex and related nutrients (nicotinic acid/niacin, pantothenic acid, folic acid, thiamine, riboflavin, pyridoxine, biotin, choline, and inositol), and pectin are some examples of bioactive compounds of *Citrus* fruits, useful not only for maintaining human health but also in food industry [7].

The presence and/or the amounts of the bioactive compounds present in citrus fruits depend on the species and cultivars [sweet oranges (*Citrus sinensis* (L.) Osbeck) including acidless, navel and blood oranges, mandarins such as satsumas (*C. unshiu* Marcov.), tangerines (*C. tangerine* Yu.Tanaka), 'Ponkan' (*C. reticulata* Blanco), and clementines (*C. clementina* hort. ex Tanaka), sour/bitter oranges (*C. aurantium* L.) such as 'Seville', lemons (*C. limon* (L.) Burm.f.), limes (*C. aurantifolia* (Cristm.) Swingle and *C. latifolia* Tanaka), grapefruit (*C. paradisi* Macfad) and pummelos (*C. maxima* Burm. f.), hybrids (*e.g.*, tangelos, tangors and limequats), citrons (*C. medica* L.), and many others], part of the fruit (juice, pulp, peel, seeds), developmental stages of fruits, among other factors [7 - 10].

In this chapter, we will focus on the chemistry and biological properties of ascorbic acid, carotenoids, flavonoids, limonoids, coumarins, and essential oils of citrus origin as well as the factors affecting the level of each of these compounds in the fruit.

ASCORBIC ACID

Chemistry

Ascorbic acid (vitamin C) is a water-soluble ketolactone with two ionizable hydroxyl groups (lactone 2, 3, - dienol- L- gluconic acid) having the chemical formula $C_6H_8O_6$. It has two pK_a's, pK_1 is 4.2 and pK_2 is 11.6, being the ascorbate monoanion, AscH−, the dominant form at physiological pH [11].

Ascorbic acid is a six-carbon compound, related to glucose. Vitamin C presents a simple structure (Fig. **1**), resembling a monosaccharide [11].

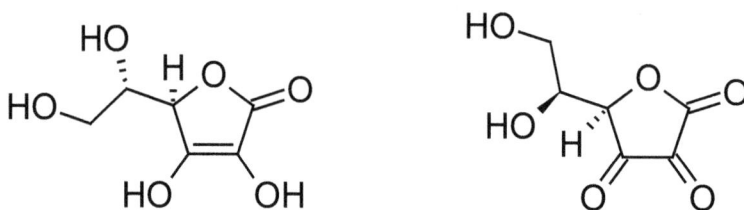

Fig. (1). Molecular structure of vitamin C.

Ascorbate is a reducing agent very powerful that it is oxidized to ascorbate radical (Ascradical dot−) and dehydroascorbic acid (DHA). The ascorbate radical it becomes easily in ascorbate and DHA ($K_{obs} = 2 \times 10^5$ M^{-1} s^{-1}, pH 7.0). In this way ascorbate radical is a very effective antioxidant [12, 13].

Ascorbic acid is a white crystalline powder, extremely soluble in water and their solution has a white-yellow color. The molar extinction coefficient of ascorbate solution at 265 nm is 14,500 M^{-1} cm^{-1} at pH values between 6 and 7.8 [14], but values ranging from 7,500 to 20,400 M^{-1} cm^{-1} have been reported.

Ascorbate is very susceptible to oxidation, depending on the pH of the solution and is accelerated by catalytic metals [15]. Nevertheless, its oxidation is slow at pH 7.0, in the absence of catalytic metals [14].

Biological Properties

Vitamin C (ascorbic acid and dehydroascorbic acid) has many biological activities and is a very important nutritional quality factor of citrus fruits.

Humans and other primates are unable of synthesize ascorbate, because the gene encoding l-gulonolactone oxidase (GLO), the enzyme required for the last step in ascorbate synthesis, is not functional and so must intake it from plants and animals that synthesize ascorbate from glucose [11, 16].

Vitamin C, because of its role in the formation of collagen, the major building protein of humans, is a essential vitamin for the growth and maintenance of healthy bones, teeth, gums, ligaments, and blood vessels, having antiscorbutic properties. It also acts in many different metabolic pathways as a reducing agent and coenzyme [17].

Ascorbate is responsible for the activity of collagen hydroxylases keeping iron in the ferrous state by being an electron-donor. It also affects the expression of a wide array of genes by reacting with different dioxygenases, for example *via* the Hypoxia-Inducible Transcription Factor (HIF) system, and by *via* the epigenetic landscape of cells and tissues. All known physiological and biochemical functions of ascorbate are due to its function as an electron donor. The ability to donate one or two electrons makes AscH− an excellent reducing and antioxidant agent. The reaction of vitamin C to produce hydrogen peroxide (H_2O_2) depends of pH and catalytic metals, acting this way as antioxidant and prooxidant [11].

In addition, ascorbic acid has also been reported for helping in the healing of wounds and burns, promoting calcium absorption, lowering the risk of ischemic heart disease, and preventing cancer [18]. However, the positive effect of this vitamin in the prevention of cancer needs a combination with other vitamins, such as vitamin E, and carotenoids (*e.g.* β-carotene), present in fruits and vegetables [18]. The prevention of cancer cannot be attributed to a sole compound and the respective amounts but to the combination of these compounds which are present in fruits and vegetables [18, 19].

Although, a vitamin C intake of 5 mg per day is enough to prevent the symptoms of scurvy in an adult [20], for ensuring a normal growth and good health a dose of 30 to 60 mg per day, equivalent to one orange intake per day, is advisable [21].

The Recommended Dietary Allowance (RDA) setted by the U.S. Department of Agriculture and U.S. Department of Health and Human Services, for non-smoking adults is of 75 to 90 mg [22]. Smoking-adults and pregnant or lactating women should take higher daily doses.

Citrus fruits are particularly rich in vitamin C, providing an average of 20 to 83 mg/100 g fresh weight [23, 24]. A medium-sized orange or grapefruit contains approximately 56 to 70 mg of ascorbic acid, and an average 225-mL of a normal orange juice contains 125 mg of ascorbic acid [25].

Content of Ascorbic Acid in Different Parts of the Plant/Fruit

The vitamin C is found naturally in citrus fruits and is its most important antioxidant compound, although other substances, like flavonoids, also have antioxidant activity. The total antioxidant capacity of vitamin C ranged from 26.9% to 45.9% in juice vesicles and segments in *Citrus unshiu*, *C. reticulata*, *C. sinensis* and *C. changshanensis* species [26].

In citrus fruit, the rind, usually not edible, is richer in ascorbic acid than the pulp. Flavedo tissues have higher ascorbic acid content than albedo, being four times superior to juice content [27].

Concentration of vitamin C is higher in unripe citrus fruits than in ripe ones [27], but total ascorbic acid content per fruit increases with fruit size during the same period.

Content of Ascorbic Acid in Different Species and Cultivars

The concentration of ascorbic acid in citrus fruits largely depends on the species and cultivars [24, 28]. Ascorbic acid concentration is generally higher in oranges than in grapefruits, mandarins, tangerines, lemons or limes [10, 23, 29, 30]. According to Cano *et al.*, [24], oranges and clementines have higher amounts of vitamin C than other citrus groups (Table **1**). Total vitamin C concentration in sweet orange (*C. sinensis* Osb.) is 1.5 times higher than in satsuma mandarin (*C. unshiu* Marc.) during fruit growth and ripening [31]. Duarte *et al.*, [30] also found the highest concentration of vitamin C in oranges and lowest values in the hybrids 'Encore', 'Wilking', 'Fortune', and 'Ortanique' and in lemon 'Lisbon'.

According to Cano *et al.*, [24], orange cultivars (navel, common and blood groups) presented the higher concentration in total vitamin C, having 'Newhall' and 'Navelate' (navel group) the highest concentrations (64.2 and 47.8 mg/100 g, respectively) and 'Lanelate' and 'Fukumoto' the lowest values (37.5 and 39.5 mg/100 g, respectively) (Table **2**). Duarte *et al.*, [30] found higher concentration of vitamin C in common orange 'Valencia Late', compared with five cultivars of navel group ('Dalmau', 'Baía', 'Newhall', 'Lanelate' and 'Rohde').

Table 1. Total vitamin C (mg/100 g fresh weight) in the pulp of different citrus groups in Valencia, Spain[a]. Adapted from Cano *et al.*, [24].

Citrus groups	Vitamin C
Clementines	41.1 c
Satsumes	22.9 a
Hybrids (tangors and mandarin hybrids)	32.7 b
Navel oranges	45.9 c
Common oranges	42.0 c
Pigmented oranges	46.0 c

[a] Data are the means of different number of cultivars of each group. Different letters indicate significant differences ($P < 0.01$).

Among mandarin cultivars, the clementines 'Oronules' and 'Hernandina' displayed the highest concentrations (47.4 and 46.0, mg/100 g, respectively) whereas mandarins of the satsuma group ('Owari', 'Okitsu', 'Avasa Pri-10' and 'Avasa Pri-19') presented the lower values (20.0, 21.3, 24.9 and 25.5 mg/100 g, respectively) [24] (Table **2**).

The rootstock also affects the concentration of ascorbic acid in the fruit. In Florida, the fruit of 'Murcott' trees grafted on rough lemon are poorer in ascorbic acid than those that were grafted on sour orange [32]. Generally, the vitamin C content is higher in fruits from trees grafted on Troyer citrange, compared with that from trees grafted on 'Cleopatra' mandarin [33].

Table 2. Total vitamin C (mg/100 g fresh weight) in the pulp of different citrus cultivars in Valencia, Spain. Adapted from Cano *et al.*, [24].[a]

Groups and cultivars	Vitamin C	Groups and cultivars	Vitamin C
Common oranges		**Clementines**	
Valencia Late	40.5 ± 0.6	Arrufatina	39.7 ± 1.7
Salustiana	43.4 ± 0.6	Clemenpons	43.6 ± 2.5
		Nules	38.7 ± 2.5
Navel oranges		Fina	41.7 ± 2.9
Fukumoto	39.5 ± 1.0	Hernandina	46.0 ± 2.2
Lanelate	37.5 ± 0.6	Loretina	39.8 ± 0.8
Navelate	47.8 ± 2.2	Marisol	37.0 ± 1.1

(Table 2) contd.....

Groups and cultivars	Vitamin C	Groups and cultivars	Vitamin C
Navelina	45.0 ± 0.9	Orogrande	39.2 ± 1.3
Navel Foyos	45.2 ± 1.4	Oronules	47.4 ± 3.1
Newhall	64.2 ± 4.2		
Powell	42.2 ± 1.4	**Satsumas**	
		Avasa Pri-10	24.9 ± 1.8
Pigmented oranges		Avasa Pri-19	25.5 ± 2.0
Sanguinelli	46.0 ± 1.0	Okitsu	21.3 ± 1.2
		Owari	20.0 ± 2.0
Tangors			
Ortanique	28.3 ± 1.2	**Mandarin hybrids**	
		Fortune	26.2 ± 0.9
		Nova	43.4 ± 1.8

[a] Data are the mean ± SEM (standard error of the mean) (n = 6).

Effect of Environmental Factors and Cultural Practices

Vitamin C content in fruits and vegetables can be influenced by various factors such as pre-harvest climatic conditions (especially temperature), cultural practices, position of fruit on the tree, fruit maturity, harvesting methods, postharvest handling procedures, type of container and processing factors [23, 27].

Shrestha *et al.*, [21], studying the variation of physicochemical composition of acid lime (*C. aurantifolia* Swingle) fruits from different sides of the tree and different locations in Nepal, observed that the percentage of ascorbic acid varies with the agro-ecological zone and the position of the fruit on the tree. South side fruits presented the highest ascorbic acid content (79.6 mg and 69.9 mg) whereas the lowest was observed in centre fruits in the High and Mid Hills zone (62.8 mg and 55.1 mg, respectively). Nevertheless, in Terai, highest ascorbic acid was observed in fruit at north side (58.7 mg) and lowest was observed in centre (41.8 mg).

Vitamin C levels in citrus fruis depends of many environmental factors like night temperatures during fruit growth (hot tropical regions have lower concentrations of ascorbic acid) [34] , environmental conditions that increase the acidity of the fruits (fruits with higher acidity have higher vitamin C levels) [17], intensity of light during the growing season (fruits with more light exposure have greater vitamin C content) [35], and hydric stress (lower irrigation reduces vitamin C content of the fruit).

High levels of nitrogen fertilizer reduced ascorbic acid content in juices of oranges, mandarins, grapefruits, and lemons, while increased potassium ferti-lization increased vitamin C content [27].

Ascorbic acid concentration in the juice is generally higher in fruits produced in the organic orchards, although this difference was not detected in all citrus cultivars [28]. The fruits produced by organic farming also have a higher concentration of citric, malic, tartaric and malonic acids and have lower levels of oxalic acid [30].

CAROTENOIDS

Chemistry

The carotenoids are tetraterpenes that give plants their colour (red, yellow and orange) with fat soluble nature and are closely related to sterols, ubiquinones, vitamin E (tocopherol), terpenes, vitamin K1 (phylloquinone) and a host of other terpenoid compounds [36, 37]. Hundreds of carotenoids are known, nevertheless only very few are of interest in human nutrition, and all of these pigments are present in citrus fruits (α-carotene, β-carotene, lycopene, lutein, zeaxanthin and cryptoxanthin) [36 - 38]. Fig. (**2**) presents the chemical structures of these compounds. However, much more carotenoid structures may be found in *Citrus*.

Generally, carotenoids are derived from a 40-carbon polyene chain. This chain may be terminated by cyclic end-groups (rings) and may be complemented with functional groups containing oxygen (Fig. **2**). According to their chemical structure, carotenoids can be classified into two categories: carotenes (hydrocarbons) and xanthophylls (oxygenated derivatives of these hydrocarbons) (Fig. **2**). There are other classification systems. For example, those based on the cyclization (acyclic, monocyclic and bicyclic) (Fig. **2**) or based on the function, which ultimately depend on the structure of a carotenoid. In this case they may be classified as primary if required for photosynthetic process (*e.g.* β-carotene, lutein, zeaxanthin, violaxanthin…) and as secondary if the presence is not directly related to plant survival (*e.g.* β-carotene, lycopene, capsaxanthin, astaxanthin, bixin) [39].

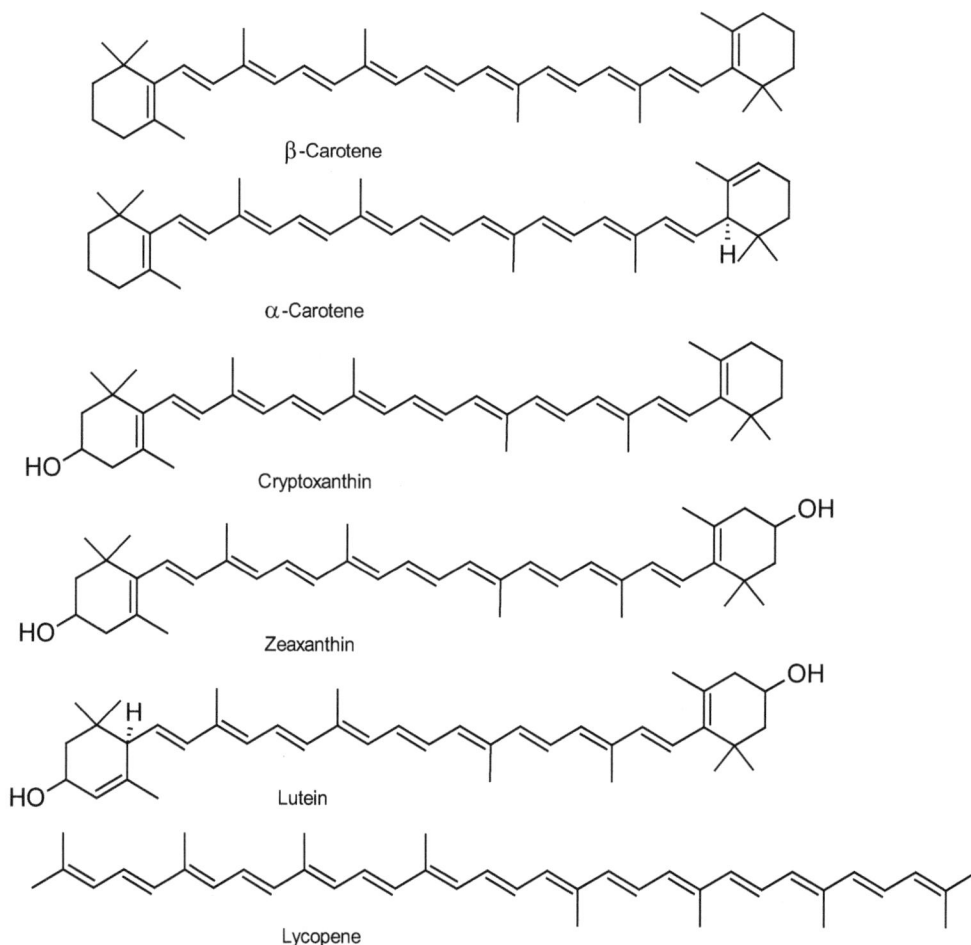

Fig. (2). Some *Citrus* carotenoids.

Carotenoids are hydrocarbons composed by eight isoprenoid units (5 atoms of carbon), joined in a head-to-tail pattern, except at the center, to give symmetry to the molecule so that the two central methyl groups are in a 1,6-positional relationship and the remaining non-terminal methyl groups are in a 1,5-positional relationship [39]. The molecules may vary in chain length, but the majority of carotenoids contain forty carbon atoms (C40). They are highly conjugated polyene chromophores (9-13 double bonds, the most common appearing as all-*trans*), and this conjugation gives carotenoids much of their functionality. The large number of possible carotenoid isomers, considering *cis/trans* configurations, in conjunction with chiral centers, can lead to tens of thousands of potential combinations.

In higher plants, carotenoids are synthesized in plastids. Both mevalonate dependent and independent pathway for the formation of isopentenyl diphosphate are well known. Isopentenyl diphosphate undergoes a series of addition and condensation reactions to form phytoene, which gets transformed into lycopene. Cyclization of lycopene either leads to the formation of β-carotene and its derivative β-cryptoxanthin, xanthophylls, antheraxanthin, zeaxanthin and violaxanthin or α-carotene, and lutein [39].

Two types of carotenoids have been reported, one with vitamin A activity (*α*-carotene, *β*-carotene and *β*-cryptoxanthin) and other without vitamin A activity (lycopene, lutein and zeaxanthin) [40, 41].

Biological Properties

Carotenoids are indispensable plant secondary metabolites that are involved in antioxidation, photosynthesis and biosynthesis of phytohormones. Carotenoids may be involved in other biological functions that have yet to be discovered. In citrus plants, altered carotenoid accumulation resulted in dramatic effects on metabolic processes involved in flavonoid/anthocyanin biosynthesis, redox modification, and starch degradation [42].

They are photoprotective pigments, antioxidants, hormone precursors, and attractants for pollinators in plant growth, development, and reproduction [43] and they also act as modulators of the physical properties of lipid membranes [44 - 47].

Carotenoids that act as precursors for vitamin A, are essential to human and animal diets. Carotenoids can also serve as antioxidants, which play an important role in reducing the risk of certain forms of cancer [48].

β-Carotene and lycopene present antioxidant activity [49]. β-Carotene also ameliorates the immune response according to the results reported by Imamura *et al.*, [50]. Cryptoxanthin, abundant in mandarins, has preventive effects on cancer [51, 52], counteracts induced bone loss *in vivo* [53], and has a positive effect on serum lipid levels [36, 54]. β-Cryptoxanthin alone or in combination with dietary phytosterols exhibited an apoptosis mediated antiproliferative action triggered by increased Ca^+ influx and reactive oxygen nitrogen species (RONS) in human colon adenocarcinoma Caco-2 cells [52].

In vitro studies have demonstrated that lycopene has the highest antioxidant activity of all the carotenoids [55, 56]. It has the ability to quench singlet oxygen (more than that of β-carotene), to trap peroxyl radicals, to inhibit the oxidation of DNA and the peroxidation of lipids [57] and in some studies to inhibit the oxidation of low density lipoprotein (LDL) [58].

Carotenoids along with vitamin E, flavonoids, limonoids, fiber, lignin, poly-saccharides, phenolic compounds and essential oils greatly contribute to the supply of anticancer agents and other nutraceutical compounds with anti-oxidant, hypocholesterolemic, anti-allergic and anti-inflammatory activities, all these essential to prevent thrombosis, cancer, cardiovascular and degenerative diseases, atherosclerosis, and obesity. Despite these beneficial characteristics, there is still a great need to improve fruit quality to meet the demands of today's consumer [59].

Carotenoids in Different Parts of the Plant/Fruit

Citrus are a complex source of carotenoids with more than 110 different carotenoids and xanthophylls reported, although many of them may be isomers. Their specific patterns of accumulation are responsible for the wide range of colors displayed by citrus fruits [60].

Carotenoid profiles varied with tissue types, citrus species, and cultivars obtained by mutations. It was observed that β-citraurin occurred in the peel of oranges but not in juice vesicles, whereas the reverse was found for violaxanthin, 9-*cis*-violaxanthin, and luteoxanthin [61]. Lutein was observed in the peel of pummelos and grapefruits and in the juice vesicles of pummelos, but not in orange tissues. The biosynthesis of carotenoids in these two tissues seems to be independent [61].

Satsuma mandarins (*C. unshiu* Marc.) accumulate β-cryptoxanthin predominantly in the flavedo and juice vesicles in mature fruit [62, 63]. Mature sweet orange (*C. sinensis* Osbeck) accumulates violaxanthin isomers predominantly in the fruit [64], in which 9-*cis*-violaxanthin was found to be the principal carotenoid [65]. Because β-cryptoxanthin was detected to a minor extent in sweet orange cultivars, the difference in its concentration can be used as a discriminating factor among orange, mandarin and their hybrids [62]. Mature lemon (*C. limon* Burm.f.) showed light yellow color in the flavedo and juice vesicles. 'Lisbon' lemon accumulates small amounts of carotenoids.

Liu *et al.*, [66] found large differences in constituents of carotenoids and limonoids in juice vesicles and flavedo of two pummelos (*C. maxima*). 'Chuhong' pummelo contained 57 times the amount of carotenoids in juice vesicles compared to 'Feicui' pummelo, mainly all-*trans*-lycopene and phytoene, while in flavedo it contained only 25% of that in 'Feicui' pummelo. This reveals that the carotenoid levels can vary greatly from one to another part of the fruit and this variation may be different in two cultivars of the same species. 'Feicui' pummelo showed a high proportion of β-carotene and other chloroplastic carotenoids.

Carotenoids in Different Species and Cultivars

Most of the carotenoids are present in almost all citrus, although in different concentrations. According to Gross [60], carotenoid concentration and composition vary greatly among citrus cultivars and depends on the growing conditions. Types of carotenoids and their concentration are also dependent on the ripening phase [67]. Ripen orange is characterised by the orange colour of the flavedo and genetic factors are responsible for the different intensity of colouration. The differences in colour of citrus peel are due to two main factors: a) carotenoid content in the peel and b) proportion of each carotenoid [67]. For example, the higher intense pigmentation of the oranges of Navel group when compared to 'Valencia Late' has been attributed to a higher ratio violaxanthin/β-citraurin in the last cultivar [68].

Alquézar *et al.*, [69] made an extensive review about content and composition of carotenoids in peel and pulp of mature fruits of selected oranges (*C. sinensis* L.

Osbeck), mandarin and mandarin hybrids, lemon (*C. limon* Burm F.), pummelo (*C. maxima* Osbeck), and grapefruit (*C. paradisi* Macf.) cultivars.

Holden and others [70] determined the major dietary carotenoid contents of selected citrus fruits and juice products (Table **3**).

Table 3. Major dietary carotenoid contents of selected citrus fruits and juice products.

Fruit description	Carotenoid concentration, µg/100 g fresh weight, edible portion			
	α-Carotene	β-Carotene	β-Cryptoxanthin	Lutein + Zeaxanthin
Grapefruit, raw, white	8	14	NA[a]	NA
Grapefruit, raw, pink and red	5	603	12	13
Orange, blood, raw	ND[a]	120	69	NA
Orange, raw, commercial varieties	16	51	122	187
Orange juice, raw	2	4	15	36
Orange juice, raw, hybrid varieties	8	39	324	105
Orange juice, frozen concentrate, unsweetened, diluted	2	24	99	138
Tangerines, raw (mandarin oranges)	14	71	485	243
Tangerine juice, raw	9	21	115	166
Tangerine juice, frozen concentrate, sweetened, undiluted	NA	227	2767	NA

Note: All values listed are µg of carotenoid/100 g fresh weight (edible portion), presented in weighted means.
[a]NA, data not available. ND, values not detected or below the detection limit.
Source: Holden and others [70].

In investigations of carotenoid-accumulating citrus mutants, some biological processes associated with carbohydrate metabolism and oxidative stress were found to differ from the wild types [71, 72].

Kato *et al.*, [73] identify differences among satsuma mandarin (*C. unshiu* Marc.), 'Valencia Late' orange (*C. sinensis* Osbeck), and 'Lisbon' lemon (*C. limon* Burm.f.) in the profiles of gene expression of carotenoid biosynthetic enzymes. Lycopene appears only in the pigmented grapefruits and pummelos and in the navel orange 'Caracara'.

To understand the origin of the diversity of carotenoid compositions of citrus fruit, Fanciullino *et al.*, [74] analyzed the carotenoid compositions of fruits from 25 genotypes that belong to 8 cultivated *Citrus* species. Statistical analyses showed a strong impact of genotype on carotenoid composition. The 25 citrus genotypes presented different carotenoid profiles with 25 distinct compounds isolated. The interspecific variability in carotenoid compositions was higher than intraspecific variability. Two carotenoids, *cis*-violaxanthin and β-cryptoxanthin, strongly determined the classification on qualitative data. These results point to the idea that, as for other phenotypical traits, the general evolution of cultivated citrus is the main factor of the organization of carotenoid diversity among citrus cultivars [74].

Matsumoto *et al.*, [75] analyzed the seasonal changes of carotenoids in the flavedo and juice vesicles of 39 citrus cultivars. Concerning the carotenoid content in flavedo, the 39 cultivars were classified into 5 clusters, in which the carotenoid profiles were carotenoid-poor, phytoene-abundant, violaxanthin-abundant, violaxanthin- and β-cryptoxanthin-abundant, and phytoene-, violaxanthin-, and β-cryptoxanthin-abundant. Regarding the content of carotenoids in the juice vesicles, the 39 cultivars were classified into 4 clusters, in which the carotenoid profiles were carotenoid-poor, violaxanthin-abundant, violaxanthin- and phytoene-abundant, and violaxanthin-, phytoene-, and β-cryptoxanthin-abundant. In flavedo, many citrus cultivars, except for the carotenoid-poor and phytoene-abundant groups, massively accumulated β,ε-carotenoids (*e.g.*, lutein), β,β-carotenoids (*e.g.*, β-cryptoxanthin and violaxanthin), and phytoene, in that order. In juice vesicles, the accumulation order among β,β-carotenoids was observed. Violaxanthin accumulation preceded β-cryptoxanthin accumulation in violaxanthin-, phytoene-, and β-cryptoxanthin-abundant cultivars.

Xu *et al.*, [61], studied the differences in the carotenoid content of ordinary "non-pigmented" citrus and lycopene-accumulating ("pigmented") mutants. They observed accumulation of lycopene and β-carotene in red mutant citrus, except for the peel of 'Cara Cara', the only cultivar of sweet orange that produces fruit with navel and pigmented. Additionally, phytoene accumulated in all tissues except for the peel of 'Chuzhou Early Red' pummelo. No obvious change in the total content of xanthophylls was observed in the 'Cara Cara' navel pigmented orange. Non-pigmented grapefruit ('Marsh') tissues and pummelo ('Yuhuan') juice vesicles were almost devoid of carotenoids, but in pigmented cultivars, the content of total carotenoids dramatically increased up to 790-fold. These results suggest that the underlying mechanisms for the mutations might be different in 'Cara Cara' sweet orange and in 'Chuzhou Early Red' pummelo.

Effect of Environmental Factors and Cultural Practices

Citrus fruit composition varies significantly due to effects from rootstock, cultivar, fruit size, maturity, storage, horticultural conditions, and climate, suggesting that nutrient and constituent analysis of the fruits only give important information about the fruits analyzed, but any estimates and general conclusions are difficult to make [76].

According to Alquézar *et al.*, [69], carotenoid synthesis is majorly influenced by environmental, nutritional and hormonal factors. High accumulation of lycopene in grapefruits growing under warm climatic conditions or the inability to develop full color of tropical-grown fruits, are two classical examples of the climatic influence in citrus fruit color.

Carotenoid biosynthesis and accumulation often occur in parallel with the ripening process [73, 77]. It is quite certain that the ripening process involves cellular activities related to metabolic networks and to organelle modification, which eventually determines product quality.

Chromoplast formation is one of the most important cellular changes during the ripening of carotenoid-rich plant tissues; it involves significant carotenoid sequestration and the use of other metabolic pathways, which are all essential for the nutritional and sensory quality of agricultural products [78].

The process of external ripening of citrus fruit consists of the conversion of chloro- to chromoplasts and involves the progressive loss of chlorophylls and the gain of carotenoids, thus changing peel color from green to orange [79]. The decline in rind chlorophyll can last for several months and the onset of carotenoid accumulation is almost coincident with the decrease of chlorophyll content [60]. In citrus, the understanding of the processes involved in the chromoplast biogenesis is of particular interest and agronomic importance, since unlike ripening processes of other fruit species, the chloro- to chromoplast conversion is reversible, even from fully differentiated chromoplasts [80].

The changes in carotenoids content during the ripening process in citrus fruits are influenced by environmental conditions, availability of nutrients and hormone levels [80, 81].

Temperature is one of the most important factors affecting synthesis and accumulation of carotenoids, influencing fruit coloration. In general, synthesis and accumulation of carotenoids is enhanced by low temperatures and inhibited by high temperatures [82].

FLAVONOIDS

Chemistry

Four types of flavonoids can be found in citrus species: flavanones, flavones, flavonols and anthocyanins. This last group of polyphenols only occurs in blood oranges. Flavanones are the most abundant citrus flavonoids [8].

Flavonoid aglycones (the forms lacking the sugar moieties) and their glycosyl derivatives are present in *Citrus* species. Glycosyl derivatives of flavonoids are more frequent in citrus juices because such compounds are more hydrophilic than the respective aglycones [9]. More than 60 flavonoids have been isolated and identified in *Citrus* species [8]. This relative high number of flavonoids in citrus species is due to the possible combinations between the polyhydroxylated aglycones and some mono- and disaccharides. The most common sugar moieties are D-glucose and L-rhamnose, generally bonded to the hydroxyl group at C-7 or at C-3 of aglycones. However, it is also possible to find C-glycosides in citrus fruit or juices [9].

Hesperetin, naringenin, isosakuranetin and eriodictiol are the flavanone aglycones present in citrus species, whereas taxifolin is the sole flavanol found in citrus juices [9]. All of these compounds are hydroxylated at C-5 and C-7 of ring A (Fig. 3). The differences among them result on the hydroxylation or methoxylation of ring B at C-3' and C-4' (Fig. 3). The flavanol taxifolin is the sole which is hydroxylated at C-3 (Fig. **3**).

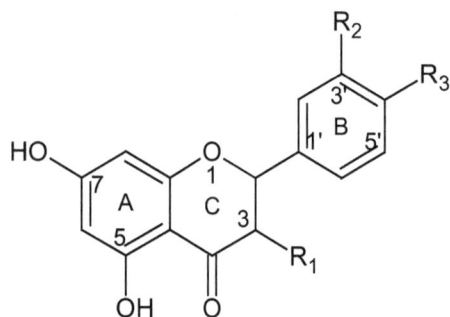

Compound name	R$_1$	R$_2$	R$_3$
Hesperetin	H	OH	OCH$_3$
Naringenin	H	H	OH
Isosakuranetin	H	H	OCH$_3$
Eriodictiol	H	OH	OH
Taxifolin	OH	OH	OH

Fig. (3). Flavanone aglycones.

Citrus flavanones may be as glycoside or aglycone forms, although usually present under diglycoside form, conferring the typical citrus taste to the fruits. Among the aglycone forms, hesperetin and naringenin are the most important flavanones, and among the glycoside forms, rutinosides and neohesperidosides prevailed. In neohesperidosides, the disaccharide rhamnosyl-α-1,2-glucose (neohesperidose) is attached to the aglycone form at hydroxyl of C-7 of naringenin, hesperetin, eriodictiol, and isosakuranetin giving arise to the glycosyl derivatives naringin, neohesperidin, neoeriocitrin, and poncirin, respectively (Fig. **4**). These heterosides have a bitter taste. In rutinosides, the aglycones hesperetin, naringenin, eriodictiol and isosakuranetin are attached through their hydroxyl groups of C-7 to the disaccharide rutinose (rhamnosyl-α-1,6-glucose), originating the heterosides hesperidin, narirutin, eriocitrin and didymin, respectively (Fig. **4**) [83].

Acacetin, isoscutellarein, luteolin, apigenin, diosmetin, and chrysoeriol are the flavone aglycones detected in citrus fruits, and kaempferol and quercetin the main flavonol aglycones present in these fruits (Fig. **5**) [9].

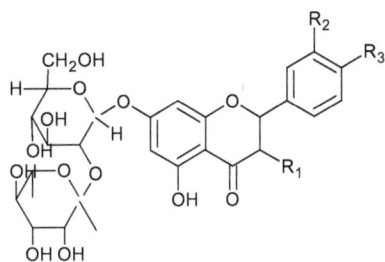

Compound name	R$_2$	R$_3$
Naringin	H	OH
Neohesperidin	OH	OCH$_3$
Neoeriocitrin	OH	OH
Poncirin	H	OCH$_3$

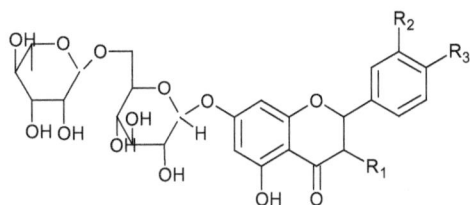

Compound name	R$_2$	R$_3$
Hesperidin	OH	OCH$_3$
Narirutin	H	OH
Eriocitrin	OH	OH
Didymin	H	OCH$_3$

Fig. (4). Flavanone neohesperidose and rutinoside forms.

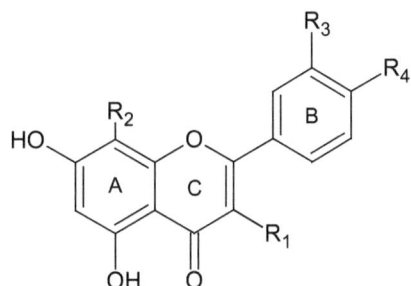

Compound name	R$_1$	R$_2$	R$_3$	R$_4$
Acacetin	H	H	H	OH$_3$
Isoscutellarein	H	OH	H	OH
Luteolin	H	H	OH	OH
Apigenin	H	H	H	OH
Diosmetin	H	H	OH	OCH$_3$
Chrysoeriol	H	H	OCH$_3$	OH
Kampferol	OH	H	H	OH
Quercetin	OH	H	OH	OH

Fig. (5). Flavones and flavonols aglycones.

7-*O*-Rutinosides or 7-*O*-neohesperidosides are the main flavone *O*-glycosides found in citrus juices, with the exception of rutin which is a 3-*O*-rutinoside [9]. *C*-diglucosides have been also found in appreciable amounts of citrus juices, along with smaller amounts of mono-*C*-glucosides, in C-6 and/or C-8 positions [9]. Lucenin-2, vicenin-2 and stellarin-2 are examples of flavone 6,8-di-*C*-glucosides of luteolin, apigenin, and chrysoeriol, respectively. An example of a mono-*C*-substituted flavone is isovitexin which is apigenin 6-*C*-glucoside [9].

Rhoifolin and isorhoifolin are examples of flavone *O*-glycosides (apigenin 7-*O*-neohesperidoside and apigenin 7-*O*-rutinoside, respectively) detected in citrus juices [9]. Nevertheless much more flavone glycosides can be found in citrus juice (Fig. **6** and Table **4**).

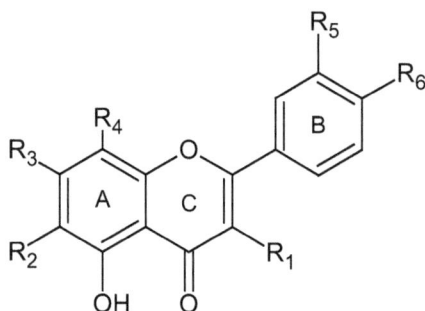

Fig. (6). Flavone-C-glucosides, flavone-O-glycosides and flavonol-O-glycosides.

Table 4. Flavone-C-glucosides, flavone-O-glycosides and flavonol-O-glycosides.

Compound name	R_1	R_2	R_3	R_4	R_5	R_6
Luteolin 6,8-di-*C*-glucoside (Lucenin-2)	H	Glu	OH	Glu	OH	OH
Apigenin 6,8-di-*C*-glucoside (Vicenin-2)	H	Glu	OH	Glu	H	OH
Chrysoeriol 6,8-di-*C*-glucoside (Stellarin-2)	H	Glu	OH	Glu	OCH_3	OH
Diosmetin 6,8-di-*C*-glucoside (Lucenin-2,4'-methyl ether)	H	Glu	OH	Glu	OH	OCH_3
Apigenin 7-*O*-neohesperidoside-4'-glucoside (Rhoifolin 4'-glucoside)	H	H	*O*-Nh	H	OH	*O*-Glu
Chrysoeriol 7-*O*-neohesperidoside-4'-glucoside	H	H	*O*-Nh	H	OCH_3	OH
Apigenin 6-*C*-glucoside (Isovitexin)	H	Glu	*OH*	H	H	OH
Luteolin 7-*O*-rutinoside	H	H	*O*-Ru	H	OH	OH
Chrysoeriol 8-*C*-glucoside (Scoparin)	H	H	OH	Glu	OCH_3	OH
Diosmetin 8-*C*-glucoside (Orientin 4'-methyl ether)	H	H	OH	Glu	OH	OCH_3
Quercetin 3-*O*-rutinoside (Rutin)	*O*-Ru	H	OH	H	OH	OH
Apigenin 7-*O*-neohesperidoside (Rhoifolin)	H	H	*O*-Nh	H	OH	OH
Apigenin 7-*O*-rutinoside (Isohoifolin)	H	H	*O*-Ru	H	OH	OH
Chrysoeriol 7-*O*-neohesperidoside	H	H	*O*-Nh	H	OCH_3	OH
Diosmetin 7-*O*-rutinoside (Diosmin)	H	H	*O*-Ru	H	OH	OCH_3
Diosmetin 7-*O*-neohesperidoside (Neodiosmin)	H	H	*O*-Nh	H	OH	OCH_3

Glu: Glucose; Nh: Neohesperidose; Ru: Rutinose

Polymethoxyflavones can be found in the essential oils present in the peel of the citrus fruits as well as in commercial juices because they are contaminated with peel constituents, in contrast to those juices obtained by hand-squeezing in which only traces may be detected [9, 84]. These components are concentrated in the skin of unripe fruit and are the constituents of bitter citrus oils [85]. Quercetogetin, nobiletin, tetra-*O*-methylisoscutellarein, heptamethoxyflavone, nobiletin, and tangeretin are examples of polymethoxyflavones detected in orange juices [84].

A review made by Hwang *et al.*, [86] cited that the rutinosides such as hesperidin, eriocitrin, narirutin, diosmin, and isorhoifolin are mainly found in oranges (*C. sinensis*), mandarins (*C. reticulata* L.), and lemons (*C. limon* L.), whereas neohesperidosides such as neohesperidin, naringin, neodiosmin, and neoeriocitrin are found in grapefruit hybrids (*C. paradisi* Macf.) and pummelos (*C. maxima* L.). The polymethoxyflavones such as tangeretin, natsudaidain, nobiletin, and heptamethoxyflavone are abundant in oranges, mandarins, and lemons, but less abundant in some grapefruits.

Fig. (**3**) depicts the flavonoids present in juices of diverse species and varieties of citrus according to the compilation made by Gattuso *et al.*, [9]. According to this document, several juices did not present polymethoxyflavones, such as *C. clementina*, *C. deliciosa*, *C. limon* and *C. bergamia* juices. *C. reticulata* juice was the sole which flavones or flavonoids were not reported.

Anthocyanins can also be found in some citrus fruits conferring them a red colour. The presence of anthocyanins is restricted to a specific group of sweet oranges (*C. sinensis*), denominated as "blood oranges" or "red oranges" and their hybrids. These pigments can accumulate in the flesh as well as in the peel [67]. 'Tarocco', 'Moro' and 'Sanguinelli' are cultivars of red or blood orange, due to the presence of anthocyanins in both the rind and fruit juice vesicles [87]. Previous studies demonstrated that the two major anthocyanins of blood oranges are cyanidin 3-glucoside and cyanidin 3-(6''-malonylglucoside) [88 - 90]. However, the amount and composition of anthocyanins in the blood orange cultivars may vary depending on several factors including the variety, maturity, region of production, environmental factors, mainly temperature, and storage conditions [91 - 93]. Low storage temperatures promoted the production of anthocyanins in blood oranges as well as low night temperatures [67, 91, 92]. The production of anthocyanins by blood orange varieties requires a wide day-night thermal range [94].

Piero [93] made a review upon the structural diversity of anthocyanins in red oranges. In such compilation, beyond the main anthocyanins reported above cyanidin 3-glucoside and cyanidin 3-(6''-malonylglucoside) the authors also cited delphinidin 3-glucoside, petunidin 3-glucoside, pelargonidin 3-glucoside, cyanidin 3-(acetyl)glucoside, some of them found as hydroxycinnamic acids derivatives of cyanidin 3-glucoside cyanidin 3-(ferulyl)glucoside, cyanidin 3-(coumarylferulyl)-glucoside, cyanidin 3-(sinapyl)glucoside, peonidin 3-(coumaryl) glucoside as well as the diglucosylated forms (3,5) of delphinidin, cyanidin, petunidin, and pelargonidin as constituting the anthocyanin fraction of red oranges. As minor anthocyanins, the author also refers peonidin 3-glucoside, cyanidin 3-rutinoside, and the 6''-malonylglucose esters of delphinidin, peonidin, and petunidin. During prolonged storage of the red orange juice, there are direct reactions between anthocyanins and hydroxycinnamic acids giving rise to 4-vinylphenol, 4-vinylcatechol, 4-vinylguaiacol, and 4-vinylsyringol adducts of cyanidin 3-glucoside. Cyanidin 3-(3''-malonyl) glucoside and cyanidin 3-(6''-dioxalyl) glucoside were also reported by Shao-Quian *et al.*, [95] for red oranges.

Biological Properties

Several biological properties have been attributed to flavonoids: antioxidant activity, free radical scavenging capacity, coronary heart disease prevention, hepatoprotective, anti-inflammatory, and anticancer activities. Antiviral activity has also been reported for flavonoids [96].

The beneficial effect of citrus fruits was reported in 1936 by Rusznyak and Szent-Gyorgyi [97]. According to the authors, lemon fruit had favourable effects on capillary permeability and fragility, along with ascorbic acid.

The effectiveness of the citrus flavonoids in infections reported by Biskind and Martin [98] was due to their ability to restore to normal impaired capillary permeability and fragility. Some of these properties might be attributed to the antioxidant activity of some flavonoids, particularly to eriocitrin which possessed the highest antioxidant activity among the remaining flavonoids present in lemon and lime fruits. However and although this evidence reported in several works [99, 100], Miyake *et al.*, [97] draw attention for the metabolization of these compounds after swallowing food, not only by the host (*e.g.* liver) but also by intestinal microorganisms, which can change the biological properties of the parent compound. Whether eriocitrin is a good antioxidant, the same cannot be reported for naringenin, according to the results described by Abeysinghe *et al.*, [26]. In this case, the antioxidant ability of ascorbic acid superposes to that of naringenin. A correlation between ascorbic acid and antioxidant activity was also reported by Miguel *et al.*, [101], and not between this activity and the levels of flavanones (hesperidin + narirutin) in juices of different cultivars of oranges and mandarins collected in multiple orchards of Algarve (Portugal).

Citrus flavonoids have been reported as having capacity to inhibit lipid peroxidation and to scavenge benzoylperoxide radicals, methyl methacrylate radicals, 2,2-diphenyl-1-picrylhydrazyl (DPPH), alkyperoxyl radicals and peroxynitrite, neverthtlees such ability is highly dependent on their structural chemistry. For example, kaempferol, luteolin, rutin, scutellarein and neoeriocitrin exert strong antioxidant activities, while hesperidin, hesperetin, neohesperidin, naringenin, and naringin exhibit moderate antioxidant activities. Generally the aglycones present higher activity than the respective glycosides [86, 102].

The antioxidant activity has also been attributed to the anthocyanins present in red oranges. The antioxidant activity of several fresh orange juices, obtained from different Italian blood oranges ('Moro', 'Sanguinello' and 'Tarocco') strongly correlated with the anthocyanin content and only a low correlation was detected between the amounts of ascorbic acid and antioxidant activity [103]. This finding was even corroborated by other authors with commercial juices of red oranges in which long shelf life leads to a diminution of anthocyanin content. Such decrease was accompanied by a remarkable loss of antioxidant activity [104]. The protective effect of blood orange juice on mononuclear blood cell against oxidative DNA damage was due to several constituents, including anthocyanins [105].

A standardized extract of blood orange juice was demonstrated to inhibit proliferation of fibroblast and epithelial prostate cells, without cytotoxic effect, so it may be useful in the management of benign prostatic hyperplasia [106].

Oxidative stress and inflammation are associated because the former can activate diverse transcription factors which can lead to the expression of a large number of diverse genes, including those for proinflammatory proteins. All of these cascades of events trigger various life-threatening diseases such as cancer, neurodegenerative diseases, and cardiovascular diseases [107]. Hesperidin with capacity for scavenging free radicals has also capacity for augmenting the antioxidant cellular defences *via* the ERK/Nrf2 signaling pathway [107]. Therefore, hesperidin and hesperetin exert their antioxidant activity *via* two ways: direct radical scavenging and increasing cellular antioxidant defence [107]. With the direct radical scavenging effect of those citroflavonoids, they are able to protect DNA and proteins (and whole tissue) against damage that is caused by intrinsic (*e.g.* oncogenes) and extrinsic (*e.g.* radiation, inflammation and toxin) factors [107]. The cellular antioxidant defence induced by hesperidin and hesperetin occur through the activation of the gene expression of Nrf2 and ERK1/2 which led to the upregulation of HO-1 expression, and consequently to a decline in intracellular pro-oxidants and an increment in bilirubin as an internal antioxidant. Nrf2 also increases the level of antioxidant enzymes (catalase, superoxide dismutase, glutathione reductase and glutathione (GSH) peroxide S-transferase and the GSH/GSSG ratio as well) [86, 107]. Hesperetin also protects cortical neurons from oxidative damage by activating prosurvival Akt and ERK1/2 signaling pathways, which inhibits the activation of proapoptotic proteins, such as apoptosis signal-regulating kinase-1 (ASK-1), Bad, caspase-9, and caspase-3 [86]. However, the neuroprotective effect of hesperetin can also be attributed to the activation of the estrogen receptor, because physiological levels of estrogen have neuroprotective effects through both estrogen receptor and tyrosine kinase receptor (Trks)-mediated signalling [86]. Contrasting with conventional anti-cancer drugs, hesperidin and hesperetin inhibit tumor development by targeting multiple cellular protein targets at the same time, including caspases, B-cell lymphoma 2 (Bcl-2) and Bax (Bcl-2 associated X protein) for the induction of apoptosis, and COX-2 (cyclooxygenase-2), MMP-2 (matrix metalloproteinase-2) and MMP-9 for the inhibition of angiogenesis and metastasis

[108]. Tumor angiogenesis is very important in the development and metastatic spread of tumors. Citrus flavonoids have showed that are able to inhibit several key events in the angiogenic processes, including multiplication and migration of endothelial cells and vascular smooth muscle cells and the expression of two main proangiogenic factors: vascular endothelial growth factor (VEGF) and matrix metalloproteinase-2 (MMP-2) [109].

The signalling actions of citrus flavonoids (hesperetin, isorhamnetin and isosakuranetin) in oxidatively stressed PC12 cells revealed that they differentially activated Akt/protein kinase B, p38 mitogen-activated protein kinase, and inhibited the activation of c-jun N-terminal kinase, which triggers apoptosis. Such signalling actions, which are involved in neuroprotection against oxidative stress were dependent on the chemical structure of the citrus flavonoids studied. In addition, with such results the authors also concluded that such molecules acted more as signalling molecules than as antioxidants [110].

Mandarin peel (*Citrus reticulata*) extract had potent anti-neuroinflammatory capacity evaluated by a lipopolysaccharide (LPS)-activated BV2 microglia culture system. The authors found that among the eight tested polymethoxy flavones and flavanone glycosides, only nobiletin displayed a capacity of > 50% to inhibit LPS-induced proinflammatory NO, TNF-α, IL-1b and IL-6 secretion, nevertheless the combination of hesperidin, nobiletin, and tangeretin showed higher activity [111].

Citrus polymethoxyflavones, such as tangeretin and nobiletin have anti-proliferative and antimutagenic properties. For example, tangeretin was demonstrated to be antimutagenic against all indirectly-acting mutagens tested; and nobiletin was antimutagenic against benzo[a] pyrene and 2-aminofluorene. These findings allow concluding that those two polymethoxyflavones might be important in the chemoprevention of cancer [112]. In addition, some authors showed that methoxylated flavones had higher antiproliferative effect in SCC-9 human oral squamous carcinoma cells than hydroxylated flavones [113].

Citrus flavonoids have demonstrated to possess anticancer activity against several human cancer cell lines, either alone or in combination either with other natural compounds, such as tocotrienols, or with synthetic ones (*e.g.* tamoxifen). In some *in vitro* studies, citrus flavonoids demonstrated to be effective either in estrogen receptor negative MDA-MB-435 or estrogen receptor-positive MCF-7 human breast cancer lines [114, 115]. Studies have also demonstrated that flavones generally possess greater activities than flavanones. In addition, glycosilation of these flavones removed their activity [112].

The mechanisms involved in the antiproliferative activity of flavonoids are mediated by the inhibition of some kinases involved in cell-cycle arrest and apoptosis. Such mechanisms are dependent on the structure of each flavonoid, showing Tripoli *et al.*, [83] that flavones and polymethoxylated flavones are more effective than flavanones due to the presence of a C_2-C_3 double bond. In addition, and within these groups, the planar structure and the two hydroxyl or methoxyl substituents in the A or B rings are essential for their potential anti-proliferative activity [36, 37].

Although the *in vitro* antiproliferative effect of the *Citrus* polymethoxyflavones, such as nobiletin and tangeritin, care must be taken into account when working *in vivo* conditions. When comparing the absorption and metabolism of nobiletin with those of luteolin in male rats, nobiletin showed a metabolic property distinct from that of luteolin, because of its wide distribution and accumulation in tissues. Walle [113] refers that compounds containing only one or two methoxy groups may be metabolically more stable than the polymethoxylated flavones.

In case of a moderate or high consumption of orange juice, favanones hesperidin and narirutin may represent an important part of the total polyphenols present in the plasma [116]. Hesperidin and hesperetin have been reported as possessing antiarrhythmic effects by prolonging the QTc interval in an experimental model, by improving both the infarct size and the edema, possibly through a reduction in the oxidative stress and inflammation. Some cardiac function biomarkers, such as aspartate aminotransferase, serum creatine kinase, and lactate dehydrogenase activities decrease significantly in diabetic rats in the presence of hesperidin [108]. Flavanones (especially hesperidin) protect against bone loss in senescent male rats [117, 118].

The reduction in the oxidative stress and inflammation with decreased concentrations of inflammatory markers, such as interleukins and tumour necrosis factor-α (TNF-α) may be influenced by a diet. In this context, orange juice consumption, due to the presence of hesperidin and naringenin, can play an important role in modulation of inflammatory markers. A review made by Coelho *et al.*, [119] reported that orange juice could prevent and treat some chronic diseases, although needing much more studies for corroborating such finding.

Such as for flavonoids, anthocyanins also present antioxidant, anti-inflammatory, antiproliferative and anticancer activities detected either *in vitro* or *in vivo* assays. The blood orange juice has an important antioxidant activity by modulating diverse antioxidant enzymes that counteract the oxidative damage. These effects may play a significant role in the etiology of several diseases, including atherosclerosis, diabetes, and cancer with the advantages that in the same juice there are flavonoids and anthocyanins, which may act synergistically [120].

Content of Flavonoids in Different Species and Cultivars

The 7-O-glycosylflavanones are the most abundant flavonoids in all citrus fruits (Tripoli *et al.*, [83] and references therein). Nevertheless, their amounts differ significantly depending on the species, variety (Table **5**) and part of the fruit [83, 121]. Among the neohesperidoside flavanones, neohesperidin, naringin and neoeriocitrin are mainly present in grapefruit, bergamot and bitter orange juices and less common in such species like *C. changshanensis* [26]. Among rutinoside flavanones, narirutin, hesperidin and didymin, are present in orange, mandarin, bergamot and lemon juices [9, 83]. Hesperidin accounted for 18.5–38.5% of the total phenolics in satsumas (*C. unshiu*), mandarins (*C. reticulata*), and oranges (*C. sinensis*), while naringin was not found in these species [26]. Highest total phenolics, total flavonoids, naringin, and TAC were found in segment membrane of *C. changshanensis* [26]. However, the content of each of these compounds varies with cultivar. The juices of the common late oranges 'Valencia Late' and 'Dom João' are relatively rich in hesperidin and poor in narirutin, compared with the juices of navel oranges, mandarins and tangors [101]. According to Bermejo *et al.*, [122], satsuma mandarin 'Owari' presented the highest amounts on the flavanone glycosides hesperidin and narirutin, as well as the highest amounts of the carotenoid and β cryptoxanthin, comparing with clementines ('Fina', 'Marisol' and 'Loretina'), navel oranges ('Navelina' and 'Navelate') and common orange ('Valencia Late'). Both studied mandarin and orange cultivars showed similar tendencies concerning to other phenolic compounds concentrations. The main flavanones detected in lemon juice are eriocitrin and hesperidin [123].

Table 5. Flavonoids found in Citrus juices (Gattuso *et al.*, [9] and references therein, [135])

Compound name	*C. sinensis*	*C. reticulata*	*C. clementina*	*C. deliciosa*	*C. aurantium*	*C. limon*	*C. aurantifolia*	*C. bergamia*	*C. paradisi*
Flavanones									
Didymin	+	+	-	-	-	-	-	-	+
Eriocitrin	+	+	-	-	-	++	+	+	+
Hesperidin	++	++	++	+	-	++	+	-	+
Naringin	-	-	+	-	+	+	-	++	++
Narirutin	+	+	+	+	-	-	-	-	+
Neoeriocitrin	+	+	-	-	-	-	+	++	+
Neohesperidin	-	-	-	-	+	+	-	++	+
Poncirin	+	-	-	-	+	-	-	++	+
Naringin-di-oxalate	-	-	-	-	-	-	-	+	-
Neoeriocitrin-di-oxalate	-	-	-	-	-	-	-	+	+
Flavones and flavonols									
6,8-di-*C*-Glu-Apigenin	+	-	+	+	-	+	-	+	-
6,8-di-*C*-Glu-Diosmetin	+	-	+	+	-	+	-	+	-
Rhoifolin	+	-	-	-	-	-	-	+	+
Isorhoifolin	+	-	-	-	-	-	-		-
Diosmin	+	-	+	-	+	+	+	+	-
Neodiosmin	+	-	-	-	-	-	-	+	-
7-*O*-Ru-Luteolin	+	-	-	-	-	+	-		-
6,8-di-*C*-Glu-Luteolin	-	-	-	-	-	-	-	+	-
6,8-di-*C*-Glu-Chrysoeriol	-	-	-	-	-	-	-	+	-
4'-*O*-Glu-Rhoifolin	-	-	-	-	-	-	-	+	-
7-*O*-Neohesp-4'--lu-Chrysoeriol	-	-	-	-	-	-	-	+	-
6-*C*-Glu-Apigenin	-	-	-	-	-	-	-	+	-
8-*C*-Glu-Chrysoeriol (Scoparin)	-	-	-	-	-	-	-	+	-
8-*C*-Glu-Diosmetin	-	-	-	-	-	-	-	+	-
7-*O*-Neohesp-Chrysoeriol	-	-	-	-	-	-	-	+	-
Rutin	-	-	-	-	-	-	-	-	+

(Table 5) contd.....

Compound name	C. sinensis	C. reticulata	C. clementina	C. deliciosa	C. aurantium	C. limon	C. aurantifolia	C. bergamia	C. paradisi
Polymethoxyflavones									
Heptamethoxyflavone	+	+	-	-	-	-	+	-	+
Nobiletin	+	+	-	-	+	-	+	-	+
Sinensetin	+	+	-	-	-	-	-	-	-
Tangeretin	+	+	-	-	+	-	+	-	+
Quercetogetin	+	+	-	-	-	-	-	-	-
Natsudaidain	-	-	-	-	-	-	+	-	-
Aglycones									
Taxifolin	+	-	-	-	-	-	+	-	+
Acacetin	+	+	-	-	-	-	-	-	-
Naringenin	+	-	-	-	-	-	-	-	+
Isoscutellarein	+	-	-	-	-	-	-	-	-
Kampferol	-	-	-	-	+	-	-	-	-
Luteolin	-	-	-	-	-	+	+	-	-
Quercetin									+
Hesperetin									+

Cultivars of clementine mandarin ('Fina', 'Nules', 'Arrufatina', 'Clemenpons', 'Hernandina', 'Loretina', 'Marisol', 'Oroval' 'Oronules' and 'Orogrande'), satsuma mandarin ('Owari', 'Okitsu', 'Avasa Pri-10' and 'Avasa Pri-19',), hybrid mandarin ('Fortune', 'Ortanique' and 'Nova'), common orange ('Salustiana' and 'Valencia Late'), navel orange ('Fukumoto', 'Lanelate', 'Navelate', 'Navelina', 'Navel Foyos', 'Newhall' and 'Powell') and pigmented orange ('Sanguinelli') groups have been analysed for its content in narirutin and hesperidin by Cano *et al.*, [24]. Sweet orange groups (common, navel and pigmented oranges) presented the higher concentration in flavonoids.

The bioactive constituents can contribute to taxonomy at the cultivar level. Dhuique-Mayer *et al.*, [30] reported the influence of species and cultivar on the content of the main antioxidant microconstituents (vitamin C, carotenoids and flavonoids) present in *Citrus* sp. Nogata *et al.*, [124] studied the concentrations of flavonoids in 42 species and cultivars of the *Citrus* genus and those in two *Fortunella* species and in *Poncirus trifoliata* according to Tanaka's system of citrus classification [125], finding coincidences in flavonoid composition within each section of this Tanaka's system, except for the species in the *Aurantium* section and those with a peculiar flavonoid composition such as bergamot (*C.*

bergamia Rissoand Poit.), sour orange (*C. aurantium* L.), 'Marsh' grapefruit (*C. paradisi* Macf.) and shunkokan (*C. shunkokan* Hort. ex Tan.). The *Aurantium* section included both hesperidin- and naringin-rich species [124].

Milella *et al.*, [126] reported for the first time the presence of rutin in clementine fruit juice. The results indicated that all chemical parameters statistically differentiated each cultivar. 'Etna' hybrid (*C. unshiu* × *C. clementina*) showed the highest narirutin content, SRA 89 cultivar the highest rutin and hesperidin content, and 'Esbal' cultivar (*C. clementina*) for naringin, hesperidin and rutin.

Auraptene, quercetin, umbelliferone, myricetin, catechin, epicatechin, vanillin, caffeic acid and ferulic acid can be considered good markers for the quality monitoring of the entire citrus production chain, from fresh fruits to the commercial beverages and to investigate possible product adulteration, sophistication or fraud in Italian Chinotto (*Citrus* × *myrtifolia*) [127].

Content of Flavonoids in Different Parts of the Plant/Fruit

Flavonoids are widely distributed in citrus fruits and most of them accumulate substantial quantities of flavonoids during the development of their different organs.

The peels are richer in flavonoids than are the seeds. For example, naringin is present in the peel of lemon or in the seeds of mandarin and bergamot, but not in the juice of lemon and never present in sweet orange juice. This compound is even a marker of adulteration of this juice [83].

The distribution of the flavanone heterosides in different parts of the fruits is also different. Only one example in lemons fruits: the lemon seed mainly contains hesperidin and eriocitrin, while the peel is rich in naringin, neoeriocitrin and neohesperidin [83]. Therefore, the peel of citrus fruits is a rich source of flavanones and many polymethoxylated flavones, which are very rare in other plant genera [128].

Abeysinghe *et al.*, [26] observed different flavonoids content and antioxidant capacity in different parts of fruit of four citrus species (*Citrus unshiu, C. reticulata, C. sinensis* and *C. changshanensis*). In segment membrane of all studied species, the content of phenolic compounds and the total antioxidant capacities were significantly higher than those in juice vesicles and segments. In segment membrane, however, a high contribution of hesperidin was observed in *C. unshiu* (54.0%), *C. sinensis* (46.7%), and *C. reticulata* (30.0%).

Gorinstein *et al.*, [129] reported total polyphenols contents for peeled lemons, oranges and grapefruits of 164±10.3; 154±10.2 and 135±10.1 and their peels 190±10.6; 179±10.5 and 155±10.3 mg/100 g, respectively. The content of total polyphenols in peeled lemons (*C. limon*) and their peels was significantly higher than in peeled oranges (*C. sinensis*) and grapefruits (*C. paradisi*) and their peels, respectively. The content of total polyphenols was significantly higher in the peels, compared with peeled fruits. Lemons (*C. limon*) possess the highest antioxidant potential among the studied citrus fruits studied by this author and are preferable for dietary prevention of cardiovascular and other diseases.

Singh *et al.*, [130], found phenols, vitamin C, carotenoids, flavonoids, hesperidin and naringin in *Citrus karna* peel extracts with maximum yield of (3.91% w/w).

Ghasemi *et al.*, [131], evaluated methanolic extracts of 13 commercially available *Citrus* species (*C. sinensis* 'Washington Navel', *C. sinensis* 'Sanguinello', *C. sinensis* 'Valencia late', *C. reticulata* 'Ponkan', *C. reticulata* 'Page', *C. unshiu* 'Sugiyama', *C. unshiu* 'Ishikawa', *C. limon*, *C. paradisi*, *C. aurantium* and *C. aurantium* 'Khosheii'). Peels and tissues growing in Iran were studied for their antioxidant activity by DPPH method. IC_{50} for antioxidant activity ranged from 0.6 to 3.8 mg ml^{-1}. Total phenolic content of the *Citrus* species samples (based on folin Ciocalteu method) varied from 66.5 to 396.8 mg gallic acid equivalent/g of extract and flavonoids content (based on colorimetric $AlCl_3$ method) varied from 0.3 to 31.1 mg quercetin equivalent/g of extract. The authors did not observe correlation between the total phenolic and/or flavonoids contents and antioxidant activity in peels and/or tissues.

Effect of Environmental Factors and Cultural Practices

According to Cano *et al.*, [24], there are some evidences of the effects of rootstock and environmental conditions on secondary metabolite production in citrus, although flavonoids content in fruits mainly depends on genetic characteristics of the cultivar.

Gil-Izquierdo *et al.*, [132] found that lemon trees 'Verna' grafted on sour orange (*C. aurantium* L.) rootstocks produced fruits with higher levels of flavonoids in the juice, compared with citrus trees grafted on *Citrus macrophyla* L. Regarding the individual flavonoids, the 6,8-di-*C*-glucosyl diosmetin was the most affected flavonoid by the rootstock used [132].

The use of an interstock (between the rootstock and the cultivar) did not increase the flavonoid content of the lemon juice, but 'Cleopatra' mandarin interstock (between 'Verna'lemon and *Citrus macrophylla* L. rootstock) decreased the phenolic content of the lemon juice [132].

The iron deficiency leads to a significant change in the phenol content in lemons. Total phenolics are higher (~ 33% average) in lemon juice from trees with iron deficiency and, in particular, flavanones, flavones and flavonols are affected in the same way [133].

Auxinic phytoregulators decreased the concentration of naringin and narirutin in the juice of 4 grapefruit cultivars ('Star Ruby', 'Marsh', 'Red Blush' and 'Shambars'). However, pulp and peel flavanones levels were mainly equal or exceed those of controls [134].

LIMONOIDS

Chemistry

Limonoids have been isolated especially from *Meliaceae* and *Rutaceae*, and much less frequently from *Cneoraceae* and *Simaroubaceae* [136]. These secondary metabolites have anticancer, antimicrobial, antiviral and antimalarial activities [137].

These compunds comprise a group of oxygenated tetracyclic triterpenes that are widely distributed in plants of the families *Rutaceae* and *Meliaceae*. In fact there are about 50 aglycones and 17 glycosides reported so far, of which about 36 aglycones were isolated from plants of the genus *Citrus* and related genera [138].

The aglycones are insoluble in water and are often responsible for the bitter taste of citrus juices, unlike glycosides which have no taste and are soluble in water. In fact, the presence of limonoid aglycones in concentrations above 6 ppm reduces the acceptability of citrus juices, which requires a reduction in the juice content through mixing, dilution or removal operations [139]. Content greater than 4 ppm classified citrus juice with lower quality, between 4 and 2 ppm as acceptable and less than 2 ppm as good quality [139].

Limonoids are modified triterpenes heavily oxygenated with 4,4,8-trimethyl-17-furanylsteroid as structural precursor [137]. Two main nucleus structures may occur in citrus limonoids: the structure of limonin with five rings; and the structure of nomilin with four rings (Fig. 7). All citrus limonoids have a furan ring attached to the D-ring, at C-17, an oxygen containing functional groups at C-3, C-4, C-7, C-16 and C-17 [136, 140]. This basic structure of intact limonoids is often referred as tetranortriterpenoids [141].

More than 50 limonoids aglycones and glucosides (44 aglycones and 18 glucosides) have been isolated, identified and quantified in different *Citrus* species: *C. aurantium*, *C. limon*, *C. reticulata*, *C. sudachi*, *C. unshiu*, *C. paradisi*, *C. sinensis*, *C. jambhiri*, and *C. pyriformis*, using diverse techniques (capillary electrophoresis, high-pressure liquid chromatography, liquid chromatography coupled to mass spectrometry, thin layer chromatography, and nuclear magnetic resonance). Limonoid glucosides are water-soluble and tasteless, whereas limonoid aglycones are water-insoluble and possessing a bitter taste [37, 136, 137, 142]. The glucosidic forms of limonoids are formed by rearrangement of D ring to accommodate a glucose moiety (Fig. 7) [37, 142]. The inter-conversion of aglycones to glucosides is carried out by uridine diphosphoglucose-limonoid glycosyl transferase and limonoid D-ring lactone hydrolase and occurs during the ripening process of citrus fruits [143].

Limonin and nomilin are the most abundant bitter limonoids occurring in *Citrus* juices and seeds, but obacunoic acid, ichangin, deoxylimonoic acid, and nomilinic acid (Fig. 7) are other limonoid aglycones found in *Citrus* seeds [137, 139]. All citrus limonoid bitterness perception has a closed D-ring, a keto group at the C-7 position, and a C_{14}-C_{15} epoxide. For 7-member A-ring limonoids, an acetyl ester group at the C-1 position is also required [137, 144].

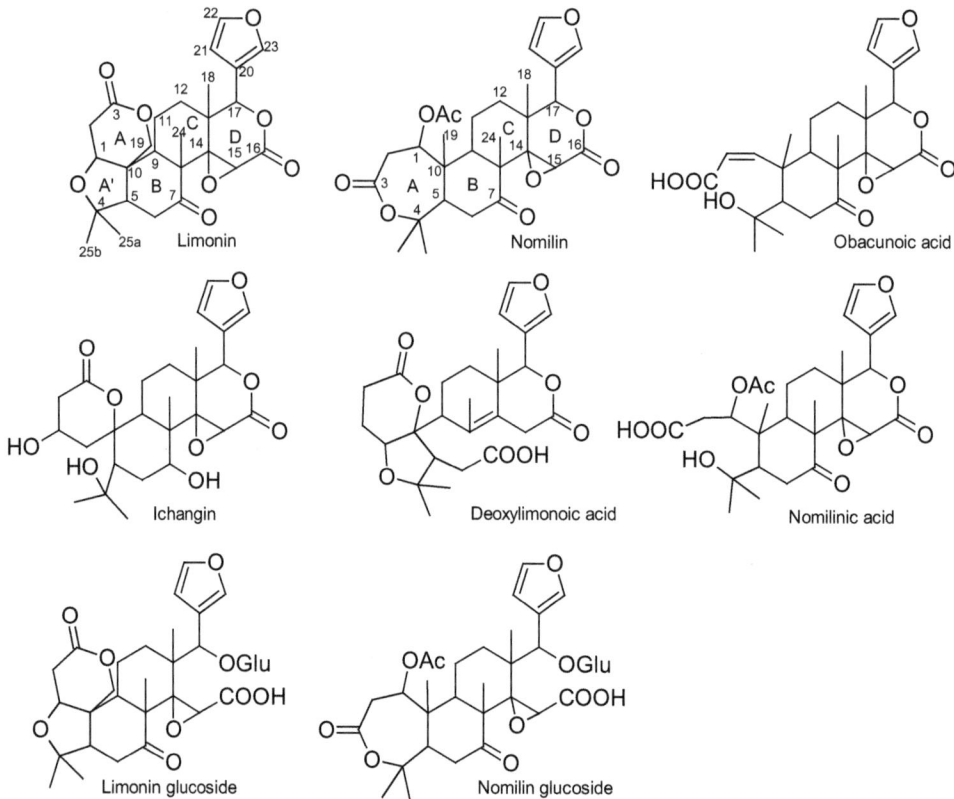

Fig. (7). Some citrus limonoids aglycones and limonoid heterosides.

The major limonoid in citrus juices is limonin glucoside, being twice the concentration respect other limonoid glucosides combined, whereas in citrus seeds predominates nomilin glucoside. Citrus seeds are the only natural reservoire of citrus limonoid aglycones, with a capacity of more than 1% of their fresh weight [137].

Citrus limonoids produce bitterness in juices at room temperature, lowering the quality and price of the commercial juice [137]. Physical disruption of the juice vesicles in the citrus fruit or freeze transforme the precursors of taste *e.g.* limonate A-ring lactone (LARL) or nomilinate A-ring lactone (NARL) to a bitter limonoid aglycone (limonin or nomilin, respectively). The development of these bitter limonoids is dependent on the pool of LARL or NARL [137].

Biological Properties

Citrus limonoids have gained importance due to their benefits on human health, particularly their role in the inhibition of several chronic diseases. Diverse cancer cell lines have been used for evaluating the antiproliferative activity of limonoids of various species and cultivars of citrus: important growth arrest of HT-29 colon cancer cells in the presence of relatively low concentrations of isolimonic and ichanexic acids from *C. aurantium* [145]; limonin, nomilin, deacetylnomilin obacunone and their glucosides isolated from *Citrus* fruits had particular capacity for inhibiting the growth of SH-SY5Y neuroblastoma, mainly glucoside ones which induced a more rapid cell death [143, 146]; limonexic acid, limonin and its glucoside, and limonexic acid from *C. aurantium* had inhibitory effect on the human pancreatic Panc-28 cells growth [147] as well as obacunone [148]; limonin, nomilin, obacunone and nomilinic acid glucosides from *C. reticulata* Blanco were able to inhibit the breast cancer MCF-7 cells [149] as well as obacunone and its glucoside derivative obtained from lemon seeds [150]. Some of these studies suggest that apoptosis mediated antiproliferative action is the major mechanism behind limonoid's chemoprotective actions along with the down regulated expression of inflammatory molecules such as nuclear factor-kappa B (NF-κB) and cyclooxigenase-2 (COX-2) [37, 148, 150].

Limonoids have also revealed to be effective to reverse the resistance of some cancer cell lines to some general anticancer drugs, including doxorubicin. P-glycoprotein (P-gp) is a membrane transporter which is encoded by the multidrug resistant 1 gene (*MDR1* gene) in human cells that mediates drug efflux from cells and plays an important role in causing multidrug resistance. This is one mechanism that explains the failure of chemotherapy in cancer treatment. Some authors found that citrus limonin was a potent P-glycoprotein inhibitor when applied at a non-toxic concentration of 20 microM, because significantly increased doxorubicin cytotoxicity 2.98-fold and 2.2-fold in Caco-2 and CEM/ADR5000 cells (multidrug-resistant human leukaemia cell line), respectively [151].

In addition, limonin, its glucoside derivative and other citrus limonoids have the capacity for inducing detoxification enzymes of phase II, including glutathione-S-transferase (GST) and NADH:QR (quinone reductase) [152, 153]. These enzymes are involved in the detoxification of toxic xenobiotics, metabolite/carcinogens. The induction of these enzymes by citrus limonoids is indicative of a possible mechanism for the cancer prevention by aiding in the detoxification of xenobiotics [152].

In vivo assays for evaluating the anticancer activity of citrus limonoids have also been performed and for that, mouse and hamster model systems have been used [154 - 158]. These studies have shown the potential of citrus limonoids as inhibitors of chemically induced carcinogenesis.

Beyond the anticancer activity of citrus limonoids, these secondary components can also present immunomodulatory activity [159], antiviral activity [160], inhibition of biofilm formation [161], hypocholesterolemic activity [162]. Antioxidant activity has been reported for citrus limonoids [146, 163, 164], nevertheless other studies using purified limonoids did not reveal such ability [165], which is consistent with limonoid chemical structure (lack of stabilizing conjugated unsaturation and electron delocalization potential) [133].

Nevertheless, more recently, some authors reported that limonin had anti-inflammatory and antioxidant activities after inducing hepatic ischemia reperfusion (I/R) injury in rats. The authors verified that limonin was able to ameliorate the deleterious effects of I/R. Limonin induced significant downregulation in the diverse elements of the toll like receptor (TLR) pathway and it also enhanced the anti-inflammatory cytokine IL-10 and decreased the activity of the apoptotic marker, caspase-3. Therefore, and according to the authors, the mechanism of the hepatoprotective effects seems to be related to the anti-inflammatory and antioxidant potential of limonin mediated by the downregulation of TLR-signaling pathway [166].

Limonoids have a wide range of biological activities, especially their potential antitumor effect. Limonoids inhibit chemically induced tumorigenesis in the mouth, stomach, small intestine, colon, lung, and skin of animals. Furthermore they inhibit the proliferation of breast cancer *in vitro*. Specifically obacunonae and limonene, the best known of citrus limonoids inhibit colon carcinogenesis in rats [167, 168].

Content of Limonoids in Different Parts of the Plant/Fruit

Citrus limonoids are present in citrus juice and tissues as water-soluble limonoid glucosides or in seeds as water-insoluble limonoid aglycones [169].

In grapefruits and pummelos, the limonoids are situated mainly in the albedo and segment membranes. In the flesh, segment membrane and albedo part of 'Rio Red' grapefruit, limonoate A-ring lactone (LARL) decrease, while limonin 17-β-D-glucopyranoside (LG) increase during late stages of fruit growth and maturation, suggesting that LARL is converted to LG as fruit maturation progresses. In the seeds, both LARL and LG increase steadily throughout fruit maturation [170].

In lemons, nomilin is synthesized from acetate in the stem, although epicotyl, hypocotyl and root tissues were also capable of biosynthesizing nomilin from acetate. Nevertheless, leaves, fruits and seeds were not capable of this synthesis. The plant was capable of translocating nomilin from the stem to other sites [171].

Content of Limonoids in Different Species and Cultivars

The limonoids are accumulated in significant quantities in grapefruit and pummelos. The segments of navel oranges membranes also have a high content of the precursor of limonin.

ESSENTIAL OILS

Chemistry

Citrus essential oils (CEO) obtained from fruits of diverse species of *Citrus*, are the most popular natural essential oils (EO) with wide application in food and beverages, cosmetics, pharmaceuticals, as natural flavors and fragrances [172]. CEO represents 2-3% of dry citrus peel. EO are complex matrices constituted up to hundreds of compounds with diverse structures and functions. The main constituent of these essential oils is limonene (monoterpene) although with different percentages depending on the varieties, seasonality, geographic origin, and fruit ripening [7, 173]. For example, (R)-(+)-limonene concentration in the essential oils may vary between 32% and 98%, depending on the cultivar: 32–45% in bergamot, 45–76% in lemon and 68–98% in sweet orange [174]. The fragrance of limonene differs depending on the isomer: (S)-(-)-limonene has a turpentine fragrance and (R)-(+)-limonene has an orange fragrance [175].

Essential oils of citrus fruits are obtained by mechanical processes (cold press), and not by hydrodistillation and steam distillation as done for obtaining the great majority of EO from plant origin, or "dry" distillation in the case of some woods [172]. Cold extraction process occurs in three steps: a) mechanical action for the rupture of the peel utricles and oil release; b) a stream of water to transport the EO; c) EO and water are separated by centrifugation. For the residue of the cold extraction it is still possible to obtain oils of lower grade by distillation [176].

Essential oils can be terpene-less, sesquiterpene-less, corrected, or deprived of a substance by partial removal, such as furocoumarines in CEO [172]. In this case, such reveals of great importance because CEO obtained by cold-extraction process present psoralens (furocoumarines) which bind to DNA under ultraviolet A light exposure producing mono- and biadducts which are cytotoxic and highly mutagenic [177].

In some cases, there is adulteration of CEO by sweet orange EO, which is the cheapest CEO. An example, is mandarin EO, characterized by the presence of α-sinensal, the major odour impact compound, the aromatic ester methyl N-methyl anthranilate and thymol, which may be adulterated with the addition of sweet orange EO, the addition of distilled mandarin oil or the addition of reconstituted mandarin oil to the natural cold-pressed one [178]. Other example is the addition of sweet orange EO to cold-pressed lemon oil [176].

Biological Properties

The physical, chemical and biological (*e.g.* antimicrobial) properties of CEO may almost be attributed to limonene due to its relative high percentage [173]. Limonene is used in the industry for producing carvone and α-terpineol, but it is also used in food products and cosmetics as a fragrance, and as a component in aromatherapy and industrial solvents [173, 177, 179]. In addition, CEO are also generally recognised as safe (GRAS) [180].

Although the importance of limonene and CEO in food and pharmaceutical industries and cosmetics, used as flavour and fragrance, the CEO and limonene also present some biological properties, including antimicrobial [180]; anti-inflammatory activity [181, 182]; antiangiogenic activity [183]; and also showed to reduce triacylglycerols and blood glucose levels in high fat diet fed C57BL/6 mice [184]. CEO and limonene were also reported as being able to inhibit colon cancer (SW480) cell proliferation [185], MCF-7 breast tumour cells [186], a lymphoma cell line [187], and THP-1 leukemia cell line [188], as well as acting against carcinogen-induced mammary tumours in rats [189].

The mechanisms that may be responsible for the anticancer activity of limonene and/or CEO include cell cycle arrest, alteration in signal transduction, apoptosis and inhibition of metastasis through suppression of VEGF expressions [190], as well as induction of phase II enzymes, activation of ERK and caspase dependent mitochondrial death pathways [37, 191, 192].

Content of Essential Oils in Different Parts of the Plant/Fruit

Citrus floral aroma is well known and used in perfumery. Hansen and Seufert [193] showed that the total terpenoid carbon emission of a flowering branch is 7.8 fold the emission of a non-flowering branch and that the presence of flowers is linked to a considerable emission of linalool. Submitting orange and clementine flowers to a hydrodistillation process, we can obtain yellowish oil in a yield, ranging from 0.05 to 0.08% (v/w) [194].

Studies of the floral orange aroma using solvent extracts, at 25 °C, demonstrated that linalool is the major component of orange (52%) and tangerine (75%) flowers, followed by sabinene (27 and 11% in orange and tangerine, respectively) [195]. Sabinene (35%), myrcene (19%), linalool (19%) and *trans*-ocimene (10%) were the main components of the solid-phase micro-extraction (SPME) of the whole flowers of Mediterranean mandarin (*C. deliciosa*) [196]. The petals seem to contribute largely to this composition since the petals volatiles include high amounts of all these compounds, whereas the other flower parts (gynoecia, stamens and pollen) show a quite different blend of compounds.

Content of Essential Oils in Different Species and Cultivars

A characterization of orange (*C. sinensis*) and clementine (*C. clementina*) flower essential oils showed that monoterpene fraction is the most abundant in both the orange and clementine oils (66-89% and 85-91%, respectively), the monoterpene hydrocarbons being the main components of this fraction (45-66% for orange flower and 50-69% for clementine flower oils) [194]. Both orange and clementine oils were dominated by sabinene (31-41% and 35-48%, respectively), linalool (15-32% and 17-29%, respectively) and limonene (4-10% and 6-10%, respectively). *trans*-Nerolidol was the major sesquiterpene component, attaining 3-10% in *C. sinensis* and 4-7% in *C. clementina* oils [194]. No significant difference were found between the composition of essential oils of the flower of clementine and common and navel orange trees [194].

COUMARINS

Chemistry

Coumarins are cinnamic acid-derived phenolic compounds composed of fused benzene and α-pyrone rings. They belong to the category of benzo-α-pyrone. Four subtypes of coumarins are reported: a) simple coumarins (hydroxylated, alkoxylated, alkylated and their glucosidic forms); b) furanocoumarins (coumarin ring attached to furan ring); c) pyranocoumarin (coumarin ring attached to pyrone ring); d) pyrone substituted coumarins (substitution on pyrone ring at 3C or 4C) [37].

Biological Properties

Auraptene and collinin are examples of citrus coumarins which are able to inhibit inflammation and obesity-related colon carcinogenesis [197]. These authors when feeding male *db/db* mice with auraptene and collinin at dose levels of 0.01% and 0.05% significantly decreased the incidence and multiplicity of colonic adenocarcinomas induced by azoxymetahene single intraperitoneal injection of 10 mg/kg body weight (bw) and dextran sodium sulphate (1% in drinking water). The authors concluded that anti-inflammatory activity of aurapene and collinin may contribute to the effects observed [197]. Auraptene also decreased the early phase preneoplastic lesions by decreasing the serum triglyceride levels in genetically altered C57BL/KsJ-*db/db* mice without any alteration in the wild mice [198]. These results along with those reported by Tanaka *et al.*, [197] suggest that auraptene can be an alternative to treat colon cancer in high risk patients, such as diabetic and obese patients [37]. Later on, studies made by Tanaka *et al.*, [199] found that auraptene included in β-cyclodextrins, when administered to male CD-1 (ICR) mice with colonic adenocarcinomas induced by azoxymetahene and dextran sodium sulphate also induced a suppressive action on colonic adenocarcinomas. Auraptene included in β-cyclodextrins inhibited colonic inflammation and also modulated proliferation, apoptosis and the expression of some proinflammatory cytokines, such as NF-κB, TNF-α, Stat3, NF-E2-related factor 2, interleukin (IL)-6 and IL-1β, which were induced in the adenocarcinomas [199].

The anti-inflammatory activity of auraptene was also reported recently by Niu *et al.*, [200]. These authors find that auraptene exhibited anti-inflammatory activity by inhibiting T cell proliferation and their inflammatory cytokine secretion after an investigation using CD3/CD28 activated lymphocytes isolated from C57BL/6 mice.

5-Geranyloxy-7-methoxycoumarin from *C. aurantifolia* was able to inhibit the growth of human colon cancer cells (SW480), which mechanism involved apoptosis [201].

The anticancer and antiproliferative activities of the antioxidant auraptene can also be attributed to its capacity to significantly increase the activities of detoxification (phase II) enzymes, such as glutathione-*S*-transferase and quinone reductase, in the colon and liver of rats. In addition, and at the same time, it has the capacity for inhibiting the expression of cell proliferation biomarkers in the colonic mucosal epithelium [202]. The elevation of those phase II enzymes by the action of auraptene may be responsible for the inhibition of the development of oral neoplasmas induced by 4-nitroquinoline-1-oxide [203].

POSTHARVEST AND INDUSTRIAL PROCESSING AS A CRITICAL PHASE FOR BIOACTIVE CONTENTS PRESERVATION IN CITRUS FRUIT

The growth and ripening stage at harvest are responsable for postharvest fruit nutritional quality [204].The ripening process involves some changes in sugars and acid content, firmness and colour, which determines the organoleptic and nutritional quality at the time of harvest. In addition, accumulation of phytochemicals (polyphenols, anthocyanins and carotenoids) and antioxidant activity occurs during on-tree ripening leading to increased functional properties. Over storage, total antioxidant activity and bioactive compounds show generally further increases, although in some cases losses of functional compounds are accompanied by losses of organoleptic quality [204].

Vitamin C of fruits and vegetables is largely affected by temperature management after harvest. High temperatures with large storage periods increases losses [205], being these more accentuated in some chilling sensitive crops.

Factors that contribute to fruit dehydration after harvest, bruising and other mechanical injuries, and trimming produce rapid losses of vitamin C. Nevertheless, irradiation at low doses (1 kGy or lower) did not show significant effects on vitamin C content in citrus fruits.

Control atmosphere storage can reduce the loss of vitamin C after harvest, but higher CO_2 levels can accelerate vitamin C loss [23]. According to Gil-Izquierdo *et al.*, [206], mild and standard juice pasteurization slightly increase total vitamin C content as a consequence of the orange solids parts, while concentration and freezing did not produce significant changes.

According to Burdurlu *et al.*, [207] ascorbic acid loss follow a first-order kinetic model with storage temperature, in citrus juice of orange, lemon, grapefruit and tangerine. Activation energy was determined in the range of 12.77 ± 0.97–25.39 ± 1.98 kcal mol^{-1}. Ascorbic acid retention after storage at 28, 37 and 45 °C was about 54.5–83.7%, 23.6–27% and 15.1–20.0%, respectively. The hydroxymethylfurfural (HMF) accumulation, one of the decomposition compounds of ascorbic acid degradation, fitted to a zero-order kinetic model and activation energy ranged from 43.41 ± 0.67 to 80.02 ± 0.07 kcal mol^{-1}. The authors observed a significant correlation between HMF accumulation and ascorbic acid loss for all storage temperatures studied in all citrus juice types.

Ascorbic acid, one of the major citrus vitamins, is also reported to be a marker for quality control of industrial processing and storage of citrus fruit [27, 208]. Orange juice produced by commercial squeezing (industrial FMC single-strength extraction) contains 25% more vitamin C and 22% more phenolics than hand domestic squeezing. [206]. Aerobic and anaerobic mechanisms demount vitamin C in processed products [27].

Vitamin C is also subject to degradation during processing and cooking. Electromagnetic energy reduces process times, energy, and water usage respect conventional heating, decreasing the loss of bioactive compounds. Blanching lower the content of vitamin C, but prevents more reductions during the frozen-storage of vegetables [23]. Frozen orange juice contains significantly smaller amounts of both soluble flavanones and insoluble chalcones [209]. Many authors studied the effect of industrial treatments to increase shelf-life of orange juice in the carotenoids content [210 - 214].

Citrus fraction of the fruit (juice, peel, pulp, mesocarp) and their by-products (peel, pulp and oil) are reach in different bioactive compounds and its maturity and the system used for juice extraction influence their composition and concentration. Álvarez *et al.*, [215] studied the effect of different juice extraction systems (plug inside fruit and rotating cylinders) on organoleptic, volatile flavor, and antioxidant quality of clementine juice. The juice extracted by the system plug inside presented higher hesperidin content and higher ORAC (Oxygen Radical Absorbance Capacity) and LDL (low-density lipoprotein) antioxidant activity. The juice organoleptic quality was not affected by the different extraction methods, but the chemical flavor profile presented significant differences. Juice extraction techniques affects chemical and functional quality of juice that must be taken into account for produce high-quality products.

Temperature and oxygen are the main factors responsible for vitamin C losses during juice processing. Fresh juices have higher levels of vitamin C than thermal processed juices, but economic factors affect the use of such methods in the citrus industry. Regarding the better packing material for preserve vitamin C in fruit juice, metal or glass containers are the best to ensure the stability of juice, whereas juice stored in plastic bottles losses more vitamin [19].

Stinco *et al.*, [216], study the impact of an industrial debittering process on nutritional and bioactive compounds in orange juice, observing that this process reduce ($p<0.001$) ascorbic acid in 26%, hydroxycinnamic acids in 32%, flavones in 28% and flavanones in 41%, while the carotenoid contents were not significantly affected by the treatment. However, the antioxidant activity of the hydrophilic fraction was significantly higher ($p<0.05$) in untreated juice than in debittered juices.

Aschoff *et al.*, [217] evaluated carotenoid, flavonoid, and vitamin C contents in fresh orange segments and a puree-like homogenate derived thereof, as well as freshly squeezed, flash-pasteurized, and pasteurized juices. Dejuicing slightly degraded Lutein and β-cryptoxanthin, whereas β-carotene levels were retained. Juice extraction did not change vitamin C levels, but decreased flavonoid levels 8-fold upon juice extraction, due to the removal of flavonoid-rich albedo and juice vesicles. The presence of such fibrous matrix compounds during *in vitro* digestion was assumed to significantly lower the total bioaccessibility of all carotenoids from fresh fruit segments (12%) as compared to juices (29–30%). The carotenoid bioaccessibility was not changed by mechanical disruption of orange segments prior to digestion, while pasteurization increased bioaccessibility by 9–11% of the freshly squeezed juice slightly.

Degreening is a common technique used in citrus, mainly mandarins, for change the color peel when the fruit is green outside and inside have reached the organoleptic maturity index, allowing early harvest and better prices.

Chaudhary *et al.*, [218], report for the first time a modulation of bioactive compounds by industrial degreening treatment. Freshly harvested 'Star Ruby' grapefruit (*Citrus paradisi* Macf.) were degreened in commercial packing shed during 60 h using 2 ppm of ethylene at a constant temperature of 20 °C and then stored at 10 °C for 21 days and later transferred to 20 °C for a period of 14 days to simulate shipment and retail store market conditions. Bioactive compounds such as carotenoids, flavonoids, limonoids and furocoumarins were analyzed. Nomilin was significantly higher ($P<0.05$) in degreened fruits at 35 days after storage. In contrast, flavonoids such as naringin, narirutin and ponciran were significantly ($P<0.05$) lower in degreened fruits at 35 days after storage. Degreening procedure reduced the levels of deacetylnomilinic acid glucoside and bergamottin after 35days of storage, although it had no significant effect on ascorbic acid, β-carotene, lycopene, limonin, neohesperidin, didymin, 6,7-dihydroxybergamottin, 5-geranyloxy-7-methoxycoumarin and radical scavenging activity. According to the authors, degreening technique could be used to enhance the grapefruit visual quality without significantly changes on nutritional quality.

Low temperature conditioning to reduce chilling injury (CI) in 'Star Ruby' grapefruit did not reduce quality and bioactive compounds [219]. Carotenoids and flavonoids were higher after 16 weeks in fruits stored at 11 °C. Storage at low temperature (2 °C) and low temperature conditioning treatment (7 days at 16 °C before cold storage at 2 °C) retained ascorbic acid for a longer period (12 weeks). Conditioned fruits had higher furocoumarins and better taste scores. Conditioning treatment can be utilized to reduce CI and to maintain taste and certain bioactive compounds of grapefruits during prolonged storage at low temperature.

Carvalho *et al.*, [220] observed an increase in total polyphenols and hesperidin content corroborated by the antioxidant activity (ORAC) in clementine juice after storage at 4 °C for 4 weeks plus 7 days at 20 °C. The cold storage did not change significantly the ascorbic acid content of juice. They also reported higher total polyphenols, hesperidin and antioxidant activity for fruit with lower maturity index (10 respect to 13).

Nowadays, food researchers are looking for alternatives to heat pasteurization to achieve high levels of food safety, preserving the nutritive quality of the products. Ultra high pressure homogenization (UHPH) is an emerging technology which has been proposed, for foods very sensitive to high temperatures, as an alternative to thermal pasteurization. UHPH consists on the dynamic application of high pressure allowing the steadily processing of fluid foods, and the equipments are designed to apply semi-industrial treatments at pressures above 350 MPa, with a better temperature control. UHPH treatments could provide minimally processed products preserving the fresh organoleptic and nutritive properties and as safe and lasting as the pasteurized ones.

Velázquez-Estrada *et al.*, [221], observed that UHPH (100, 200 and 300 MPa) preserved significantly more the vitamin C and polyphenols contents of juice respect to pasteurization. The UHPH also increased the content of flavanones of orange juice, more specifically the amounts of hesperidin, achieving its highest concentration on the samples treated at 200 and 300 MPa. The level of decrease of both l-ascorbic acid and carotenoids was significantly depended ($P<0.05$) on the pressure of the treatment and on the maximal temperature achieved during the treatment. The UHPH treatment did not affect the total polyphenol content and antioxidant capacity respect to fresh orange juice, while these contents where significantly reduced by juice pasteuriztion.

High pressure processing (HPP) controls microbial growth in grapefruit (*Citrus paradisi* Macfad) juice without significantly changes in the levels of health promoting compounds [222]. Thermal processing (TP) treatment showed significantly lower content of ascorbic acid as compared to control and HPP during 0 and 7 days of storage analysis. The HPP treatment had no significant ($P \leq 0.05$) effect on the levels of citric acid, flavonoids, limonoids and coumarins, in comparison to the control treatment. The levels of carotenoids were significantly ($P \leq 0.05$) lower among both processing treatments and control at 21 days after storage at 4 °C. The authors recommended the HPP technique for providing fresh like grapefruit juice without significantly changes in the health beneficial bioactive compounds levels.

Physiologically immature citrus fruits that are dropped have good potential as sources of different bioactive compounds and antioxidants and drying could be a good method to preserve these fruits. Sun *et al.*, [223] studied the effects of different drying methods (sun-drying, hot air-drying and freeze-drying) on phytochemical compounds and antioxidant activity of these class of citrus fruit. According to the authors freeze-drying is a good method for preserve phenolic compounds, synephrine and antioxidants, while hot air-drying is good for retaining flavonoids, and all three techniques can be used for keeping limonoids.

According to these results, the industrial production of high-quality extraction materials for bioactive compounds must take in account the drying technique for ensure high efficiency.

Carotenoid biosynthesis during postharvest in 'Cara Cara' fruits could be affected by temperature in a tissue-dependent way [224]. Storage at 20 °C rapidly increased the carotenoid content in the fruit peel, while the content remained unchanged in the pulp before 35 days of storage. Contrarily, storage at 4 °C preserved the carotenoid content in the peel before 35 days of storage, after which it slightly increased as time progressed. However, the content in the pulp increased gradually over the entire storage period. Among foods, citrus fruits are the most important source of β-cryptoxanthin. The best dietary sources are papaya, tangerine, and orange [225], but it is also found in red chilies, peaches, pumpkins [226] and guava [227]. β-cryptoxanthin bioavailability from these food sources is affected by the food matrix, processing, and storage state. The degradation of β-cryptoxanthin in its native form (before consumption) is caused by natural light and heat, which results in isomerization. Carotenoids could be more bioavailable for digestion by cooking and other thermal processing [228].

According to Escudero-López [229], juice fermentation could significantly increase total and individual flavanones and carotenoids, due to an enhanced extraction of these compounds from the pulp. A higher content of hesperetin-7-*O*-glucoside (2-fold higher at the end of the fermentation process) incresead the bioavailability of flavanone. Nevertheless, total phenolics decreased by fermentation process, while ascorbic acid did not undergo a significant change.

CONCLUDING REMARKS

Citrus fruits are rich in bioactive compounds, which all contribute to determine a high nutraceutical potential of these fruits. They have biological properties such as antioxidant, analgesic and anti-inflammatory effects, anti-fungal, anti-viral, anti-bacterial, and blood clot inhibition activities (inhibition of human platelet aggregation), up to anti-tumoral, anti-thrombotic and neuroprotective properties [230 - 238]. The intake of orange resulted also in reduced incidence of asthma [239] and can prevent coronary heart disease [240, 241] and breast cancer [242].

For all above mentioned, the consumption of citrus fruit is highly recommended. It should be taken into account the high levels of bioactive compounds in the membranes of the segments of the fruit [26], which means that we should consume all the edible portion of the fruit, rather than just consume the juice. The rind of the fruit, despite not being edible by its unpleasant taste, is rich in bioactive compounds and must be used in cooking and pastries. Aqueous extracts of fruit peel of *Citrus sinensis* (L.) showed anti typhoid activity [243] and could be used for the future therapeutic medicine [130].

Oranges, the citrus fruit most consumed worldwide, are the principal food sources of vitamin C, β-cryptoxanthin and zeaxanthin in the Mediterranean diet [244].

Vitamin C is the compound that longer is recognized as an important component of citrus fruits. Ascorbic acid is known by its importance in collagen synthesis. This vitamin is also a good antioxidant and it has also been reported to help in healing wounds and burns, promoting calcium absorption, lowering the risk of ischemic heart disease, and preventing cancer [18].

However, the positive effect of this vitamin in the prevention of cancer needs a combination with other vitamins, such as vitamin E and carotenoids (*e.g.* β-carotene), present in fruits and vegetables [18]. The prevention of cancer cannot be attributed to a sole compound and the respective amounts but to the combination of these compounds which are present in fruits and vegetables, particularly in citrus [18, 19].

Carotenoids, flavonoids, limonoids and coumarins only later they began to be considered important to human health. These compounds have been the subject of many studies over the past decades. Among these components, phenols, particularly flavonoids, are greatly studied due to their antioxidant, radical-scavenging activity and anti-inflammatory properties [102]. Due to these properties, consumption of these polyphenols is associated with lower risk of cardiovascular and degenerative diseases as well as different types of cancer [245]. Several works *in vitro* and *in vivo* have demonstrated that citrus limonoids have the capacity for inhibiting diverse types of cancer and reducing LDL cholesterol [18].

Essential oils are important for fruit flavour and fragrance, and therefore, with high value in food industry. Complex mixtures of citrus essential oil have antimicrobial and antioxidant activities. Due to these properties they can be used in food industry as natural additives [7].

Although many research already performed on citrus and the large number of compounds yet discovered, recent studies have revealed the presence of compounds, whose presence was unknown in these fruits. In 2004, two new flavonoids quercetin 3-*O*-rutinoside-7-*O*-glucoside and chrysoeriol 6,8-di-C-glucoside (stellarin-2) were detected in the lemon juice by HPLC-ESI/MS^n analyses [132]. The presence of 3-*O*-caffeoylferuoylquinic acid and two hydrated feruloylquinic acids in orange and the presence of 3,5-diferuoylquinic acid in grapefruit were detected in 2014 [246].

The richness of citrus fruits, in bioactive compounds makes these fruits promising for the development of promoting health drinks [247 - 249], general food-based neuroprotection and brain foods [86] and other health promoting foods.

The content of bioactive compounds depends on the species and cultivar, as well as the environmental conditions and cultural practices. The health concerns may cause in the future, we begin to prefer to use, and to plant, the richest cultivars in bioactive compounds. Furthermore, cultural practices can be directed not only to increase the productivity of orchards, but also to produce fruits with higher levels of bioactive compounds.

Many factors must be taken in account since preharvest to achieve harvest quality and high levels of bioactive compounds in the fruit. Postharvest and industrial processing are critical for preserve bioactive contents in citrus fruit. Factors as temperature, light, oxygen concentration, time and others are crucial to maintain the initial quality and functionality of citrus products. By other side, the concentration of these compounds and its functionality in postharvest can also be increased by some factors or techniques that induces a stress response in fruit.

CONFLICT OF INTEREST

The author confirms that author has no conflict of interest to declare for this publication.

ACKNOWLEDGEMENTS

Declared none.

REFERENCES

[1]   Lembke J. Virgil's Georgics. Yale University Press 2006.

[2]   Zargaran A, Zarshenas MM, Karimi A, Yarmohammadi H, Borhani-Haghighi A. Management of stroke as described by Ibn Sina (Avicenna) in the Canon of Medicine. Int J Cardiol 2013; 169(4): 233-7.
        [http://dx.doi.org/10.1016/j.ijcard.2013.08.115] [PMID: 24063916]

[3]   Mizrahi A, Knekt P, Montonen J, Laaksonen MA, Heliövaara M, Järvinen R. Plant foods and the risk of cerebrovascular diseases: a potential protection of fruit consumption. Br J Nutr 2009; 102(7): 1075-83.
        [http://dx.doi.org/10.1017/S0007114509359097] [PMID: 19646291]

[4]   Frada J. Contributos Portugueses do Período Expansionista e da Época Colonial para as Ciências Médicas. Acta Pediátrica Portuguesa 1996; 4(27): 721-4.

[5]   Liu YQ, Heying E, Tanumihardjo SA. History, global distribution, and nutritional importance of citrus fruits. Compr Rev Food Sci Food Saf 2012; 11(6): 530-45.
        [http://dx.doi.org/10.1111/j.1541-4337.2012.00201.x]

[6]   Kitts DD. Bioactive substances in food: identification and potential uses. Can J Physiol Pharmacol 1994; 72(4): 423-34.
        [http://dx.doi.org/10.1139/y94-062] [PMID: 7922875]

[7]   Ledesma-Escobar CA, de Castro MD. Towards a comprehensive exploitation of citrus. Trends Food Sci Technol 2014; 39: 63-75.
        [http://dx.doi.org/10.1016/j.tifs.2014.07.002]

[8]   Benavente-García O, Castillo J, Marin FR, Ortuño A, del Rio JA. Uses and properties of *Citrus* flavonoids. J Agric Food Chem 1997; 45: 4505-15.
        [http://dx.doi.org/10.1021/jf970373s] [PMID: 18593176]

[9]   Gattuso G, Barreca D, Gargiulli C, Leuzzi U, Caristi C. Flavonoid composition of *Citrus* juices. Molecules 2007; 12(8): 1641-73.
        [http://dx.doi.org/10.3390/12081641] [PMID: 17960080]

[10]   Turner T, Burri BJ. Potential nutritional benefits of current citrus consumption. Agriculture 2013; 3: 170-87.
        [http://dx.doi.org/10.3390/agriculture3010170]

[11]   Du J, Cullen JJ, Buettner GR. Ascorbic acid: Chemistry, biology and the treatment of cancer. Biochimica et Biophysica Acta (BBA). Rev Can 2012; 1826(2): 443-57.

[12]   Bielski BH. Chemistry of ascorbic acid radicals. In: Tolbert BM, Ed. Ascorbic Acid: Chemistry, Metabolism, and Uses Washington DC. Seib, PA: American Chemical Society 1982.

[http://dx.doi.org/10.1021/ba-1982-0200.ch004]

[13]　Buettner GR. The pecking order of free radicals and antioxidants: lipid peroxidation, [alpha]-tocopherol, and ascorbate. Arch Biochem Biophys 1993; 300(2): 535-43.
[http://dx.doi.org/10.1006/abbi.1993.1074] [PMID: 8434935]

[14]　Buettner GR. In the absence of catalytic metals ascorbate does not autoxidize at pH 7: ascorbate as a test for catalytic metals. J Biochem Biophys Methods 1988; 16(1): 27-40.
[http://dx.doi.org/10.1016/0165-022X(88)90100-5] [PMID: 3135299]

[15]　Buettner GR, Jurkiewicz BA. Catalytic metals, ascorbate and free radicals: combinations to avoid. Radiat Res 1996; 145(5): 532-41.
[http://dx.doi.org/10.2307/3579271] [PMID: 8619018]

[16]　Nishikimi M, Fukuyama R, Minoshima S, Shimizu N, Yagi K. Cloning and chromosomal mapping of the human nonfunctional gene for L-gulono-gamma-lactone oxidase, the enzyme for L-ascorbic acid biosynthesis missing in man. J Biol Chem 1994; 269(18): 13685-8.
[PMID: 8175804]

[17]　Njoku PC, Ayuk AA, Okoye CV. Temperature effects on vitamin C content in citrus fruits. Pak J Nutr 2011; 10(12): 1168-9.
[http://dx.doi.org/10.3923/pjn.2011.1168.1169]

[18]　Patil BS, Jayaprakasha GK, Chidambara Murthy KN, Vikram A. Bioactive compounds: historical perspectives, opportunities, and challenges. J Agric Food Chem 2009; 57(18): 8142-60.
[http://dx.doi.org/10.1021/jf9000132] [PMID: 19719126]

[19]　Martí N, Mena P, Cánovas JA, Micol V, Saura D. Vitamin C and the role of citrus juices as functional food. Nat Prod Commun 2009; 4(5): 677-700.
[PMID: 19445318]

[20]　Mapson LW. The Vitamins. London, New York: Academic Press 1987; 1.

[21]　Shrestha R, Dhakal D, Gautum DK, Shrestha S. Variation of physiochemical components of acid lime (*Citrus aurantifolia* Swingle) Fruits at Different Sides of the Tree in Nepal. Am J Plant Sci 2012; 3(12): 1688-92.
[http://dx.doi.org/10.4236/ajps.2012.312206]

[22]　United States Department of Agriculture and United States Department of Health and Human Services. Dietary Guidelines for Americans 2010; 2011 [USDA/USDHH]

[23]　Lee SK, Kader AA. Preharvest and postharvest factors influencing vitamin C content of horticultural crops. Postharvest Biol Technol 2000; 20: 207-20.
[http://dx.doi.org/10.1016/S0925-5214(00)00133-2]

[24]　Cano A, Medina A, Bermejo A. Bioactive compounds in different citrus varieties. Discrimination among cultivars. J Food Compos Anal 2008; 21: 377-81.
[http://dx.doi.org/10.1016/j.jfca.2008.03.005]

[25]　Whitney E, Whitney EN, Rolfes SR. Understanding nutrition Belmont. Calif.: Wadsworth 2009.

[26]　Abeysinghe DC, Li X, Sun CD, Zhang WS, Zhou CH, Chen KS. Bioactive compounds and antioxidant capacities in different edible tissues of citrus fruit of four species. Food Chem 2007; 104: 1338-44.
[http://dx.doi.org/10.1016/j.foodchem.2007.01.047]

[27]   Nagy S. Vitamin C contents of citrus fruit and their products: a review. J Agric Food Chem 1980; 28(1): 8-18.
[http://dx.doi.org/10.1021/jf60227a026] [PMID: 7358939]

[28]   Duarte A, Caixeirinho D, Miguel MG, *et al.* Vitamin C content of citrus from conventional versus organic farming systems. Acta Hortic 2010; 868: 389-94.
[http://dx.doi.org/10.17660/ActaHortic.2010.868.52]

[29]   Dhuique-Mayer C, Caris-Veyrat C, Ollitrault P, Curk F, Amiot MJ. Varietal and interspecific influence on micronutrient contents in citrus from the Mediterranean area. J Agric Food Chem 2005; 53(6): 2140-5.
[http://dx.doi.org/10.1021/jf0402983] [PMID: 15769147]

[30]   Duarte A, Caixeirinho D, Miguel MG, *et al.* Organic acids concentration in citrus juice from conventional versus organic farming. Acta Hortic 2012; 933: 601-6.
[http://dx.doi.org/10.17660/ActaHortic.2012.933.78]

[31]   Yang XY, Xie JX, Wang FF, *et al.* Comparison of ascorbate metabolism in fruits of two citrus species with obvious difference in ascorbate content in pulp. J Plant Physiol 2011; 168(18): 2196-205.
[http://dx.doi.org/10.1016/j.jplph.2011.07.015] [PMID: 21925761]

[32]   Long WG, Sunday MB, Harding PL. Seasonal changes in Florida Murcott honey oranges. U.S. Department of Agriculture 1962.

[33]   Bermejo A, Cano A. Analysis of nutritional constituents in twenty citrus cultivars from the mediterranean area at different stages of ripening. Food Nutr Sci 2012; 3: 639-50.
[http://dx.doi.org/10.4236/fns.2012.35088]

[34]   Padayatty SJ, Katz A, Wang Y, *et al.* Vitamin C as an antioxidant: evaluation of its role in disease prevention. J Am Coll Nutr 2003; 22(1): 18-35.
[http://dx.doi.org/10.1080/07315724.2003.10719272] [PMID: 12569111]

[35]   Lado J, Alós E, Rodrigo MJ, Zacarías L. Light avoidance reduces ascorbic acid accumulation in the peel of Citrus fruit. Plant Sci 2015; 231: 138-47.
[http://dx.doi.org/10.1016/j.plantsci.2014.12.002] [PMID: 25575999]

[36]   Codoñer-Franch P, Valls-Bellés V. Citrus as functional foods. Curr Top Nutraceutical Res 2010; 8: 173-84.

[37]   Kaur J, Kaur G. An insight into the role of Citrus bioactives in modulation of colon cancer. J Funct Foods 2015; 13: 239-61.
[http://dx.doi.org/10.1016/j.jff.2014.12.043]

[38]   Okwu DE. Citrus fruits: a rich source of phytochemicals and their roles in human health. Int J Chem Sci 2008; 6(2): 451-71.

[39]   Namitha KK, Negi PS. Chemistry and biotechnology of carotenoids. Crit Rev Food Sci Nutr 2010; 50(8): 728-60.
[http://dx.doi.org/10.1080/10408398.2010.499811] [PMID: 20830634]

[40]   Männistö S, Yaun SS, Hunter DJ, *et al.* Dietary carotenoids and risk of colorectal cancer in a pooled analysis of 11 cohort studies. Am J Epidemiol 2007; 165(3): 246-55.
[http://dx.doi.org/10.1093/aje/kwk009] [PMID: 17158857]

[41]   Tanaka T, Shnimizu M, Moriwaki H. Cancer chemoprevention by carotenoids. Molecules 2012; 17(3): 3202-42.
[http://dx.doi.org/10.3390/molecules17033202] [PMID: 22418926]

[42]   Cao H, Wang J, Dong X, *et al.* Carotenoid accumulation affects redox status, starch metabolism, and flavonoid/anthocyanin accumulation in citrus. BMC Plant Biol 2015; 15: 27.
[http://dx.doi.org/10.1186/s12870-015-0426-4] [PMID: 25644332]

[43]   Cazzonelli CI, Pogson BJ. Source to sink: regulation of carotenoid biosynthesis in plants. Trends Plant Sci 2010; 15(5): 266-74.
[http://dx.doi.org/10.1016/j.tplants.2010.02.003] [PMID: 20303820]

[44]   Gruszecki WI, Strzalka K. Carotenoids as modulators of lipid membrane physical properties. Biochim Biophys Acta 2005; 1740: 108-15.

[45]   de Saint Germain A, Bonhomme S, Boyer FD, Rameau C. Novel insights into strigolactone distribution and signalling. Curr Opin Plant Biol 2013; 16(5): 583-9.
[http://dx.doi.org/10.1016/j.pbi.2013.06.007] [PMID: 23830996]

[46]   Alder A, Jamil M, Marzorati M, *et al.* The path from β-carotene to carlactone, a strigolactone-like plant hormone. Science 2012; 335(6074): 1348-51.
[http://dx.doi.org/10.1126/science.1218094] [PMID: 22422982]

[47]   Ramel F, Mialoundama AS, Havaux M. Nonenzymic carotenoid oxidation and photooxidative stress signalling in plants. J Exp Bot 2013; 64(3): 799-805.
[http://dx.doi.org/10.1093/jxb/ers223] [PMID: 22915744]

[48]   Olson JA. Provitamin A function of carotenoids: the conversion of β-carotene into vitamin A. J Nutr 1989; 119(1): 105-8.
[PMID: 2643691]

[49]   Guimarães R, Barros L, Barreira JC, Sousa MJ, Carvalho AM, Ferreira IC. Targeting excessive free radicals with peels and juices of citrus fruits: grapefruit, lemon, lime and orange. Food Chem Toxicol 2010; 48(1): 99-106.
[http://dx.doi.org/10.1016/j.fct.2009.09.022] [PMID: 19770018]

[50]   Imamura T, Bando N, Yamanishi R. Beta-carotene modulates the immunological function of RAW264, a murine macrophage cell line, by enhancing the level of intracellular glutathione. Biosci Biotechnol Biochem 2006; 70(9): 2112-20.
[http://dx.doi.org/10.1271/bbb.60056] [PMID: 16960387]

[51]   Nishino H, Murakoshi M, Tokuda H, Satomi Y. Cancer prevention by carotenoids. Arch Biochem Biophys 2009; 483(2): 165-8.
[http://dx.doi.org/10.1016/j.abb.2008.09.011] [PMID: 18848517]

[52]   Cilla A, Attanzio A, Barberá R, Tesoriere L, Livrea MA. Anti-proliferative effect of main dietary phytosterols and β-cryptoxanthin alone or combined in human colon cancer Caco-2 cells through cytosolic Ca^{+2} and oxidative stress-induced apoptosis. J Funct Foods 2015; 12: 282-93.
[http://dx.doi.org/10.1016/j.jff.2014.12.001]

[53]   Uchiyama S, Yamaguchi M. Oral administration of β-cryptoxanthin prevents bone loss in ovariectomized rats. Int J Mol Med 2006; 17(1): 15-20.

[PMID: 16328006]

[54]   Sugiura M, Matsumoto H, Kato M, Ikoma Y, Yano M, Nagao A. Seasonal changes in the relationship between serum concentration of β-cryptoxanthin and serum lipid levels. J Nutr Sci Vitaminol (Tokyo) 2004; 50(6): 410-5.
[http://dx.doi.org/10.3177/jnsv.50.410] [PMID: 15895516]

[55]   Böhm V, Bitsch R. Intestinal absorption of lycopene from different matrices and interactions to other carotenoids, the lipid status, and the antioxidant capacity of human plasma. Eur J Nutr 1999; 38(3): 118-25.
[http://dx.doi.org/10.1007/s003940050052] [PMID: 10443333]

[56]   Sengupta A, Das S. The anti-carcinogenic role of lycopene, abundantly present in tomato. Eur J Cancer Prev 1999; 8(4): 325-30.
[http://dx.doi.org/10.1097/00008469-199908000-00009] [PMID: 10493308]

[57]   Leal M, Shimada A, Ruíz F, González de Mejía E. Effect of lycopene on lipid peroxidation and glutathione-dependent enzymes induced by T-2 toxin in vivo. Toxicol Lett 1999; 109(1-2): 1-10.
[http://dx.doi.org/10.1016/S0378-4274(99)00062-4] [PMID: 10514025]

[58]   Sonia S, Srivastava K, Saxena M, Mishva R. Tomato; A Rich Source of Lycopene and its Role in Human Health – A Review. J Med Arom Plant Sci 2007; 29: 81-9.

[59]   Iglesias DJ, Cercós M, Colmenero-Flores JM, *et al.* Physiology of citrus fruiting. Braz J Plant Physiol 2007; 19(4): 333-62.
[http://dx.doi.org/10.1590/S1677-04202007000400006]

[60]   Gross J. Pigments in Fruits. London: Academic Press 1987.

[61]   Xu CJ, Fraser PD, Wang WJ, Bramley PM. Differences in the carotenoid content of ordinary citrus and lycopene-accumulating mutants. J Agric Food Chem 2006; 54(15): 5474-81.
[http://dx.doi.org/10.1021/jf060702t] [PMID: 16848534]

[62]   Goodner KL, Rouseff RL, Hofsommer HJ. Orange, mandarin, and hybrid classification using multivariate statistics based on carotenoid profiles. J Agric Food Chem 2001; 49(3): 1146-50.
[http://dx.doi.org/10.1021/jf000866o] [PMID: 11312826]

[63]   Ikoma Y, Komatsu A, Kita M, *et al.* Expression of a phytoene synthase gene and characteristic carotenoid accumulation during citrus fruit development. Physiol Plant 2001; 111: 232-8.
[http://dx.doi.org/10.1034/j.1399-3054.2001.1110215.x]

[64]   Lee HS, Castle WS. Seasonal changes of carotenoid pigments and color in Hamlin, Earlygold, and budd blood orange juices. J Agric Food Chem 2001; 49(2): 877-82.
[http://dx.doi.org/10.1021/jf000654r] [PMID: 11262044]

[65]   Molnár P, Szabolcs J. β-Citraurin epoxide, a new carotenoid from Valencia orange peel. Phytochemistry 1980; 19: 633-7.
[http://dx.doi.org/10.1016/0031-9422(80)87029-4]

[66]   Liu C, Yan F, Gao H, *et al.* Features of citrus terpenoid production as revealed by carotenoid, limonoid and aroma profiles of two pummelos (*Citrus maxima*) with different flesh color. J Sci Food Agric 2015; 95(1): 111-9.
[http://dx.doi.org/10.1002/jsfa.6689] [PMID: 24723118]

[67]   Rodrigo MJ, Alquézar B, Alós E, Lado J, Zacarias L. Biochemical bases and molecular regulation of pigmentation in the peel of *Citrus* fruit. Sci Hortic (Amsterdam) 2013; 163: 46-62.
[http://dx.doi.org/10.1016/j.scienta.2013.08.014]

[68]   Oberholster R, Cowan AK, Molnár P, Tóth G. Biochemical basis of color as an aesthetic quality in Citrus sinensis. J Agric Food Chem 2001; 49(1): 303-7.
[http://dx.doi.org/10.1021/jf0007840] [PMID: 11170592]

[69]   Alquézar B, Rodrigo MJ, Zacarías L. Carotenoid biosynthesis and their regulation in citrus fruits. Tree For Sci Biotech 2008; 2(1): 23-37.

[70]   Holden JM, Eldridge AL, Beecher GR, *et al.* Carotenoid content of U.S. foods: an update of the database. J Food Compos Anal 1999; 12: 169-96.
[http://dx.doi.org/10.1006/jfca.1999.0827]

[71]   Liu Q, Xu J, Liu Y, *et al.* A novel bud mutation that confers abnormal patterns of lycopene accumulation in sweet orange fruit (*Citrus sinensis* L. Osbeck). J Exp Bot 2007; 58(15-16): 4161-71.
[http://dx.doi.org/10.1093/jxb/erm273] [PMID: 18182424]

[72]   Pan Z, Liu Q, Yun Z, *et al.* Comparative proteomics of a lycopene-accumulating mutant reveals the important role of oxidative stress on carotenogenesis in sweet orange (*Citrus sinensis* [L.] osbeck). Proteomics 2009; 9(24): 5455-70.
[http://dx.doi.org/10.1002/pmic.200900092] [PMID: 19834898]

[73]   Kato M, Ikoma Y, Matsumoto H, Sugiura M, Hyodo H, Yano M. Accumulation of carotenoids and expression of carotenoid biosynthetic genes during maturation in citrus fruit. Plant Physiol 2004; 134(2): 824-37.
[http://dx.doi.org/10.1104/pp.103.031104] [PMID: 14739348]

[74]   Fanciullino AL, Dhuique-Mayer C, Luro F, Casanova J, Morillon R, Ollitrault P. Carotenoid diversity in cultivated citrus is highly influenced by genetic factors. J Agric Food Chem 2006; 54(12): 4397-406.
[http://dx.doi.org/10.1021/jf0526644] [PMID: 16756373]

[75]   Matsumoto H, Ikoma Y, Kato M, Kuniga T, Nakajima N, Yoshida T. Quantification of carotenoids in citrus fruit by LC-MS and comparison of patterns of seasonal changes for carotenoids among citrus varieties. J Agric Food Chem 2007; 55(6): 2356-68.
[http://dx.doi.org/10.1021/jf062629c] [PMID: 17300198]

[76]   Kefford JF, Chandler BV. General composition of citrus fruits. In: Chichester CO, Mrak EM, Stewart GF, Eds. The chemical constituents of citrus fruits. New York: Academic Press 1970.

[77]   Barsan C, Zouine M, Maza E, *et al.* Proteomic analysis of chloroplast-to-chromoplast transition in tomato reveals metabolic shifts coupled with disrupted thylakoid biogenesis machinery and elevated energy-production components. Plant Physiol 2012; 160(2): 708-25.
[http://dx.doi.org/10.1104/pp.112.203679] [PMID: 22908117]

[78]   Wang YQ, Yang Y, Fei Z, *et al.* Proteomic analysis of chromoplasts from six crop species reveals insights into chromoplast function and development. J Exp Bot 2013; 64(4): 949-61.
[http://dx.doi.org/10.1093/jxb/ers375] [PMID: 23314817]

[79]   Huff A. Nutritional control of regreening and degreening in citrus peel segments. Plant Physiol 1983;

73(2): 243-9.
[http://dx.doi.org/10.1104/pp.73.2.243] [PMID: 16663202]

[80]   Goldschmidt EE. Regulatory aspects of chloro-chromoplast interconvensions in senescing *Citrus* fruit peel. Isr J Bot 1988; 47: 123-30.

[81]   Iglesias DJ, Tadeo FR, Legaz F, Primo-Millo E, Talon M. *In vivo* sucrose stimulation of colour change in citrus fruit epicarps: Interactions between nutritional and hormonal signals. Physiol Plant 2001; 112(2): 244-50.
[http://dx.doi.org/10.1034/j.1399-3054.2001.1120213.x] [PMID: 11454230]

[82]   Koshita Y. Effect of Temperature on Fruit Color Development. In: Kanayama Y, Kochetov A, Eds. Abiotic Stress Biology in Horticultural Plants. Springer Japan 2015; pp. 47-58.
[http://dx.doi.org/10.1007/978-4-431-55251-2_4]

[83]   Tripoli E, la Guardia M, Griammanco S, di Majo D, Griammanco M. Citrus flavonoids: molecular structure, biological activities and nutritional properties: a review. Food Chem 2007; 104: 466-79.
[http://dx.doi.org/10.1016/j.foodchem.2006.11.054]

[84]   Leuzzi U, Caristi C, Panzera V, Licandro G. Flavonoids in pigmented orange juice and second-pressure extracts. J Agric Food Chem 2000; 48(11): 5501-6.
[http://dx.doi.org/10.1021/jf000538o] [PMID: 11087509]

[85]   Drewnowski A, Gomez-Carneros C. Bitter taste, phytonutrients, and the consumer: a review. Am J Clin Nutr 2000; 72(6): 1424-35.
[PMID: 11101467]

[86]   Hwang SL, Shih PH, Yen GC. Neuroprotective effects of citrus flavonoids. J Agric Food Chem 2012; 60(4): 877-85.
[http://dx.doi.org/10.1021/jf204452y] [PMID: 22224368]

[87]   Rapisarda P, Giuffrida A. Anthocyanins level in Italian blood oranges. Proc Int Soc Citricult 1992; 3: 1130-3.

[88]   Maccarone E, Rapisarda P, Fanella F, Arena E, Mondello L. Cyanidin-3-(6"-malonyl)-β-glucoside. One of the major anthocyanins in blood orange juice. Ital J Food Sci 1998; 10: 367-72.

[89]   Lee HS. Characterization of major anthocyanins and the color of red-fleshed Budd Blood orange (Citrus sinensis). J Agric Food Chem 2002; 50(5): 1243-6.
[http://dx.doi.org/10.1021/jf011205+] [PMID: 11853511]

[90]   Hillebrand S, Schwarz M, Winterhalter P. Characterization of anthocyanins and pyranoanthocyanins from blood orange [*Citrus sinensis* (L.) Osbeck] juice. J Agric Food Chem 2004; 52(24): 7331-8.
[http://dx.doi.org/10.1021/jf0487957] [PMID: 15563216]

[91]   Cultrone A, Cotroneo PS, Recupero GR. Cloning and molecularcharacterization of R2R2-MYB and bHLH-MYC transcription factors from *Citrus sinensis*. Tree Genet Genomes 2010; 6: 101-12.
[http://dx.doi.org/10.1007/s11295-009-0232-y]

[92]   Crifò T, Puglisi I, Petrone G, Recupero GR, Lo Piero AR. Expression analysis in response to low temperature stress in blood oranges: implication of the flavonoid biosynthetic pathway. Gene 2011; 476(1-2): 1-9.
[http://dx.doi.org/10.1016/j.gene.2011.02.005] [PMID: 21349317]

[93]   Piero AR. The state of the art in biosynthesis of anthocyanins ant its regulation in pigmented sweet oranges [(*Citrus sinensis*) L. Osbeck]. J Agric Food Chem 2015; 63: 4031-41.
[http://dx.doi.org/10.1021/acs.jafc.5b01123] [PMID: 25871434]

[94]   Butelli E, Licciardello C, Zhang Y, *et al.* Retrotransposons control fruit-specific, cold-dependent accumulation of anthocyanins in blood oranges. Plant Cell 2012; 24(3): 1242-55.
[http://dx.doi.org/10.1105/tpc.111.095232] [PMID: 22427337]

[95]   Shao-Quian C, Si-yi P, Xiao-lin Y, Hong-fei F. Isolation and purification of anthocyanins from blood oranges by column chromatography. Agric Sci China 2010; 9(2): 207-15.
[http://dx.doi.org/10.1016/S1671-2927(09)60085-7]

[96]   Kumar S, Pandey AK. Chemistry and biological activities of flavonoids: an overview. Scientific World Journal 2013; 2013: 162750.
[http://dx.doi.org/10.1155/2013/162750] [PMID: 24470791]

[97]   Miyake Y, Shimoi K, Kumazawa S, Yamamoto K, Kinae N, Osawa T. Identification and antioxidant activity of flavonoid metabolites in plasma and urine of eriocitrin-treated rats. J Agric Food Chem 2000; 48(8): 3217-24.
[http://dx.doi.org/10.1021/jf990994g] [PMID: 10956094]

[98]   Biskind MS, Martin WC. The use of citrus flavonoids in infections. II. Am J Dig Dis 1955; 22(2): 41-5.
[http://dx.doi.org/10.1007/BF02886400] [PMID: 13228388]

[99]   Miyake Y, Yamamoto K, Osawa T. Isolation of eriocitrin (eriodictyol 7-rutinoside) from lemon fruit (*Citrus limon* BURM. f.) and its antioxidative activity. Food Sci Technol Int 1997; 3: 84-9.
[http://dx.doi.org/10.3136/fsti9596t9798.3.84]

[100] Miyake Y, Yamamoto K, Morimitsu Y, Osawa T. Characteristics of antioxidative flavonoid glycosides in lemon fruit. Food Sci Technol Int 1998; 4: 48-53.
[http://dx.doi.org/10.3136/fsti9596t9798.4.48]

[101] Miguel MG, Duarte A, Nunes S, Sustelo V, Martins D, Dandlen SA. Ascorbic acid and flavanone glycosides in citrus: Relationship with antioxidant activity. J Food Agric Environ 2009; 7(2): 222-7.

[102] Barreca D, Bellocco E, Caristi C, Leuzzi U, Gattuso G. Flavonoid profile and radical-scavenging activity of Mediterranean sweet lemon (*Citrus limetta* Risso) juice. Food Chem 2011; 129(2): 417-22.
[http://dx.doi.org/10.1016/j.foodchem.2011.04.093]

[103] Rapisarda P, Tomaino A, Lo Cascio R, Bonina F, De Pasquale A, Saija A. Antioxidant effectiveness as influenced by phenolic content of fresh orange juices. J Agric Food Chem 1999; 47(11): 4718-23.
[http://dx.doi.org/10.1021/jf990111l] [PMID: 10552879]

[104] Fiore A, La Fauci L, Cervellati R, *et al.* Antioxidant activity of pasteurized and sterilized commercial red orange juices. Mol Nutr Food Res 2005; 49(12): 1129-35.
[http://dx.doi.org/10.1002/mnfr.200500139] [PMID: 16254888]

[105] Guarnieri S, Riso P, Porrini M. Orange juice vs vitamin C: effect on hydrogen peroxide-induced DNA damage in mononuclear blood cells. Br J Nutr 2007; 97(4): 639-43.
[http://dx.doi.org/10.1017/S0007114507657948] [PMID: 17349075]

[106] Vitali F, Pennisi C, Tomaino A, *et al.* Effect of a standardized extract of red orange juice on proliferation of human prostate cells in vitro. Fitoterapia 2006; 77(3): 151-5.
[http://dx.doi.org/10.1016/j.fitote.2005.10.001] [PMID: 16530345]

[107] Parhiz H, Roohbakhsh A, Soltani F, Rezaee R, Iranshahi M. Antioxidant and anti-inflammatory properties of the citrus flavonoids hesperidin and hesperetin: an updated review of their molecular mechanisms and experimental models. Phytother Res 2015; 29(3): 323-31.
[http://dx.doi.org/10.1002/ptr.5256] [PMID: 25394264]

[108] Roohbakhsh A, Parhiz H, Soltani F, Rezaee R, Iranshahi M. Molecular mechanisms behind the biological effects of hesperidin and hesperetin for the prevention of cancer and cardiovascular diseases. Life Sci 2015; 124: 64-74.
[http://dx.doi.org/10.1016/j.lfs.2014.12.030] [PMID: 25625242]

[109] Schindler R, Mentlein R. Flavonoids and vitamin E reduce the release of the angiogenic peptide vascular endothelial growth factor from human tumor cells. J Nutr 2006; 136(6): 1477-82.
[PMID: 16702307]

[110] Hwang SL, Yen GC. Modulation of Akt, JNK, and p38 activation is involved in citrus flavonoid-mediated cytoprotection of PC12 cells challenged by hydrogen peroxide. J Agric Food Chem 2009; 57(6): 2576-82.
[http://dx.doi.org/10.1021/jf8033607] [PMID: 19222219]

[111] Ho SC, Kuo CT. Hesperidin, nobiletin, and tangeretin are collectively responsible for the anti-neuroinflammatory capacity of tangerine peel (*Citri reticulatae* pericarpium). Food Chem Toxicol 2014; 71: 176-82.
[http://dx.doi.org/10.1016/j.fct.2014.06.014] [PMID: 24955543]

[112] Li Y, Fang H, Xu W. Recent advance in the research of flavonoids as anticancer agents. Mini Rev Med Chem 2007; 7(7): 663-78.
[http://dx.doi.org/10.2174/138955707781024463] [PMID: 17627579]

[113] Walle T. Methoxylated flavones, a superior cancer chemopreventive flavonoid subclass? Semin Cancer Biol 2007; 17(5): 354-62.
[http://dx.doi.org/10.1016/j.semcancer.2007.05.002] [PMID: 17574860]

[114] Guthrie N, Carroll KK. Inhibition of mammary cancer by citrus flavonoids. Adv Exp Med Biol 1998; 439: 227-36.
[http://dx.doi.org/10.1007/978-1-4615-5335-9_16] [PMID: 9781306]

[115] Manthey JA, Guthrie N. Antiproliferative activities of citrus flavonoids against six human cancer cell lines. J Agric Food Chem 2002; 50(21): 5837-43.
[http://dx.doi.org/10.1021/jf020121d] [PMID: 12358447]

[116] Manach C, Morand C, Gil-Izquierdo A, Bouteloup-Demange C, Rémésy C. Bioavailability in humans of the flavanones hesperidin and narirutin after the ingestion of two doses of orange juice. Eur J Clin Nutr 2003; 57(2): 235-42.
[http://dx.doi.org/10.1038/sj.ejcn.1601547] [PMID: 12571654]

[117] Habauzit V, Sacco SM, Gil-Izquierdo A, *et al.* Differential effects of two citrus flavanones on bone quality in senescent male rats in relation to their bioavailability and metabolism. Bone 2011; 49(5):

1108-16.
[http://dx.doi.org/10.1016/j.bone.2011.07.030] [PMID: 21820093]

[118] Habauzit V, Trzeciakiewicz A, Mardon J, *et al.* Hesperidin and naringin, two main flavonoids of citrus fruit, modulate bone metabolism in gonad-intact senescent male rats. Proc Nutr Soc 2008; 67: E189.
[http://dx.doi.org/10.1017/S0029665108008215]

[119] Coelho RC, Hermsdorff HH, Bressan J. Anti-inflammatory properties of orange juice: possible favorable molecular and metabolic effects. Plant Foods Hum Nutr 2013; 68(1): 1-10.
[http://dx.doi.org/10.1007/s11130-013-0343-3] [PMID: 23417730]

[120] Grosso G, Galvanno F, Mistretta A, *et al.* Red orange: experimental models and epidemiological evidence of its benefits on human health. Oxid Med Cell Longev 2013; 2013: 11. Article ID 157240

[121] Peterson JJ, Beecher GR, Bhagwat SA, *et al.* Flavanones in grapefruit, lemons, and limes: a compilation and review of the data from the analytical literature. J Food Compos Anal 2006; 19: 574-80.

[122] Bermejo A, Llosá MJ, Cano A. Analysis of bioactive compounds in seven citrus cultivars. Food Sci Technol Int 2011; 17(1): 55-62.
[http://dx.doi.org/10.1177/1082013210368556] [PMID: 21364046]

[123] Caristi C, Bellocco E, Panzera V, Toscano G, Vadalà R, Leuzzi U. Flavonoids detection by HPLC-DAD-MS-MS in lemon juices from Sicilian cultivars. J Agric Food Chem 2003; 51(12): 3528-34.
[http://dx.doi.org/10.1021/jf0262357] [PMID: 12769519]

[124] Nogata Y, Sakamoto K, Shiratsuchi H, Ishii T, Yano M, Ohta H. Flavonoid composition of fruit tissues of citrus species. Biosci Biotechnol Biochem 2006; 70(1): 178-92.
[http://dx.doi.org/10.1271/bbb.70.178] [PMID: 16428836]

[125] Tanaka T. Misunderstanding with regards citrus classification and nomenclature. Bulletin of the University of Osaka Prefecture 1969; Series B 21: 139-45.

[126] Milella L, Caruso M, Galgano F, Favati F, Padula MC, Martelli G. Role of the cultivar in choosing Clementine fruits with a high level of health-promoting compounds. J Agric Food Chem 2011; 59(10): 5293-8.
[http://dx.doi.org/10.1021/jf104991z] [PMID: 21504146]

[127] Protti M, Valle F, Poli F, Raggi MA, Mercolini L. Bioactive molecules as authenticity markers of Italian Chinotto (*Citrus×myrtifolia*) fruits and beverages. J Pharm Biomed Anal 2015; 104: 75-80.
[http://dx.doi.org/10.1016/j.jpba.2014.11.024] [PMID: 25482848]

[128] Ahmad MM, Rehman S, Iqbal Z, Anjum FM, Sultan JI. Genetic variability to essential oil composition in four citrus fruit species. Pak J Bot 2006; 38(2): 319-24.

[129] Gorinstein S, Martin-Belloso O, Park Y, *et al.* Comparison of some biochemical characteristics of different citrus fruits. Food Chem 2001; 74: 309-15.
[http://dx.doi.org/10.1016/S0308-8146(01)00157-1]

[130] Singh J, Sood S, Muthuraman A. *In-vitro* evaluation of bioactive compounds, anti-oxidant, lipid peroxidation and lipoxygenase inhibitory potential of *Citrus karna* L. peel extract. J Food Sci Technol 2014; 51(1): 67-74.
[http://dx.doi.org/10.1007/s13197-011-0479-9] [PMID: 24426049]

[131] Ghasemi K, Ghasemi Y, Ebrahimzadeh MA. Antioxidant activity, phenol and flavonoid contents of 13 citrus species peels and tissues. Pak J Pharm Sci 2009; 22(3): 277-81.
[PMID: 19553174]

[132] Gil-Izquierdo A, Riquelme MT, Porras I, Ferreres F. Effect of the rootstock and interstock grafted in lemon tree (*Citrus limon* (L.) Burm.) on the flavonoid content of lemon juice. J Agric Food Chem 2004; 52(2): 324-31.
[http://dx.doi.org/10.1021/jf0304775] [PMID: 14733516]

[133] Mellisho CD, González-Barrio R, Ferreres F, *et al.* Iron deficiency enhances bioactive phenolics in lemon juice. J Sci Food Agric 2011; 91(12): 2132-9.
[PMID: 21560131]

[134] Botia JM. Efecto de la aplicación de fitorreguladores auxínicos sobre los niveles de flavanonas en frutos de pomelo. Levante Agrícola 2010; 403: 355-60.

[135] Gardana C, Nalin F, Simonetti P. Evaluation of flavonoids and furanocoumarins from *Citrus bergamia* (Bergamot) juice and identification of new compounds. Molecules 2008; 13(9): 2220-8.
[http://dx.doi.org/10.3390/molecules13092220] [PMID: 18830151]

[136] Roy A, Saraf S. Limonoids: overview of significant bioactive triterpenes distributed in plants kingdom. Biol Pharm Bull 2006; 29(2): 191-201.
[http://dx.doi.org/10.1248/bpb.29.191] [PMID: 16462017]

[137] Tundis R, Loizzo MR, Menichini F. An overview on chemical aspects and potential health benefits of limonoids and their derivatives. Crit Rev Food Sci Nutr 2014; 54(2): 225-50.
[http://dx.doi.org/10.1080/10408398.2011.581400] [PMID: 24188270]

[138] Manners GD. Citrus limonoids: analysis, bioactivity, and biomedical prospects. J Agric Food Chem 2007; 55(21): 8285-94.
[http://dx.doi.org/10.1021/jf071797h] [PMID: 17892257]

[139] Manners GD, Breksa AP III, Schoch TK, Hidalgo MB. Analysis of bitter limonoids in citrus juices by atmospheric pressure chemical ionization and electrospray ionization liquid chromatography-mass spectrometry. J Agric Food Chem 2003; 51(13): 3709-14.
[http://dx.doi.org/10.1021/jf021124t] [PMID: 12797731]

[140] Khalil AT, Maatooq GT, El Sayed KA. Limonoids from *Citrus reticulata* . Z Naturforsch, C, J Biosci 2003; 58(3-4): 165-70.
[http://dx.doi.org/10.1515/znc-2003-3-403] [PMID: 12710721]

[141] Heasley B. Synthesis of limonoid natural products European J Org Chem 2011; 2011: 19-46.
[http://dx.doi.org/10.1002/ejoc.201001218]

[142] Breksa AP III, King DE, Vilches AM. Determination of citrus limonoid glucosides by high performance liquid chromatography coupled to post-column reaction with Ehrlich's reagent. Beverages 2015; 1: 70-81.
[http://dx.doi.org/10.3390/beverages1020070]

[143] Poulose SM, Harris ED, Patil BS. Antiproliferative effects of citrus limonoids against human neuroblastoma and colonic adenocarcinoma cells. Nutr Cancer 2006; 56(1): 103-12.
[http://dx.doi.org/10.1207/s15327914nc5601_14] [PMID: 17176224]

[144]   Hasegawa S. Limonin bitterness in citrus juices. In: Teranishi , Ed. Flavor Chemistry Thirty Years of Progress. New York: Kluwer Academic/Plenum Publishers 1999.
[http://dx.doi.org/10.1007/978-1-4615-4693-1_9]

[145]   Jayaprakasha GK, Mandadi KK, Poulose SM, Jadegoud Y, Nagana Gowda GA, Patil BS. Novel triterpenoid from *Citrus aurantium* L. possesses chemopreventive properties against human colon cancer cells. Bioorg Med Chem 2008; 16(11): 5939-51.
[http://dx.doi.org/10.1016/j.bmc.2008.04.063] [PMID: 18490169]

[146]   Poulose SM, Harris ED, Patil BS. Citrus limonoids induce apoptosis in human neuroblastoma cells and have radical scavenging activity. J Nutr 2005; 135(4): 870-7.
[PMID: 15795449]

[147]   Patil JR, Jayaprakasha GK, Murthy CK, Chetti MB, Patil BS. Characterization of *Citrus aurantifolia* bioactive compounds and their inhibition of human pancreatic cancer cells through apoptosis. Microchem J 2010; 94: 108-17.
[http://dx.doi.org/10.1016/j.microc.2009.09.008]

[148]   Chidambara Murthy KN, Jayaprakasha GK, Patil BS. Apoptosis mediated cytotoxicity of citrus obacunone in human pancreatic cancer cells. Toxicol In Vitro 2011; 25(4): 859-67.
[http://dx.doi.org/10.1016/j.tiv.2011.02.006] [PMID: 21333732]

[149]   Tian Q, Miller EG, Ahmad H, Tang L, Patil BS. Differential inhibition of human cancer cell proliferation by citrus limonoids. Nutr Cancer 2001; 40(2): 180-4.
[http://dx.doi.org/10.1207/S15327914NC402_15] [PMID: 11962254]

[150]   Kim J, Jayaprakasha GK, Patil BS. Obacunone exhibits anti-proliferative and anti-aromatase activity in vitro by inhibiting the p38 MAPK signaling pathway in MCF-7 human breast adenocarcinoma cells. Biochimie 2014; 105: 36-44.
[http://dx.doi.org/10.1016/j.biochi.2014.06.002] [PMID: 24927687]

[151]   El-Readi MZ, Hamdan D, Farrag N, El-Shazly A, Wink M. Inhibition of P-glycoprotein activity by limonin and other secondary metabolites from *Citrus* species in human colon and leukaemia cell lines. Eur J Pharmacol 2010; 626(2-3): 139-45.
[http://dx.doi.org/10.1016/j.ejphar.2009.09.040] [PMID: 19782062]

[152]   Perez JL, Jayaprakasha GK, Valdivia V, *et al.* Limonin methoxylation influences the induction of glutathione S-transferase and quinone reductase. J Agric Food Chem 2009; 57(12): 5279-86.
[http://dx.doi.org/10.1021/jf803712a] [PMID: 19480426]

[153]   Perez JL, Jayaprakasha GK, Cadena A, Martinez E, Ahmad H, Patil BS. In vivo induction of phase II detoxifying enzymes, glutathione transferase and quinone reductase by citrus triterpenoids. BMC Complement Altern Med 2010; 10: 51.
[http://dx.doi.org/10.1186/1472-6882-10-51] [PMID: 20846448]

[154]   Miller EG, Fanous R, Rivera-Hidalgo F, Binnie WH, Hasegawa S, Lam LK. The effect of citrus limonoids on hamster buccal pouch carcinogenesis. Carcinogenesis 1989; 10(8): 1535-7.
[http://dx.doi.org/10.1093/carcin/10.8.1535] [PMID: 2502323]

[155]   Miller EG, Gonzales-Sanders AP, Couvillon AM, Wright JM, Hasegawa S, Lam LK. Inhibition of hamster buccal pouch carcinogenesis by limonin 17-beta-D-glucopyranoside. Nutr Cancer 1992;

17(1): 1-7.
[http://dx.doi.org/10.1080/01635589209514167] [PMID: 1574440]

[156] Miller EG, Gonzales-Sanders AP, Couvillon AM, Binnie WH, Hasegawa S, Lam LK. Citrus limonoids as inhibitors of oral carcinogenesis. Food Technol 1994; 48: 110-4.

[157] Lam LK, Zhang J, Hasegawa S. Citrus limonoid reduction of chemically induced tumorgenesis. Food Technol 1994; 48: 104-8.

[158] Tanaka T, Kohno H, Tsukio Y, *et al.* Citrus limonoids obacunone and limonin inhibit azoxymethane-induced colon carcinogenesis in rats. Biofactors 2000; 13(1-4): 213-8.
[http://dx.doi.org/10.1002/biof.5520130133] [PMID: 11237184]

[159] Raphael TJ, Kuttan G. Effect of naturally occurring triterpenoids glycyrrhizic acid, ursolic acid, oleanolic acid and nomilin on the immune system. Phytomedicine 2003; 10(6-7): 483-9.
[http://dx.doi.org/10.1078/094471103322331421] [PMID: 13678231]

[160] Battinelli L, Mengoni F, Lichtner M, *et al.* Effect of limonin and nomilin on HIV-1 replication on infected human mononuclear cells. Planta Med 2003; 69(10): 910-3.
[http://dx.doi.org/10.1055/s-2003-45099] [PMID: 14648393]

[161] Vikram A, Jesudhasan PR, Jayaprakasha GK, Pillai SD, Patil BS. Citrus limonoids interfere with Vibrio harveyi cell-cell signalling and biofilm formation by modulating the response regulator LuxO. Microbiology 2011; 157(Pt 1): 99-110.
[http://dx.doi.org/10.1099/mic.0.041228-0] [PMID: 20864476]

[162] Kurowska EM, Borradaile NM, Spence JD, Carroll KK. Hypocholesterolemic effects of dietary citrus juices in rabbits. Nutr Res 2000; 20: 121-9.
[http://dx.doi.org/10.1016/S0271-5317(99)00144-X]

[163] Sun C, Chen K, Chen Y, Chen Q. Contents and antioxidant capacity of limonin and nomilin in different tissues of citrus fruit of four cultivars during fruit growth and maturation. Food Chem 2005; 93: 599-605.
[http://dx.doi.org/10.1016/j.foodchem.2004.10.037]

[164] Yu J, Wang L, Walzem RL, Miller EG, Pike LM, Patil BS. Antioxidant activity of citrus limonoids, flavonoids, and coumarins. J Agric Food Chem 2005; 53(6): 2009-14.
[http://dx.doi.org/10.1021/jf0484632] [PMID: 15769128]

[165] Breksa AP III, Manners GD. Evaluation of the antioxidant capacity of limonin, nomilin, and limonin glucoside. J Agric Food Chem 2006; 54(11): 3827-31.
[http://dx.doi.org/10.1021/jf060901c] [PMID: 16719503]

[166] Mahmoud MF, Gamal S, El-Fayoumi HM. Limonin attenuates hepatocellular injury following liver ischemia and reperfusion in rats via toll-like receptor dependent pathway. Eur J Pharmacol 2014; 740: 676-82.
[http://dx.doi.org/10.1016/j.ejphar.2014.06.010] [PMID: 24967531]

[167] Chidambara Murthy KN, Jayaprakasha GK, Patil BS. Obacunone and obacunone glucoside inhibit human colon cancer (SW480) cells by the induction of apoptosis. Food Chem Toxicol 2011; 49(7): 1616-25.
[http://dx.doi.org/10.1016/j.fct.2011.04.014] [PMID: 21515332]

[168] Champagne DE, Koul O, Isman MB, Scudder GG, Neil Towers GH. Biological activity of limonoids from the rutales. Phytochemistry 1992; 31(2): 377-94.
[http://dx.doi.org/10.1016/0031-9422(92)90003-9]

[169] Jayaprakasha GK. Girennavar BaPBS. Radical scavenging activity of red grapefruits and sour orange fruit extracts in different in vitro model systems. Bioresour Technol 2008; 99: 4484-94.
[http://dx.doi.org/10.1016/j.biortech.2007.07.067] [PMID: 17935981]

[170] Li J. Citrus limonoids: seasonal changes and their potential in glutathione S-transferase induction. Master's thesis 2002.

[171] Hasegawa S, Herman Z, Orme E, Ou P. Biosynthesis of limonoids in *Citrus*: Sites and translocation. Phytochemistry 1986; 25(12): 2783-5.
[http://dx.doi.org/10.1016/S0031-9422(00)83741-3]

[172] Do TK, Hadji-Minaglou F, Antoniotti S, Fernandez X. Authenticity of essential oils. Trends Analyt Chem 2015; 66: 146-57.
[http://dx.doi.org/10.1016/j.trac.2014.10.007]

[173] Ruiz B, Flotats X. Citrus essential oils and their influence on the anaerobic digestion process: an overview. Waste Manag 2014; 34(11): 2063-79.
[http://dx.doi.org/10.1016/j.wasman.2014.06.026] [PMID: 25081855]

[174] Moufida S, Marzouk B. Biochemical characterization of blood orange, sweet orange, lemon, bergamot and bitter orange. Phytochemistry 2003; 62(8): 1283-9.
[http://dx.doi.org/10.1016/S0031-9422(02)00631-3] [PMID: 12648552]

[175] de Carvalho CC, da Fonseca MM. Biotransformation of terpenes. Biotechnol Adv 2006; 24(2): 134-42.
[http://dx.doi.org/10.1016/j.biotechadv.2005.08.004] [PMID: 16169182]

[176] Tranchida PQ, Bonaccorsi I, Dugo P, Mondello L, Dugo G. Analysis of *Citrus* essential oils: state of the art and future perspectives. A review. Flavour Fragrance J 2012; 27: 98-123.
[http://dx.doi.org/10.1002/ffj.2089]

[177] Bakkali F, Averbeck S, Averbeck D, Idaomar M. Biological effects of essential oils--a review. Food Chem Toxicol 2008; 46(2): 446-75.
[http://dx.doi.org/10.1016/j.fct.2007.09.106] [PMID: 17996351]

[178] Schipilliti L, Tranchida PQ, Sciarrone D, *et al.* Genuineness assessment of mandarin essential oils employing gas chromatography-combustion-isotope ratio MS (GC-C-IRMS). J Sep Sci 2010; 33(4-5): 617-25.
[http://dx.doi.org/10.1002/jssc.200900504] [PMID: 20112303]

[179] Badee AZ, Helmy SA, Morsy NF. Utilisation of orange peel in the production of -terpineol by Penicillium digitatum (NRRL 1202). Food Chem 2011; 126: 849-54.
[http://dx.doi.org/10.1016/j.foodchem.2010.11.046]

[180] Fisher K, Phillips C. Potential antimicrobial uses of essential oils in food: is citrus the answer? Trends Food Sci Technol 2008; 19: 156-64.
[http://dx.doi.org/10.1016/j.tifs.2007.11.006]

[181] Kummer R, Fachini-Queiroz FC, Estevão-Silva CF, *et al.* Evaluation of anti-inflammatory activity of citrus latifolia tanaka essential oil and limonene in experimental mouse models. Evid Based Complement Alternat Med 2013; 2013: 1-8.
[http://dx.doi.org/10.1155/2013/859083]

[182] Darland GK, Lukaczer DO, Liska DJ, Irving TA, Bland JS. Inventors Patent No 6,210,701 2001.

[183] Salminen A, Lehtonen M, Suuronen T, Kaarniranta K, Huuskonen J. Terpenoids: natural inhibitors of NF-kappaB signaling with anti-inflammatory and anticancer potential. Cell Mol Life Sci 2008; 65(19): 2979-99.
[http://dx.doi.org/10.1007/s00018-008-8103-5] [PMID: 18516495]

[184] Jing L, Zhang Y, Fan S, *et al.* Preventive and ameliorating effects of citrus D-limonene on dyslipidemia and hyperglycemia in mice with high-fat diet-induced obesity. Eur J Pharmacol 2013; 715(1-3): 46-55.
[http://dx.doi.org/10.1016/j.ejphar.2013.06.022] [PMID: 23838456]

[185] Jayaprakasha GK, Murthy KN, Uckoo RM, Patil BS. Chemical composition of volatile oil from *Citrus limettioides* and their inhibition of colon cancer cell proliferation. Ind Crops Prod 2013; 45: 200-7.
[http://dx.doi.org/10.1016/j.indcrop.2012.12.020]

[186] Monzote L, Hill GM, Cuellar A, Scull R, Setzer WN. Chemical composition and anti-proliferative properties of Bursera graveolens essential oil. Nat Prod Commun 2012; 7(11): 1531-4.
[PMID: 23285824]

[187] Roberto D, Micucci P, Sebastian T, Graciela F, Anesini C. Antioxidant activity of limonene on normal murine lymphocytes: relation to H2O2 modulation and cell proliferation. Basic Clin Pharmacol Toxicol 2010; 106(1): 38-44.
[PMID: 19796276]

[188] Aazza S, Lyoussi B, Megías C, *et al.* Anti-oxidant, anti-inflammatory and anti-proliferative activities of Moroccan commercial essential oils. Nat Prod Commun 2014; 9(4): 587-94.
[PMID: 24868891]

[189] Monajemi R, Oryan S, Haeri-Roohani A, Ghannadi A, Jafarian A. Cytotoxic effects of essential oils of some Iranian Citrus peels. Iran J Pharm Res 2005; 3: 183-7.

[190] Lu XG, Zhan LB, Feng BA, Qu MY, Yu LH, Xie JH. Inhibition of growth and metastasis of human gastric cancer implanted in nude mice by d-limonene. World J Gastroenterol 2004; 10(14): 2140-4.
[http://dx.doi.org/10.3748/wjg.v10.i14.2140] [PMID: 15237454]

[191] Chen J, Lu M, Jing Y, Dong J. The synthesis of L-carvone and limonene derivatives with increased antiproliferative effect and activation of ERK pathway in prostate cancer cells. Bioorg Med Chem 2006; 14(19): 6539-47.
[http://dx.doi.org/10.1016/j.bmc.2006.06.013] [PMID: 16806947]

[192] Ji J, Zhang L, Wu YY, Zhu XY, Lv SQ, Sun XZ. Induction of apoptosis by d-limonene is mediated by a caspase-dependent mitochondrial death pathway in human leukemia cells. Leuk Lymphoma 2006; 47(12): 2617-24.
[http://dx.doi.org/10.1080/00268970600909205] [PMID: 17169807]

[193] Hansen U, Seufert G. Terpenoid emission from *Citrus sinensis* (L.) Osbeck under drought stress. Phys

Chem Earth (B) 1999; 42: 681-7.
[http://dx.doi.org/10.1016/S1464-1909(99)00065-9]

[194] Miguel MG, Dandlen S, Figueiredo AC, *et al.* Essential oils of flowers of *Citrus sinensis* and *Citrus clementina* cultivated in Algarve, Portugal. Acta Hortic 2008; 773: 89-94.
[http://dx.doi.org/10.17660/ActaHortic.2008.773.12]

[195] Alissandrakis E, Daferera D, Tarantilis PA, Polissiou M, Harizanis PC. Ultrasound-assisted extraction of volatile compounds from citrus flowers and citrus honey. Food Chem 2003; 82: 575-82.
[http://dx.doi.org/10.1016/S0308-8146(03)00013-X]

[196] Flamini G, Cioni PL, Morelli I. Use of solid-phase micro-extraction as a sampling technique in the determination of volatiles emitted by flowers, isolated flower parts and pollen. J Chromatogr A 2003; 998(1-2): 229-33.
[http://dx.doi.org/10.1016/S0021-9673(03)00641-1] [PMID: 12862387]

[197] Tanaka T, Yasui Y, Ishigamori-Suzuki R, Oyama T. Citrus compounds inhibit inflammation- and obesity-related colon carcinogenesis in mice. Nutr Cancer 2008; 60 (Suppl. 1): 70-80.
[http://dx.doi.org/10.1080/01635580802381253] [PMID: 19003583]

[198] Hayashi K, Suzuki R, Miyamoto S, *et al.* Citrus auraptene suppresses azoxymethane-induced colonic preneoplastic lesions in C57BL/KsJ-db/db mice. Nutr Cancer 2007; 58(1): 75-84.
[http://dx.doi.org/10.1080/01635580701308216] [PMID: 17571970]

[199] Tanaka T, de Azevedo MB, Durán N, *et al.* Colorectal cancer chemoprevention by 2 β-cyclodextrin inclusion compounds of auraptene and 4′-geranyloxyferulic acid. Int J Cancer 2010; 126(4): 830-40.
[PMID: 19688830]

[200] Niu X, Huang Z, Zhang L, Ren X, Wang J. Auraptene has the inhibitory property on murine T lymphocyte activation. Eur J Pharmacol 2015; 750: 8-13.
[http://dx.doi.org/10.1016/j.ejphar.2015.01.017] [PMID: 25620131]

[201] Patil JR, Jayaprakasha GK, Kim J, *et al.* 5-Geranyloxy-7-methoxycoumarin inhibits colon cancer (SW480) cells growth by inducing apoptosis. Planta Med 2013; 79(3-4): 219-26.
[PMID: 23345169]

[202] Kostova I. Synthetic and natural coumarins as antioxidants. Mini Rev Med Chem 2006; 6(4): 365-74.
[http://dx.doi.org/10.2174/138955706776361457] [PMID: 16613573]

[203] Wu L, Wang X, Xu W, Farzaneh F, Xu R. The structure and pharmacological functions of coumarins and their derivatives. Curr Med Chem 2009; 16(32): 4236-60.
[http://dx.doi.org/10.2174/092986709789578187] [PMID: 19754420]

[204] Valero D, Serrano M. Growth and ripening stage at harvest modulates postharvest quality and bioactive compounds with antioxidant activity. Stewart Postharvest Review 2013; 9(3): 1-8.
[http://dx.doi.org/10.2212/spr.2013.3.7]

[205] Murcia MA, Lopez-Ayerra BL, Martinez-Tomé M, Vera AM, García-Carmona F. Evolution of ascorbic acid and peroxidase during industrial processing of broccoli. J Sci Food Agric 2000; 80: 1882-6.
[http://dx.doi.org/10.1002/1097-0010(200010)80:13<1882::AID-JSFA729>3.0.CO;2-B]

[206] Gil-Izquierdo A, Gil MI, Ferreres F. Effect of processing techniques at industrial scale on orange juice

antioxidant and beneficial health compounds. J Agric Food Chem 2002; 50(18): 5107-14.
[http://dx.doi.org/10.1021/jf020162+] [PMID: 12188615]

[207] Burdurlu HS, Koca N, Karadeniz F. Degradation of vitamin C in citrus juice concentrates during storage. J Food Eng 2006; 74(2): 211-6.
[http://dx.doi.org/10.1016/j.jfoodeng.2005.03.026]

[208] Lee HS, Coates GA. Vitamin C contents in processed Florida citrus juice products from 1986–1995 survey. J Agric Food Chem 1997; 45: 2550-5.
[http://dx.doi.org/10.1021/jf960851j]

[209] Gil-Izquierdo A, Gil MI, Tomás-Barberan FA, Ferreres F. Influence of industrial processing on orange juice flavanone solubility and transformation to chalcones under gastrointestinal conditions. J Agric Food Chem 2003; 51(10): 3024-8.
[http://dx.doi.org/10.1021/jf020986r] [PMID: 12720386]

[210] Lee HS, Coates GA. Effect of thermal pasteurization on Valencia orange juice color and pigments. Lebensm Wiss Technol -. Food Sci Technol (Campinas) 2003; 36: 153-6.

[211] Sánchez-Moreno C, Plaza L, Elez-Martínez P, De Ancos B, Martín-Belloso O, Cano MP. Impact of high pressure and pulsed electric fields on bioactive compounds and antioxidant activity of orange juice in comparison with traditional thermal processing. J Agric Food Chem 2005; 53(11): 4403-9.
[http://dx.doi.org/10.1021/jf048839b] [PMID: 15913302]

[212] Cortés C, Torregrosa F, Esteve MJ, Frígola A. Carotenoid profile modification during refrigerated storage in untreated and pasteurized orange juice and orange juice treated with high-intensity pulsed electric fields. J Agric Food Chem 2006; 54(17): 6247-54.
[http://dx.doi.org/10.1021/jf060995q] [PMID: 16910715]

[213] Gama JJ, de Sylos CM. Effect of thermal pasteurization and concentration on carotenoid composition of Brazilian Valencia orange juice. Food Chem 2007; 100: 1686-90.
[http://dx.doi.org/10.1016/j.foodchem.2005.01.062]

[214] Meléndez-Martínez AJ, Britton G, Vicario IM, Heredia FJ. The complex carotenoid pattern of orange juices from concentrate. Food Chem 2008; 109: 546-53.
[http://dx.doi.org/10.1016/j.foodchem.2008.01.003]

[215] Álvarez R, Carvalho CP, Sierra J, Lara O, Cardona D, Londoño-Londoño J. Citrus juice extraction systems: effect on chemical composition and antioxidant activity of clementine juice. J Agric Food Chem 2012; 60(3): 774-81.
[http://dx.doi.org/10.1021/jf203353h] [PMID: 22225414]

[216] Stinco CM, Fernández-Vázquez R, Hernanz D, Heredia FJ, Meléndez-Martínez AJ, Vicario IM. Industrial orange juice debittering: Impact on bioactive compounds and nutritional value. J Food Eng 2013; 116(1): 155-61.
[http://dx.doi.org/10.1016/j.jfoodeng.2012.11.009]

[217] Aschoff JK, Kaufmann S, Kalkan O, Neidhart S, Carle R, Schweiggert RM. In vitro bioaccessibility of carotenoids, flavonoids, and vitamin C from differently processed oranges and orange juices [*Citrus sinensis* (L.) Osbeck]. J Agric Food Chem 2015; 63(2): 578-87. [*Citrus sinensis* (L.) Osbeck].
[http://dx.doi.org/10.1021/jf505297t] [PMID: 25539394]

[218] Chaudhary P, Jayaprakasha GK, Porat R, Patil BS. Degreening and postharvest storage influences 'Star Ruby' grapefruit (*Citrus paradisi* Macf.) bioactive compounds. Food Chem 2012; 135(3): 1667-75.
[http://dx.doi.org/10.1016/j.foodchem.2012.05.095] [PMID: 22953908]

[219] Chaudhary PR, Jayaprakasha GK, Porat R, Patil BS. Low temperature conditioning reduces chilling injury while maintaining quality and certain bioactive compounds of 'Star Ruby' grapefruit. Food Chem 2014; 153: 243-9.
[http://dx.doi.org/10.1016/j.foodchem.2013.12.043] [PMID: 24491726]

[220] Carvalho CP, Navarro P, Salvador A. Poscosecha. In: Garcés LF, Carvalho CP, Eds. Cítricos: cultivo, poscosecha e industrialización: Corporación Universitaria Lasallista. 2012.

[221] Velázquez-Estrada RM, Hernández-Herrero MM, Rüferb CE, Guamis-López B, Roig-Sagués AX. Influence of ultra high pressure homogenization processing on bioactive compounds and antioxidant activity of orange juice. Innov Food Sci Emerg Technol 2013; 18: 89-94.
[http://dx.doi.org/10.1016/j.ifset.2013.02.005]

[222] Uckoo RM, Jayaprakasha GK, Somerville JA, Balasubramaniam VM, Pinarte M, Patil BS. High pressure processing controls microbial growth and minimally alters the levels of health promoting compounds in grapefruit (*Citrus paradisi* Macfad) juice. Innov Food Sci Emerg Technol 2013; 18: 7-14.
[http://dx.doi.org/10.1016/j.ifset.2012.11.010]

[223] Sun Y, Shen Y, Liu D, Ye X. Effects of drying methods on phytochemical compounds and antioxidant activity of physiologically dropped un-matured citrus fruits. LWT -. Food Sci Technol (Campinas) 2015; 60(2): 1269-75.

[224] Tao N, Wang C, Xu J, Cheng Y. Carotenoid accumulation in postharvest "Cara Cara" navel orange (*Citrus sinensis* Osbeck) fruits stored at different temperatures was transcriptionally regulated in a tissue-dependent manner. Plant Cell Rep 2012; 31(9): 1667-76.
[http://dx.doi.org/10.1007/s00299-012-1279-z] [PMID: 22562781]

[225] Arscott SA, Howe JA, Davis CR, Tanumihardjo SA. Carotenoid profiles in provitamin A-containing fruits and vegetables affect the bioefficacy in Mongolian gerbils. Exp Biol Med (Maywood) 2010; 235(7): 839-48.
[http://dx.doi.org/10.1258/ebm.2010.009216] [PMID: 20558838]

[226] Burri BJ, Chang JS, Neidlinger TR. β-Cryptoxanthin- and α-carotene-rich foods have greater apparent bioavailability than β-carotene-rich foods in Western diets. Br J Nutr 2011; 105(2): 212-9.
[http://dx.doi.org/10.1017/S0007114510003260] [PMID: 20807466]

[227] Maiani G, Castón MJ, Catasta G, *et al.* Carotenoids: actual knowledge on food sources, intakes, stability and bioavailability and their protective role in humans. Mol Nutr Food Res 2009; 53 (Suppl. 2): S194-218.
[http://dx.doi.org/10.1002/mnfr.200800053] [PMID: 19035552]

[228] Shardell MD, Alley DE, Hicks GE, *et al.* Low-serum carotenoid concentrations and carotenoid interactions predict mortality in US adults: the Third National Health and Nutrition Examination Survey. Nutr Res 2011; 31(3): 178-89.
[http://dx.doi.org/10.1016/j.nutres.2011.03.003] [PMID: 21481711]

[229] Escudero-López B, Cerrillo I, Herrero-Martín G, *et al.* Fermented orange juice: source of higher carotenoid and flavanone contents. J Agric Food Chem 2013; 61(37): 8773-82.
[http://dx.doi.org/10.1021/jf401240p] [PMID: 24004007]

[230] García-Lafuente A, Guillamón E, Villares A, Rostagno MA, Martínez JA. Flavonoids as anti-inflammatory agents: implications in cancer and cardiovascular disease. Inflamm Res 2009; 58(9): 537-52.
[http://dx.doi.org/10.1007/s00011-009-0037-3] [PMID: 19381780]

[231] Stojković D, Petrović J, Soković M, Glamočlija J, Kukić-Marković J, Petrović S. *In situ* antioxidant and antimicrobial activities of naturally occurring caffeic acid, p-coumaric acid and rutin, using food systems. J Sci Food Agric 2013; 93(13): 3205-8.
[http://dx.doi.org/10.1002/jsfa.6156] [PMID: 23553578]

[232] Darvesh AS, Carroll RT, Bishayee A, Geldenhuys WJ, Van der Schyf CJ. Oxidative stress and Alzheimer's disease: dietary polyphenols as potential therapeutic agents. Expert Rev Neurother 2010; 10(5): 729-45.
[http://dx.doi.org/10.1586/ern.10.42] [PMID: 20420493]

[233] Joshi UJ, Gadge AS, D'Mello P, Sinha R, Srivastava S, Govil G. Anti-inflammatory, antioxidant and anticancer activity of quercetin and its analogues. Int J Res Pharm Biomed Sci 2011; 2: 1756-66.

[234] Middleton EJ, Kandaswami C. Potential health promoting properties of citrus flavonoids. Food Technol 1994; 18: 115-20.

[235] Yehoshua SB, Rodov V, Fang DQ, Kim JJ. Preformed antifungal compounds of citrus fruit: Effect of postharvest treatments with heat and growth regulators. J Agric Food Chem 1995; 43: 1062-6.
[http://dx.doi.org/10.1021/jf00052a040]

[236] Braddock RJ. By-products of citrus fruit. Food Technol 1995; 49: 74-7.

[237] Craig WJ. Phytochemicals: guardians of our health. J Am Diet Assoc 1997; 97(10) (Suppl. 2): S199-204.
[http://dx.doi.org/10.1016/S0002-8223(97)00765-7] [PMID: 9336591]

[238] Okwu DE. Phytochemicals and vitamins content of indigenous spices of south eastern nigeria. J Sustainable Agric Environm 2004; 6(1): 30-7.

[239] Knekt P, Kumpulainen J, Järvinen R, *et al.* Flavonoid intake and risk of chronic diseases. Am J Clin Nutr 2002; 76(3): 560-8.
[PMID: 12198000]

[240] Bang G, Kristoffersen T. Dental caries and diet in an Alaskan Eskimo population. Scand J Dent Res 1972; 80(5): 440-4.
[PMID: 4403921]

[241] Olsson B. Dental health situation in privileged children in Addis Ababa, Ethiopia. Community Dent Oral Epidemiol 1979; 7(1): 37-41.
[http://dx.doi.org/10.1111/j.1600-0528.1979.tb01183.x] [PMID: 282955]

[242] Song JK, Bae JM. Citrus fruit intake and breast cancer risk: a quantitative systematic review. J Breast Cancer 2013; 16(1): 72-6.
[http://dx.doi.org/10.4048/jbc.2013.16.1.72] [PMID: 23593085]

[243] Kumar V, Nandini SS, Anitha S. Anti typhoid activity of aqueous extract of fruit peel *Citrus sinensis* (L.). Int J Pharm Res Develop 2010; 2(9): 217-21.

[244] García-Closas R, Berenguer A, José Tormo M, *et al.* Dietary sources of vitamin C, vitamin E and specific carotenoids in Spain. Br J Nutr 2004; 91(6): 1005-11.
[http://dx.doi.org/10.1079/BJN20041130] [PMID: 15182404]

[245] Kris-Etherton PM, Hecker KD, Bonanome A, *et al.* Bioactive compounds in foods: their role in the prevention of cardiovascular disease and cancer. Am J Med 2002; 113 (Suppl. 9B): 71S-88S.
[http://dx.doi.org/10.1016/S0002-9343(01)00995-0] [PMID: 12566142]

[246] Gironés-Vilaplana A, Moreno DA, García-Viguera C. Phytochemistry and biological activity of Spanish Citrus fruits. Food Funct 2014; 5(4): 764-72.
[http://dx.doi.org/10.1039/c3fo60700c] [PMID: 24563112]

[247] Gironés-Vilaplana A, Valentão P, Andrade PB, Ferreres F, Moreno DA, García-Viguera C. Phytochemical profile of a blend of black chokeberry and lemon juice with cholinesterase inhibitory effect and antioxidant potential. Food Chem 2012; 134(4): 2090-6.
[http://dx.doi.org/10.1016/j.foodchem.2012.04.010] [PMID: 23442660]

[248] Gironés-Vilaplana A, Mena P, García-Viguera C, Moreno DA. A novel beverage rich in antioxidant phenolics: Maqui berry (*Aristotelia chilensis*) and lemon juice. Food Sci Technol (Campinas) 2012; 47: 279-86.

[249] Gironés-Vilaplana A, Villaño D, Moreno DA, García-Viguera C. New isotonic drinks with antioxidant and biological capacities from berries (maqui, açaí and blackthorn) and lemon juice. Int J Food Sci Nutr 2013; 64(7): 897-906.
[http://dx.doi.org/10.3109/09637486.2013.809406] [PMID: 23815554]

CHAPTER 3

Bioactive Compounds of Apples and Pears as Health Promoters

Andrea Catalina Galvis-Sánchez [1,*], **Ada Rocha** [2,3]

[1] *REQUIMTE, Department of Chemical Engineering, Faculty of Engineering, University of Porto, Rua Dr. Roberto Frias, 4200-465, Porto, Portugal*

[2] *Faculty of Nutrition and Food Sciences, University of Porto, Rua Dr. Roberto Frias, 4200-465, Porto, Portugal*

[3] *LAQV@REQUIMTE, Porto, Portugal*

Abstract: Contemporaneous dietary patterns could be best described as in need of improvement. Obesity, diabetes and metabolic syndrome prevalence increased dramatically in the recent years likely due to unbalanced dietary patterns and sedentary lifestyle. "Epidemiological studies have shown that dietary patterns are significantly associated with the prevention of chronic diseases such as heart disease, cancer, diabetes and Alzheimer's disease".

"Fruits and vegetables" have improved the human diet for centuries, enriching it nutritionally and sensorially. A significant amount of vitamins and minerals in the diet come from fruits and vegetables. "Approximately half of the vitamin A, in the form of carotene, over 90% of vitamin C and 40% of folacin come from this food group". Fruits contribute with "considerable amounts of vitamins A, C, B$_6$, thiamin, niacin and minerals (*i.e.* magnesium and iron)" to our diet. Furthermore, "they supply proteins, starch and sugars, and they are important sources of dietary and crude fiber".

"Epidemiological and clinical investigations demonstrate significant decrease in morbidity and mortality from cardiovascular and other diseases among fruit and vegetables consumers". These benefits have been associated to their content on dietary fiber and different bioactive compounds with anti-atherosclerotic and anticancer effects.

[*] **Corresponding author Andrea C. Galvis-Sánchez:** REQUIMTE, Department of Chemical Engineering, Faculty of Engineering, University of Porto, Porto, Portugal; Email: agalvissanchez@gmail.com.

Apples and pears are some of the most "common and frequently consumed fruits in the world and the most widely consumed fruits by Western populations". A revision of the type and content of bioactive compounds present in these types of fruits, as the main methodologies used for the assessment of their antioxidant potential are presented in this chapter.

Keywords: Antioxidant-potential-indicators, Bioactive-compounds, Cancer, Chronic-diseases, Diabetes, Dietary-fiber, Flavonoids, Heart-disease, *In vitro*assay, *In vivo*assay, Lipid-peroxidation, Low-density-lipoprotein cholesterol, Obesity, Peel, Phenolic compounds, Pulp, Vitamins.

INTRODUCTION

Fruits and vegetables are essential components of a balanced and healthy diet since they are important sources of some essential nutrients and contain phytochemicals, which may lower the risk of cardiovascular diseases [1, 2].

The link between dietary factors and cardiovascular disease (CVD) has been mainly and almost only correlated to lipids, especially saturated fat and cholesterol, namely "low density lipoprotein (LDL) levels resulting in atherosclerotic vascular changes" [3]. CVD have been reported by several international health organizations as one of the main causes of death worldwide [4], and several "epidemiological, clinical and biochemical studies have proved that increased lipid concentrations, namely triglycerides, total cholesterol, low density lipoprotein cholesterol are risk factors to the development of CVD and atherosclerosis" [3].

Some studies associated the ingestion of fruits and vegetables with "the reduced risk of chronic diseases like cancer prevention" [5] and with a general decrease in all-cause mortality [6]. An inverse association has been established "between the consumption of fruits and vegetables and the prevalence of CVD". On the other hand, inversely low consumption of fruits and vegetables produce less favorable plasma biomarker profiles predictive of CVD, bone diseases and higher mortality rates (Khaw *et al.*, 2001) [7].

Phytochemicals like phenolics, present in fruit and vegetables, have been "suggested to be the major bioactive compounds for human health benefits" [8, 9]. "The protective effect of fruits and vegetables has generally been attributed to their antioxidant constituents, including vitamin C (ascorbic acid), vitamin E (α tocopherol), carotenoids, glutathione, flavonoids, and phenolic acids" [10 - 13]. The antioxidant capacity of pears and apples has been associated with their total phenolic content; having a positive influence on lipids metabolism (hypocholesteromic effect and plasma antioxidant potential) [14, 15].

In general pears present lower antioxidant capacity in comparison with apples; such difference was associated to pears' low percentage (76%) of free phenolics in comparison then apples (91.8% of free phenolics) [8]. On the other hand, the percentage of bound phenolics was higher in pears than in apples, 24% and 8.2%, respectively. Nevertheless, "the effect of bound phytochemicals to human health is not clear yet", but some research in this area suggests that bound phytochemical might "survive stomach and small intestine digestion to reach the colon and be digested by bacteria flora to release phytochemicals locally to provide health benefits" [16].

This chapter focuses on the differences in terms of bioactive compounds (phenolic composition, vitamins, fiber) between apples and pears and how these differences may be associated to antiproliferative activities like tumor cancer cell growth and also how the changes that occur during the gastric digestion can influence the bioavailability of apple and pear constituents.

Apples

"Apples are one of the most popular and frequently consumed fruits in the world". This type of fruit is a "good source of polyphenolic compounds such as flavonoids and phenolic acids". In general, the phenolic composition of apple skin has been demonstrated to be significantly higher when compared with the phenolic content of apple flesh extracts [15]. At least twenty-nine phytochemical compounds were identified in apple peels; from those a significant part corresponds to flavonoids (*e.g.* quercetin-3-O-β-D-glucopyranoside and quercentin) which showed potent antiproliferative activity against tumor cell proliferation [17]. The high content of flavonoids together with the concentration of phenolic acids found in apple peels confers to apple fruit an important antioxidant activity [17]. However, diversity and concentration of apple polyphenols was found to be dependent on apple fruit variety.

Flavonoid compounds extracted from "Jonagold" apples were tested in terms of their antihypertensive properties by assaying their capacity to inhibit the "angiotensin converting enzyme (ACE) pointed as the main therapeutic target in controlling blood pressure" [18]. Inhibition of ACE was tested by "biochemical assay and by human umbilical vein endothelial cell (HUVEC) model" using selected quercetin metabolites. Enzymatic assays show that "quercetin metabolites" *e.g.* quercetin-3-O-glucoronic acid and quercetin-3-O-sulfate, are competitive inhibitors of ACE; then are effective for inhibit ACE at low concentration levels, presenting good inhibition capacity of ACE [18]. Quercetin "QAE" and triterpene "TAE" compounds were extracted from "Jonagold" and "Idared" apples peels, respectively [19]; these extracts were tested in terms of their "ability to inhibit human low density lipoprotein cholesterol (LDL-C) oxidation by *in vitro* tests" [19]. Overall the two extracts inhibited LDL-C oxidation during *in vitro* tests; however, "QAE" metabolites were more effective to inhibit LDL-C oxidation by their optimum concentrations at physiological conditions. In the case of the "TAE" constituents, ursolic acid was detected at higher concentration and performs better than corosolic and oleanolic acids [19]. This study confirms that "an antioxidant compound will not be effective at concentrations lower or higher than its optimal concentration range. At low levels, antioxidant compounds cannot provide enough protection while at high concentrations they can act as pro-oxidants" [19]. Another work in the area of triterpene compounds in this case extracted from "Gold-red" apples revealed that apple peel is the richest part in triterpene compounds and from them ursolic acid presents the highest concentration [20]. Some of the isolated triterpenes; namely, ursolic, oleanolic and betulinic acids have the highest toxicity inhibition and most of the isolated triterpenes presented an inhibitory activity of the α-Glucosidase enzyme activity, an enzyme linked to the final breakdown of carbohydrates to absorbable monosaccharide units [20]. Triterpenoids were also isolated from red "Fuji" apple peels. The ethanolic extracts (AP), the total triterpenoids (ATT), the ursolic acid (UA) and the oleanolic acid (OA) isolates from "Fuji" apple peels were tested against four cancer cell lines proliferation: "human breast cancer (MCF-7 and MDA-MB-231)", "human colon cancer (Caco-2)", and "human liver cancer (HepG2)" [21]. In general all triterpenoids extracts presented "antiproliferative activities against all four cancer cell lines", especially to "human breast cancer" [21]. However, "ATT showed the highest antiproliferative activity toward MCG-7 and MDA-MB-231 cells" and only "ATT was evaluated for *in vivo* antimammary tumor activity" in rat model [21]. It was observed that "ATT could reduce mammary tumor in a rat model as the total cumulative weight of observable tumors in the ATT supplemented rat group was significantly lower when compared to the control group" [21]. Furthermore, this study investigated the related mechanism behind "ATT-induced MDA-MB-231 cell death" [21]. They find out that apoptosis was the biochemical mechanism associated with cancer cell death [21]. In the same line of the triterpenes bioactive compounds a new triterpene "identified as 3β-trans-cinnamiyloxy-2α-hydroxy-urs-12-en-28-oic acid (CHUA) was isolated" from "Fuji" apples [22]. CHUA antitumor activity was investigated by *in vitro* and *in vivo* assays. *In vivo* assays were conducted to test the inhibitory efficiency of different CHUA doses against human tumor cell lines [22]. As it was reported previously [21], the best results were obtained for the inhibition of breast cancer (MDA-MB-231). Then, *"in-vivo"* antitumor activity of CHUA was tested for antimammary tumor activity using the "nude mouse xenograft model" [22]. The results indicated a significant reduction in the weight of tumors in mice after 10 days of CHUA treatment. Antitumor mechanism of CHUA was explained by apoptotic process which was confirmed by the "depletion of the mitochondria membrane potential ($\Delta\psi m$)" [22]. Besides, the increase in the ratio of the apoptosis-related proteins, Bax/Bcl-XL, accompanied by the liberation of "cytochrome c from the mitochondria" it was also associated to antitumor mechanism [22]. A remark of this study is that "triterpenoids from 3 kg of apple peels could contribute with enough amount of CHUA for chemoprevention for a 100 lb human" according to *in vivo* experiments [22].

Pears

Together with apples, "pears are one of the most frequently consumed fruits in the world". Pears are a good source of nutrients including bioactive compounds.

Besides, they also contribute to provide sensorial attributes, such as "aroma, sweetness and crispness" to the human diet [13]. The edible parts (just the flesh) of eight Chinese pears were analyzed in terms of "sugars, organic acids, amino acids, fatty acids and minerals" [23]. It was observed that "fructose was the dominant sugar in the eight varieties followed by glucose and sucrose, while malic acid was the principal organic acid" [23]. Linoleic acids and palmitic were "the principal fatty acids, constituting 70-80% of the total fatty acids in the pear fruit" and in terms of amino acids were asparagine and serine the principal ones [23]. Finally pears were rich in "potassium followed by magnesium and calcium" and some other micronutrients were also detected like iron, copper, zinc, manganese and boron [23].

Research in the area of pears phytochemicals reported that the phenolic profile in pears also differs significantly between flesh and peel; and the phenolic composition differs and depends on pear's cultivar [24]. A comparative study using six pears cultivars showed that the peels of pears presented "higher concentration of chlorogenic acid, flavonols and arbutin than the flesh, where only chlorogenic acid was detected" [24]. In terms of vitamins, "ascorbic acid and dehydroascorbic acid were detected in the peel of pears, whereas only dehydroascorbic acid was present in the flesh" [24]. Arbutin, a characteristic phenolic constituent of pears, was systematically evaluated in 17 Oriental cultivars [25]. The interest in arbutin quantification comes as the Oriental medicine identified this phenolic compound as an important antibacterial and anti-inflammatory agent. This study reported that the greatest concentration of arbutin was present in the peel of the pears. The authors also compared Oriental pears with 5 Occidental pear cultivars; and they observed significant differences between the two groups of pears concluding that arbutin was two-fold higher in Oriental pears than in Occidental pears. On the other hand, chologenic acid was three times higher in Occidental pears than in Oriental pears [25]. Therefore, as it was mentioned before, fruits constituents can vary and their concentration can depend on different factors; those including: fruit origin, variety, part of the fruit and physiological stage of the fruit [26].

In vitro assay models were used to assess the correlation of different varieties of pear phenolic extracts to hyperglycemia and hypertension diseases [27]. It was observed that flesh and peel pear extracts inhibited enzymes involved in the carbohydrate digestion (alpha-glucoside and alpha-amylase enzymes). Then, pear extracts were associated with the regulation of carbohydrate metabolism helping in the control of postprandial hyperglycemia. In this case "only a small correlation was found between the total phenolic content and alpha-glucosidase inhibitory activity" [27]. On the other hand, no correlation was established between the total phenolic content and the alpha-amylase inhibitory activity. In this study just two pear extracts from the "Green D' Anjou" and "Bartle" varieties presented bioactivity "to inhibit angiotensin I-converting enzyme (ACE), which plays an important role in the regulation of blood pressure" [27].

Bioaccessibility and Bioavailability of Bioactive Compounds

Results from chemical extractions have been often used to estimate the amounts of "available nutrients and phytochemicals (*e.g.* polyphenols) for human daily diets or portions", but according with research in the area of phytochemicals those recommendations does not take into account the changes that might happen during gastro-intestinal (GI) digestion [28]. The term "bioaccessibility" was defined as "the fraction of a compound that is released from the food matrix in the gastrointestinal tract and thus becomes available for the intestinal absorption"; and "bioavailability" is used to "describe the proportion of the ingested compound that reaches the systemic circulation". In the case of apple bioactive compounds, a study that simulated an *in vitro* model of the GI digestion of apple phytochemical shows that most of the polyphenols release takes place during the "gastric phase (ca. 65% of phenolics and flavonoids) and only a small release ($\approx$ 10%) occurs during intestinal digestion" [28]. This study also showed that only 55% and 44% from the initial concentrations of polyphenols and flavonoids respectively were soluble dialyzable [28]. Then, from this study it is possible to see that maybe a huge part of "apple polyphenols are bounded to macromolecular compounds that are non-dyalisable" and probably some of the research in the area of phytochemicals quantification overestimates polyphenol availability. Moreover, it is known that "the transition from the acidic gastric to the mild alkaline intestinal environment caused a decrease in the amount of bioaccessible total polyphenols, flavonoids and anthocyanins" [28]. By the same approach, using an "*in vitro* model of the gastrointestinal (GI) digestion", researchers observed that effectively there is a drastic reduction in "the bioaccessibility of individual polyphenols" following the gastric and intestinal phases [29]. Bioactive compounds like "flavan-3-ols (catequin, epicatechin and procyanidin B1 and B2) were completely unstable in the intestinal medium", due to their pH sensitivity. Chlorogenic acid, an abundant hydroxycinnamic acid in apple fruits, was "significantly debased during intestinal digestion with its partial isomerization to cryoptochlorogenic and neochlorogenic acids" [29]. In the terms of polyphenols abundance detected in final phase of the *in vitro* model GI digestion, quercetin-3-O-galactoside was in general present at higher concentration in the intestinal juice followed by phloridzin and quercetin-3-O-rhamnoside compounds [29]. A study that compares by *in vitro* assays the "bioaccessibility, bioavailability and plasma protein interaction of apple polyphenols from different apple varieties" shows that important changes take place in terms of polyphenols metabolites concentration along the digestion process [30]. This study reported that apple polyphenols varied significantly among apple cultivars and those differences were also detected in terms of fruit part (flesh or skin). In this particular case the 'Anurca' apple flesh was significantly rich in procyanidin compounds in comparison with the other apple varieties or in relation to its skin part. Concerning the digestion process and starting from the salivary digestion, it was found significant amounts of procyanidins (35.0%), rutin (35.1%) and phloridzin (27.3%) from both peel and flesh during this part of the digestion. Salivary digestion contributes to the release of important amounts of "nutrients and non-nutrients from the food matrix" by the mastication process and by the action of the digestive enzymes. Also, salivary digestion allows that large part of food constituents and some bioactive compounds, that are "gastro-sensitive and/or poorly absorbed in the intestinal tract" can be absorbed through the oral mucosal epithelium [30]. Continuing the digestion process, after the salivary step, "an increase" in the concentration of "procyanidins, rutin and phloridzin" was detected in the gastric phase [30]; indicating that the process of polyphenol extraction continues and flavonoids glycosides such as "rutin and phloridzin" are stable at gastric acidic conditions. Concerning the bioaccessibility of apple flesh and peel polyphenols after intestinal digestion, "both procyanidins and phloridzin showed an increase in their concentration, while rutin levels decrease" [30]. As "approximately 50% of native procyanidins were not absorbed they are expected to accumulate in the intestinal lumen where a potential inhibition capacity of cellular cholesterol uptake could be assumed". "At least 18% of permeated procyanidins significantly bound to plasma HDLs suggesting their major participation in cholesterol metabolism" [30].

Some of the benefits of "apple bioactive compounds are also associated to the compounds that are not absorbed and metabolized during gastrointestinal digestion which can induce changes in microbial population faeces" [16]. A study that involved seven different apple cultivars was conducted in which the objective was to characterize apple varieties in terms of their extractable phenolic concentration, non-extractable proanthocyanidins, dietary fibre and carbohydrates [16]. Among these varieties the 'Granny Smith' cultivar has "the highest content in extractable fibre and phenolics, and low in available carbohydrates" [16]. 'Granny Smith' extractable and non-extractable compounds were quantified after *in vitro* digestion test; it was observed that extractable phenolics were reduced to almost 50% of the initial value. On the other hand, the proanthocyanidins content was the same after *in vitro* digestion test. This result indicates that proanthocyanidins could be at the colon and be metabolized by faecal bacteria. Then, faecal bacteria collected from diet-induced obese mice and lean mice were cultured with medium containing dried apples after *in vitro* digestion. The results show that the bacterial population in faeces from obese mice tended to be the same to the lean control, suggesting that "apple non-digestible compounds might help to re-establish microbiota balance in obesity cases" [16].

CONCLUDING REMARKS

Evaluation of the total, subclasses and individual dietary intake of specific pear and apple polyphenolic compounds would be crucial to determine their influence on chronic diseases prevention. In the case of apples there is the necessity to study the biochemical mechanisms behind apple bioactive compounds and disease prevention. Pears bioactive compounds remain unexplored in terms of their bioaccessibility and bioavailability; so, they should be study in more detail.

CONFLICT OF INTEREST

The author confirms that author has no conflict of interest to declare for this publication.

ACKNOWLEDGEMENTS

Declare none.

REFERENCES

[1]   Darmon N, Darmon M, Maillot M, Drewnowski A. A nutrient density standard for vegetables and fruits: nutrients per calorie and nutrients per unit cost. J Am Diet Assoc 2005; 105(12): 1881-7.
       [http://dx.doi.org/10.1016/j.jada.2005.09.005] [PMID: 16321593]

[2]   Pennington JA, Fisher RA. Classification of fruits and vegetables. J Food Compos Anal 2009; S23-1.
       [http://dx.doi.org/10.1016/j.jfca.2008.11.012]

[3]   Serra AT, Rocha J, Sepodes B, *et al*. Evaluation of cardiovascular protective effect of different apple varieties - Correlation of response with composition. Food Chem 2012; 135(4): 2378-86.
       [http://dx.doi.org/10.1016/j.foodchem.2012.07.067] [PMID: 22980816]

[4]   Global Atlas on Cardiovascular Disease Prevention and Control. Geneva 2011.

[5]   Terry P, Terry JB, Wolk A. Fruit and vegetable consumption in the prevention of cancer: an update. J Intern Med 2001; 250(4): 280-90.
       [http://dx.doi.org/10.1046/j.1365-2796.2001.00886.x] [PMID: 11576316]

[6]   Bazzano LA, He J, Ogden LG, *et al*. Fruit and vegetable intake and risk of cardiovascular disease in US adults: the first National Health and Nutrition Examination Survey Epidemiologic Follow-up Study. Am J Clin Nutr 2002; 76(1): 93-9.
       [PMID: 12081821]

[7]   Khaw KT, Bingham S, Welch A, *et al*. Relation between plasma ascorbic acid and mortality in men and women in EPIC-Norfolk prospective study: a prospective population study. European Prospective Investigation into Cancer and Nutrition. Lancet 2001; 357(9257): 657-63.
       [http://dx.doi.org/10.1016/S0140-6736(00)04128-3] [PMID: 11247548]

[8]   Sun J, Chu Y-F, Wu X, Liu RH. Antioxidant and antiproliferative activities of common fruits. J Agric Food Chem 2002; 50(25): 7449-54.
       [http://dx.doi.org/10.1021/jf0207530] [PMID: 12452674]

[9]   Upadhyay R, Mohan Rao LJ. An outlook on chlorogenic acids-occurrence, chemistry, technology, and biological activities. Crit Rev Food Sci Nutr 2013; 53(9): 968-84.
       [http://dx.doi.org/10.1080/10408398.2011.576319] [PMID: 23768188]

[10]   Sies H, Stahl W. Vitamins E and C, beta-carotene, and other carotenoids as antioxidants. Am J Clin Nutr 1995; 62(6) (Suppl.): 1315S-21S.
       [PMID: 7495226]

[11]   Liu S, Willett WC, Manson JE, Hu FB, Rosner B, Colditz G. Relation between changes in intakes of dietary fiber and grain products and changes in weight and development of obesity among middle-aged women. Am J Clin Nutr 2003; 78(5): 920-7.
       [PMID: 14594777]

[12]   Hegedús A, Engel R, Abrankó L, *et al*. Antioxidant and antiradical capacities in apricot (Prunus armeniaca L.) fruits: variations from genotypes, years, and analytical methods. J Food Sci 2010; 75(9):

C722-30.
[http://dx.doi.org/10.1111/j.1750-3841.2010.01826.x] [PMID: 21535583]

[13]   Li X, Gao WY, Huang LJ, Zhang JY, Guo XH. Antioxidant and antiinflammation capacities of some pear cultivars. J Food Sci 2011; 76(7): C985-90.
[http://dx.doi.org/10.1111/j.1750-3841.2011.02302.x] [PMID: 21824135]

[14]   Leontowicz H, Gorinstein S, Lojek A, *et al*. Comparative content of some bioactive compounds in apples, peaches and pears and their influence on lipids and antioxidant capacity in rats. J Nutr Biochem 2002; 13: 603-10.
[http://dx.doi.org/10.1016/S0955-2863(02)00206-1] [PMID: 12550072]

[15]   Leontowicz M, Gorinstein S, Leontowicz H, *et al*. Apple and pear peel and pulp and their influence on plasma lipids and antioxidant potentials in rats fed cholesterol-containing diets. J Agric Food Chem 2003; 51(19): 5780-5.
[http://dx.doi.org/10.1021/jf030137j] [PMID: 12952433]

[16]   Condezo-Hoyos L, Mohanty IP, Noratto GD. Assessing non-digestible compounds in apple cultivars and their potential as modulators of obese faecal microbiota *in vitro*. Food Chem 2014; 161: 208-15.
[http://dx.doi.org/10.1016/j.foodchem.2014.03.122] [PMID: 24837942]

[17]   He X, Liu RH. Phytochemicals of apple peels: isolation, structure elucidation, and their antiproliferative and antioxidant activities. J Agric Food Chem 2008; 56(21): 9905-10.
[http://dx.doi.org/10.1021/jf8015255] [PMID: 18828600]

[18]   Balasuriya N, Rupasinghe HP. Antihypertensive properties of flavonoid-rich apple peel extract. Food Chem 2012; 135(4): 2320-5.
[http://dx.doi.org/10.1016/j.foodchem.2012.07.023] [PMID: 22980808]

[19]   Thilakarathna SH, Rupasinghe HP, Needs PW. Apple peel bioactive rich extracts effectively inhibit *in vitro* human LDL cholesterol oxidation. Food Chem 2013; 138(1): 463-70.
[http://dx.doi.org/10.1016/j.foodchem.2012.09.121] [PMID: 23265512]

[20]   He Q-Q, Yang L, Zhang J-Y, Ma J-N, Ma C-M. Chemical constituents of gold-red apple and their α-glucosidase inhibitory activities. J Food Sci 2014; 79(10): C1970-83.
[http://dx.doi.org/10.1111/1750-3841.12599] [PMID: 25227714]

[21]   He X, Wang Y, Hu H, Zhang Z. *In vitro* and *in vivo* antimammary tumor activities and mechanisms of the apple total triterpenoids. J Agric Food Chem 2012; 60(37): 9430-6.
[http://dx.doi.org/10.1021/jf3026925] [PMID: 22924395]

[22]   Qiao A, Wang Y, Xiang L, Wang C, He X. A novel triterpenoid isolated from apple functions as an anti-mammary tumor agent via a mitochondrial and caspase-independent apoptosis pathway. J Agric Food Chem 2015; 63(1): 185-91.
[http://dx.doi.org/10.1021/jf5053546] [PMID: 25521501]

[23]   Chen J, Wang Z, Wu J, Wang Q, Hu X. Chemical compositional characterization of eight pear cultivars grown in China. Food Chem 2007; 104: 268-75.
[http://dx.doi.org/10.1016/j.foodchem.2006.11.038]

[24]   Galvis-Sánchez AC, Gil-Izquierdo A, Gil MI. Comparative study of six pear cultivars in terms of their phenolic and vitamin C contents and antioxidant capacity. J Sci Food Agric 2003; 83: 995-1003.

[http://dx.doi.org/10.1002/jsfa.1436]

[25]   Cui T, Nakamura K, Ma L, Li J-Z, Kayahara H. Analyses of arbutin and chlorogenic acid, the major phenolic constituents in Oriental pear. J Agric Food Chem 2005; 53(10): 3882-7.
[http://dx.doi.org/10.1021/jf047878k] [PMID: 15884812]

[26]   Galvis-Sánchez AC, Fonseca SC, Gil-Izquierdo A, Gil MI, Malcata FX. Effect of different levels of CO_2 on the antioxidant content and the polyphenol oxidase activity of 'Rocha' pears during cold storage. J Sci Food Agric 2006; 86: 509-17.
[http://dx.doi.org/10.1002/jsfa.2359]

[27]   Barbosa AC, Sarkar D, Pinto MD, Ankolekar C, Greene D, Shetty K. Type 2 diabetes relevant bioactive potential of freshly harvested and long-term stored pears using *in vitro* assay models. J Food Biochem 2013; 37: 677-86.
[http://dx.doi.org/10.1111/j.1745-4514.2012.00665.x]

[28]   Bouayed J, Hoffmann L, Bohn T. Total phenolics, flavonoids, anthocyanins and antioxidant activity following simulated gastro-intestinal digestion and dialysis of apple varieties: Bioaccessibility and potential uptake. Food Chem 2011; 128(1): 14-21.
[http://dx.doi.org/10.1016/j.foodchem.2011.02.052] [PMID: 25214323]

[29]   Bouayer J, Deußer H, Hoffmann L, Bohn T. Bioaccessible and dialyzable polyphenols in selected apple varieties following *in vitro* digestion *vs.* their native patterns. Food Chem 2012; 131: 1466-72.
[http://dx.doi.org/10.1016/j.foodchem.2011.10.030]

[30]   Tenore GC, Campiglia P, Ritieni A, Novellino E. *In vitro* bioaccessibility, bioavailability and plasma protein interaction of polyphenols from Annurca apple *(M. pumila Miller cv Annurca)*. Food Chem 2013; 141(4): 3519-24.
[http://dx.doi.org/10.1016/j.foodchem.2013.06.051] [PMID: 23993515]

CHAPTER 4

Stone Fruits as a Source of Bioactive Compounds

Juliana Vinholes[1,*], Daniel Pens Gelain[2], Márcia Vizzotto[1,*]

[1] *Embrapa Temperate Agriculture, Pelotas, Brazil*

[2] *Center of Oxidative Stress Research, Department of Biochemistry, Federal University of Rio Grande do Sul, Porto Alegre, Brazil*

Abstract: Fruits constitute one of the most important sources of phytochemicals in human diet. Stone fruits, such as peaches, plums, almonds, apricots and cherries have been investigated concerning their therapeutic effects in the prevention of a range of diseases. The consumption of these fruits is related with the lower prevalence of diabetes, overweight or general obesity, lower risk for estrogen receptor-negative tumors and cardiovascular protection among others. Phenolic compounds, predominantly flavonoids and phenolic acids, are the main phytochemicals in stone fruits. Considering the importance of stone fruits as a source of biologically active compounds the present chapter aims to provide the current findings in this field and the main implications to human health associated with its consumption.

Keywords: Almond, Anticancer, Antidiabetic, Antiinflammatory, Antioxidant, Apricot, Bioactive Compounds, Cardiovascular protection, Cherry, Flavonoids, Nectarine, Obesity, Peach, Phenolic acids, Phenolic compounds, Plum, Prunus, Stone fruits.

INTRODUCTION

Different studies have demonstrated that a diet rich in fruits and vegetables may decrease the risk of diabetes, cancer, cardiovascular and neurodegenerative disea-ses (*i.e.* Alzheimer and Parkinson) [1 - 4]. This beneficial effect is associated to

* **Corresponding author Juliana Vinholes:** Embrapa Temperate Agriculture, Pelotas, Brazil; Email: julianarochavinholes@gmail.com and **Márcia Vizzotto:** Embrapa Temperate Agriculture, Pelotas, Brazil; Email: marcia.vizzotto@embrapa.br

the nutrients they contain such as fibers, minerals, vitamins and the presence of phytochemicals. Phytochemicals are secondary metabolites produced by plants, in relatively small amounts, where they are responsible for a variety of functions. The ecological role of secondary metabolites production can be related to the potential medicinal effect observed in humans. For example, secondary metabolites in plant defense through cytotoxicity towards microbial pathogens could be useful as antimicrobial drugs in humans, if not very toxic [5].

Phytochemicals compounds have been extensively investigated since they possess a range of activities, which may be involved in the protection against chronic diseases (Fig. **1**). They may also regulate inflammatory and immune responses, inhibit cancer cell proliferation, and protect cells against oxidative damages, caused by free radicals and reactive oxygen species, to macromolecules such as lipids, proteins, and DNA [6]. Examples of phytochemicals present in fruits are phenolic, terpenoids, alkaloids and organosulfur compounds.

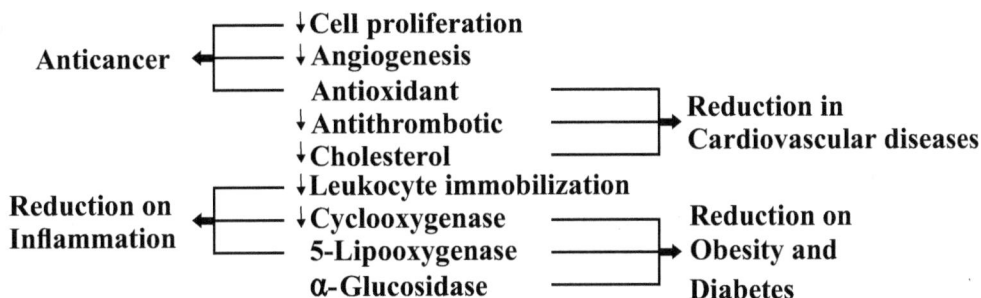

Anticancer
- ↓ **Cell proliferation**
- ↓ **Angiogenesis**
- **Antioxidant**
- ↓ **Antithrombotic** → **Reduction in Cardiovascular diseases**
- ↓ **Cholesterol**

Reduction on Inflammation
- ↓ **Leukocyte immobilization**
- ↓ **Cyclooxygenase**
- **5-Lipooxygenase** → **Reduction on Obesity and Diabetes**
- **α-Glucosidase**

Fig. (1). Relation of phytochemicals actions on different diseases (adapted from [7]).

Stone fruits, also known as drupes, are trees and shrubs members of Prunus genus. Peaches (*Prunus persica* (L.) Batsch), nectarines (*P. persica*, var. nectarine), European plum (*P. domestica* L.), Japanese plum (*P. salicina*), apricot (*P. armeniaca* L.), mume or Japanese apricot (*P. mume*), sweet cherry (*P. avium*), sour cherry (*P. cerasus*) and almond (*P. amygdalus*) are examples of stone fruits with economical interest and are highly consumed worldwide [8]. Good nutritional properties are described for these fruits which are also a good source of phytochemicals affording considerable amounts of bioactive phenolic compounds, mainly flavonoids and phenolic acids.

In the last years, the consumption of fruits has been investigated in different epidemiological studies. Stone fruits consumption was correlated with lower prevalence of diabetes and obesity [9 - 11], cardiovascular protection [12 - 17] and inversely associated with estrogen receptor-negative (ER−) breast cancer [18, 19] and risk of esophageal squamous carcinomas [20]. Moreover, *in vivo* and *in vitro* studies with stone fruits, their extracts, purified fractions and their phytochemicals have been related with these epidemiological evidences on chronic diseases. Considering these facts, we intend to present in this chapter recent research, mostly from the past 10 years, related with epidemiological evidences of the consumption of stones fruits and their health benefits as well as the relationship between these effects and the phytochemicals, mainly phenolic compounds, present in these matrices.

Stone Fruits Bioactive Compounds

Stone fruits can provide different bioactive phytochemicals such as terpenoids, mainly carotenoids, tocopherols and phenolic compounds. In this chapter greater attention will be given to phenolic compounds, since these are the most abundant compounds found in stone fruits. Nevertheless, stone fruits carotenoids and tocopherols, their amounts and their role to human health will be briefly commented.

Carotenoids

Carotenoids are C_{40} terpenoids pigments present in all stone fruits. The carotenoids α- and β-carotene, capsantine, β-cryptoxanthin, lycopene, lutein and zeaxantin, have been reported [21 - 24]. However, there is a large variation on the amounts of these compounds that is dependent of the specie and variety studied. Apricots is the most abundant stone fruit concerning carotenoids with 1512-16500 µg/100g of fresh weight, followed by peaches (up to 1160 µg/100g of fresh weight), plums (231 µg/100g of fresh weight), nectarines (162 µg/100g of fresh weight), being cherries (1.1 µg/100g of fresh weight) the poorest one [25 - 27]. Carotenoids are important to human health since they are precursors of vitamin A, that are implicated in the production of retinoids, vital for human vision [28].

Besides, different studies report carotenoids as antioxidants [29, 30], a very important activity linked with the protection of cells from the damage caused by free radicals and reactive oxygen species which can result in damage to proteins, DNA and lipids resulting in cell death and neoplasia, which are probably in the origin of many human diseases [6].

Tocopherols

Tocopherols and tocotrienols (vitamin E) are natural liposoluble compounds that exist as a mixture of 8 homologues, Alpha, Beta, Gamma and Delta. Almonds are the richest provider of tocopherols (26 mg α-TOC/100g), followed by apricots, nectarines and peaches (0.89, 0.75 and 0.73 mg α-TOC/100g, respectively), plums (0.26 mg α-TOC/100g) and cherries (0.07 mg α-TOC/100g) [8]. The main contribution of tocopherols and tocotrienols for human health are their antioxidant properties [31], that can protect cells against oxidative injury as mentioned for carotenoids. Nevertheless, others properties have been related to vitamin E, such as gene regulation and antiproliferative effects, prevention of platelet aggregation, enhancement of vasodilation and modulation of enzymes associated with the immune system as recently reviewed by [8, 32].

Phenolic Compounds

Phenolic compounds are the main phytochemical compounds found in stone fruits. They are formed *via* the shikimic acid pathway, responsible by the production of the phenylpropanoids, and the acetic acid pathway, in which the simple phenol is mainly formed [33, 34]. Phenolic compounds are structurally characterized by the presence of at least one aromatic ring with one or more hydroxyl groups attached. They represent the largest and wide group of compounds, accounting for more than 8000 phenolic structures, and are classified according with the number of carbons atoms and their basic structure [35, 36] or as flavonoids and non-flavonoids. Non-flavonoids include phenol, benzoquinones and phenolic acids, while flavonoids are the remaining ones. Flavonoids can be divided into classes including mainly flavan-3-ols, flavones, flavonols, anthocyanins, flavanones and isoflavones, and phenolic acids are represented by hydroxybenzoic and hydroxycinnamic acids. Flavonoids and phenolic acids are the two major classes of dietary polyphenols [37]. Stone fruits are a good source of phenolic compounds being peaches capable of providing up to 1260 mg/100 g of fresh weight (FW) (Table **1**).

Table 1. Stone fruits phenolic composition (mg/ 100g of fresh weight).

Total	Peach	Nectarine	Plum	Almond	Apricot	Cherry
	mg/100 g FW					
Phenolic compounds	21.0-1260.0	13.6-102.4	42.0-563.0	45.0-241.0	32.0-211.0	27.8-312.4
Flavonoids	76.4	6.5-21.7	5.4-257.5	14.0-26.7	78.5-139.0	30.2-109.2
Phenolic acids	163.0	10.0	0.1-58.4	0.3-0.5	12.3-40.4	11.0-33.6
Refs	[22, 25, 38-41]	[9, 25, 42]	[25, 40, 41, 43-45]	[46-48]	[46, 47, 49, 50]	[27, 51, 52]

Daily intake of phenolic compounds has been estimated in different studies from different countries. Total phenolic intake can reach 1193 mg/day where flavonoids contribute with 42% (506 mg/day) and phenolic acids with 53% (639 mg/day) [53]. Fruits were reported to contribute with 17% (206 mg/day) of phenolic compounds, being mostly represented by flavonoids (83%, 172 mg/day), and with small contribution of phenolic acids (15%, 32 mg/day) [53]. The median total flavonoids intake was reported to be 166.0-193.3 mg/day, where flavon-3-ols were the most representative group (85.0-161.7 mg/day), followed by flavonone (15.5-46.4 mg/day), flavonols (12.9-20.7 mg/day) and anthocyanins (2.6-7.6 mg/day) [54, 55].

Hydroxycinnamic acids were the main contributors to the total phenolic acid intake, accounting for 84-95% of intake [53, 71]. Fruits, vegetables and nuts are the main food sources of hydroxycinnamic acids, after coffee [53, 71, 72]. Concerning stone fruits, plums and cherries were found among the fruits with high contributions for phenolic compounds intake in this food group (8%) [53]. The type and amounts of phenolic compounds on stone fruits can range significantly inter and intra species since it is dependent of edaphoclimatic factors (Table **2**) [73, 74].

Table 2. Stone fruits phenolic composition (mg/ 100g of fresh weight).

	Peach	Nectarine	Plum	Almond	Apricot	Cherry
	mg / 100 g FW					
Catechin	1.0-42.0	0.3-4.7	3.3-31.8	0.5-22.7	2.8-23.2	0.6-6.8
Epicatechin	1.4-9.2	2.5	0.6-15.6	0.3-4.0	1.1-14.5	13.0-15.0
Epigallocatechin	0.3-1.5			2.6	1.3-9.6	0.05-0.2
Procyanidin B1	14.7	9.9	2.4-31.8		0.09	0.2
Procyanidin B2	2.3		0.7-10.1			2.1
Cyanidin-3-*O*-glucoside	2.4		0.1-48.4		0.8-1.2	1.43-18.7
Cyanidin-3-*O*-rutinoside			0.2-33.8			6.9-143.2
Naringenin			0.2-2.5	0.01-28.1		
Naringenin-7-*O*-glucoside				0.08-10.5		
Eriodictyol				0.03-0.5		
Quercetin		0.1	0.2-6.6	0.02-3.5	0.8-1.6	
Quercetin-3-*O*-glucoside		0.1	0.02-14.3	0.09	0.6-3.6	0.2-0.7
Quercetin-3-*O*-rutinoside	10.2	0.1	0.2-7.9	0.05-0.5	6.7-37.1	0.9-4.6
Quercetin-3-*O*-galactoside			0.3	0.2-1.2		
Quercetin-3-*O*-rhaminoside					2.0-73.0	
Kaempferol			0.3-1.1	0.003	0.6	
Kaempferol-3-*O*-rutinoside	7.0	0.12		0.2-7.0		0.3-1.39
Myricetin					0.6	
Isorhamnetin-3-O-glucoside				0.1-1.0		0.09-2.65
2,5-Dihydroxy benzoic	98.5					
Protocatechuic acid				0.1-12.2	12.6	
Vanillic acid				0.09-9.3	0.2-3.8	
Gallic acid				0.2-1.6	35.0	
p-Coumaric acid			0.2-2.6		0.2-8.8	0.8-6.84
Ferulic acid			0.5-2.2		0.3-1.6	
Caffeic acid			0.3-6.1	0.2-3.2	0.2-7.3	
3-*O*-Caffeoylquinic acid	4.0-29.3	3.9	3.5-75.8	0.03-2.2	2.4-94.0	0.13-19.1
5-*O*-Caffeoylquinic acid	2.5-25.0	6.1	1.1-43.4		3.6-18.6	2.9-53.0
Refs	[21, 22, 25, 38-40, 42, 56, 57]	[42, 58]	[24, 25, 40, 42-45, 57-62]	[42, 46-48, 63, 64]	[46, 47, 49, 57, 58, 65-68]	[42, 51, 52, 57, 58, 69, 70]

Stone fruits are responsible to provide 1% of the intake of caffeic acid and 6% of *p*-coumaric acid [71]. More specifically peaches can provide 1 mg/day of catechin and 4.5 mg/day of procyanidin B1, plums 4.2 mg/day of caffeic acid and 1.6 mg/day of cyanidin-3-*O*-rutinoside, almonds contributes with 0.02 mg/day of epigallocatechin and cherries with 2.8 mg/day of caffeic acid and 9.0 mg/day of cyanidin-3-*O*-rutinoside [53].

Health Benefits of Stone Fruits

In Vivo Studies and Their Mechanism of Action

Diabetes and Obesity

Diabetes mellitus and obesity are becoming a major public health concern with high social and health care costs both in developed and developing countries [75]. Type 2 diabetes mellitus (T2DM) is characterized by individuals with post-prandial hyperglycemia associated with low production of insulin, resistance to insulin or both. The T2DM accounts for 90% of cases of diabetes mellitus and is directly related to obesity, since 80 and 90% of obese individuals have T2DM. Concerning obesity its etiology is multifactorial, with a combination of genetic and environmental factors. This metabolic disorder is characterized by the combination of hyperglycemia, dyslipidemia, abdominal obesity and hypertension. Both, diabetes and obesity, are risk factors in several other pathologies such as coronary heart disease, stroke, diabetic ophthalmopathy, diabetic neuropathy and chronic renal failure [76].

Experiments on obese Zucker and lean rats fed with peach (47.5 mL/day) and plum (45.2 mL/day) juices revealed protective effect against a combination of obesity-induced metabolic disorders including hyperglycemia, insulin and leptin resistance, dyslipidemia and low-density lipoprotein oxidation [75]. In addition, a decrease on the expression of pro-atherogenic and pro-inflammatory biomarkers in plasma and heart tissues including intercellular cell adhesion molecule-1, monocyte chemotactic protein-1, nuclear factor-κB (NF-κB) and foam cell adherence to aortic arches was reported. A reduction on levels of angiotensin II in plasma and its receptor Agtr1 in heart tissues was also reported, suggesting a role of peach and plum polyphenols as peroxisome proliferator-activated receptor-γ agonists. Nevertheless, only plum juice prevented body weight gain and increased the ratio high-density lipoprotein cholesterol/total cholesterol in plasma [75].

Insulin-resistant obese Wistar rats fed with plum concentrated (0.25%) were reported to present a decrease in blood glucose and plasma triglyceride concentrations. Besides, a reduction on the areas under the curve for glucose and insulin during a glucose tolerance test was also reported. Authors suggest that plum treatment may increase insulin sensitivity in Wistar obese rats *via* adiponectin-related mechanisms [77].

These effects can be related to the presence of phenolic compounds that can act on diabetes and obesity by different mechanisms. Cyanidin-3-*O*-glucoside (Table **2**, Fig. **2**) reduces blood glucose and increases glucose tolerance [78], improves insulin sensitivity and reduces white adipose tissue messenger RNA levels and serum concentrations of inflammatory cytokines [79], preserves insulin sensitivity [80] and may exert insulin-like activities [81]. These properties have direct effect on reduction of free fatty acids in plasma and improvement on insulin resistance that represents a risk factor for metabolic, cardiovascular, and neoplastic disorders [82]. Cyanidin-3-*O*-rutinoside retards *in vivo* absorption of carbohydrates by inhibition of α-glucosidase [83], fact that was also observed for this compound and for cyanidin-3-*O*-glucoside and cyanidin-3-*O*-galactoside by different authors in *in vitro* studies[83 - 85]. Naringenin and quercetin-3-*O*-glucoside lower mean levels of fasting blood glucose [86, 87], glycosylated hemoglobin, and elevated serum insulin levels [87], plasma C-peptide, triglyceride, total cholesterol and blood urea nitrogen levels and improves glucose tolerance and the immunoreactive of pancreatic islets β-cells [86]. Quercetin-3-*O*-glucoside lowers the biochemical changes and delays paw and tail withdrawal latency in hyperalgesia and allodynia, showing beneficial effects in preventing the progression of early diabetic neuropathy in rats [88]. Quercetin reduces glycaemia and diminishes total cholesterol, triglycerides (TG) and low density lipoprotein (LDL) and increases high density lipoprotein (HDL) levels by inhibition of 11β-hydroxysteroid dehydrogenase type 1 [89]. Myricetin protects the diabetic nephrotoxic rats on all the biochemical parameters studied [90], alters their lipid metabolism [91] and improves carbohydrate metabolism, subsequently enhancing glucose utilization and renal function [92]. Protocatechuic acid activates the nuclear factor erythroid 2-related factor (Nrf2) system [93] *in vivo*, but this compound was also reported with *in vitro* protective effect on insulin sensitivity [80]. Ferulic acid increases insulin release and reduces hepatic glycogenolysis *in vitro*, while caffeic acid affects only the last [94]. Moreover, caffeic acid decreases the hepatic glucose output along with the increased level of adipocyte glucose disposal [94] and induced a decrease of blood glucose and glycosylated hemoglobin levels [95]. 3-*O*-Caffeoylquinic acid inhibits protein tyrosine phosphatase 1B, a negative regulator of the insulin signaling pathway [96].

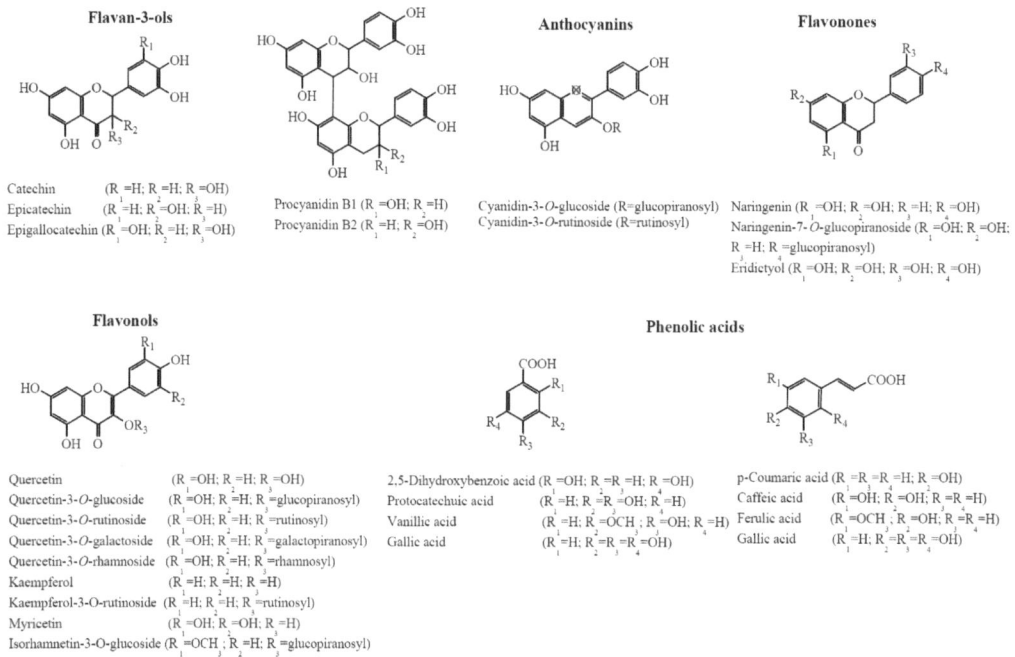

Flavan-3-ols

Catechin (R_1=H; R_2=H; R_3=OH)
Epicatechin (R_1=H; R_2=OH; R_3=H)
Epigallocatechin (R_1=OH; R_2=H; R_3=OH)

Procyanidin B1 (R_1=OH; R_2=H)
Procyanidin B2 (R_1=H; R_2=OH)

Anthocyanins

Cyanidin-3-*O*-glucoside (R=glucopiranosyl)
Cyanidin-3-*O*-rutinoside (R=rutinosyl)

Flavonones

Naringenin (R_1=OH; R_2=OH; R_3=H; R_4=OH)
Naringenin-7-*O*-glucopiranoside (R_1=OH; R_2=OH; R_3=H; R_4=glucopiranosyl)
Eridictyol (R_1=OH; R_2=OH; R_3=OH; R_4=OH)

Flavonols

Quercetin (R_1=OH; R_2=H; R_3=OH)
Quercetin-3-*O*-glucoside (R_1=OH; R_2=H; R_3=glucopiranosyl)
Quercetin-3-*O*-rutinoside (R_1=OH; R_2=H; R_3=rutinosyl)
Quercetin-3-*O*-galactoside (R_1=OH; R_2=H; R_3=galactopiranosyl)
Quercetin-3-*O*-rhamnoside (R_1=OH; R_2=H; R_3=rhamnosyl)
Kaempferol (R_1=H; R_2=H; R_3=H)
Kaempferol-3-O-rutinoside (R_1=H; R_2=H; R_3=rutinosyl)
Myricetin (R_1=OH; R_2=OH; R_3=H)
Isorhamnetin-3-O-glucoside (R_1=OCH$_3$; R_2=H; R_3=glucopiranosyl)

Phenolic acids

2.5-Dihydroxybenzoic acid (R_1=OH; R_2=R$_3$=H; R_4=OH)
Protocatechuic acid (R_1=H; R_2=R$_3$=OH; R_4=H)
Vanillic acid (R_1=H; R_2=OCH$_3$; R_3=OH; R_4=H)
Gallic acid (R_1=H; R_2=R$_3$=R$_4$=OH)

p-Coumaric acid (R_1=R$_2$=H; R_3=OH)
Caffeic acid (R_1=OH; R_2=OH; R_3=R$_4$=H)
Ferulic acid (R_1=OCH$_3$; R_2=OH; R_3=R$_4$=H)
Gallic acid (R_1=H; R_2=R$_3$=R$_4$=OH)

Fig. (2). Chemical structures of biological active phenolic compounds present in stone fruits.

Concerning obesity consumption of catechins and phenolic acids (Fig. **2**), Table **2**) has been associated with a variety of beneficial effects including increased plasma antioxidant activity, blood vessel expansion, fat oxidation, and resistance of LDL to oxidation [97 - 100]. Epicatechin and epigallocatechin lower lymphatic cholesterol absorption in rats, and decrease lymphatic absorption of triacylglycerols [101]. Cyanidin-3-*O*-glucoside reduced obesity, accumulation of fat in tissues, and plasma triglyceride levels, by inhibiting lipoprotein lipase [82], activating of protein kinase phosphorylation [102], and upregulating hepatic cholesterol 7α-hydroxylase expression [103]. Quercetin can protect LDL against oxidation [104]. Quercetin and kaempferol could significantly improve insulin-stimulated glucose uptake [105] and regulates hepatic apolipoprotein A-I (apo A-I) and HDL synthesis by inducing apo A-I gene expression in HepG2 and Caco-2 cells [106].

Cardiovascular Disease

Cardiovascular diseases are a group of disorders of the heart and blood vessels and is the number 1 cause of death globally [107]. Cardiovascular disease is usually resulting from a vascular dysfunction, such as atherosclerosis, thrombosis and hypertension, which can compromises heart function.

The cardio-protective potential of apricot-feeding in the ischemia-reperfusion model of rats *in vivo* was evaluated by [108]. Rats were fed with 10% or 20% dried apricot during 3 months and ischemia-reperfusion produced by occlusion of the left main coronary artery for 30 min, followed by 120 min reperfusion, in anesthetized rats. Significant and similar decrease on infarct sizes were observed in 10% (55%) and 20% (57%) apricot-fed groups compared to control group (68%). Besides, Cu, Zn superoxide dismutase (SOD) and catalase (CAT) activities were increased, and lipid peroxidation was decreased in the hearts of 20% apricot-fed group after ischemia-reperfusion [108].

Phenolic compounds present in stone fruits can have *in vivo* protective effect on cardiovascular diseases by exerting for instance anti-atherosclerotic activity, as reported for catechin [109]. Other example is preventive effect of cyanidin-3-*O*-glucoside on formation of glycated-LDL products on aortic endothelial cells induced by NADPH, the impairment of mitochondrial electron transport chain enzymes and cell viability in cultured vascular endothelial cells [110]. Naringenin-7-*O*-glucoside, prevent cardiomyocytes from doxorubicin-induced toxicity by induction of endogenous antioxidant enzymes *via* phosphorylation of extracellular signal-regulated kinases 1 and 2 (ERK1/2) and nuclear translocation of Nrf2 [111] and by stabilizing the cell membrane and reducing reactive oxygen species generation [112]. Quercetin has been shown to induce a progressive reduction in blood pressure when given chronically in several rat models of hypertension. A high dose of quercetin also reduced blood pressure in stage 1 hypertensive patients in a randomized, double-blind, placebo-controlled, crossover study [113]. In addition, it can protect isolated vessels by activation of adenosine monophosphate-activated protein kinase and nitric oxide synthase (eNOS) in human aortic endothelial cells [114]. Quercetin-3-*O*-glucoside improve cell survival in the oxygen-glucose deprivation model of ischemia and increase neurite outgrowth in differentiated PC12 cells subjected to ischemic insult [115].

Cancer

Cancer is a multifactorial disease and a progressive process involving gene-environment interactions which can cause dysfunction in multiple systems, including DNA repair, apoptotic and immune functions.

Peaches and plums extracts have been reported with antiproliferative effects in estrogen-receptor negative breast cancer cells but not in estrogen positive breast cancer line or the normal breast cell line [18, 19, 116, 117]. Crude extracts and fractions rich in flavonoids for peach and plum [19] and hydroxycinnamic acids for peach [18] showed to be very effective against cell proliferation. Further, *in vivo* studies on peach polyphenolic extracts revealed inhibitory and anti-metastatic action on female athymic mice implanted with MDA-MB-435 breast cancer cell line. Inhibition was achieved by feeding mices with 0.8-1.6 mg/day, this effect was mediated by inhibition of metalloproteinases gene expression [118]. Sour cherry extracts were also reported with antiproliferative activity and induction of apoptosis in mammary adenocarcinoma (MCF-7) and mouse mammary tumour cell (4T1) breast cancer cells lines [119]. Moreover, peach and plum extracts and purified phenolic compounds can inhibit growth and induce differentiation of colon cancer cells [120], and plums extracts have similar beneficial effect *in vivo* [121].

A Japanese apricot extract (MK615), has shown strong antiproliferative and antitumorigenic effects in different cancer cell lines and *in vivo* model [122 - 126]. MK615 also inhibit A375 melanoma cells growth by reducing the mRNA- and protein expression levels of DNA binding 1, a basic helix-loop-helix transcription factor family that is essential for DNA binding and the transcriptional regulation of various proteins that play important roles in the development, progression and invasion of tumour [122]. SK-MEL28 cell line, another melanoma cell model, has also their growth inhibited by MK615 in a dose-dependent manner, by increasing the proportion of cells in sub-G1 phase and inducing apoptosis. Inhibition of advanced glycation end products and suppression of the release of a specific cytokine (high mobility group box protein 1) indicates the MK615 for the treatment of malignant melanoma [124]. MK615 also showed strong growth suppression effect mainly in human cancer cells while sparing normal cells such as human umbilical vein endothelial cells (HUVEC) and mouse bone marrow cells [123]. Moreover, MK615 induce the accumulation of reactive oxygen species in cancer cells but not in HUVEC indicating that the antiproliferative effect is due to a reactive oxygen species-dependent mechanism. *In vivo* tests showed that MK615, in both the presence and absence of gemcitabine, significantly inhibited the growth of human pancreatic cancer cells as xenografts without apparent adverse effects [123]. Japanese apricot extract combined with anticancer drugs 5-fluorouracil, irinotecan and cisplatin showed synergistic cytotoxic effects on esophageal squamous carcinoma cells. The effect was observed in both *in vitro* model using YES-2 cells and *in vivo* model by the injection of YES-2 cells into the peritoneal cavity of a severe combined immuno-deficiency mouse. The Japanese apricot extract and 5-fluorouracil induced cell cycle arrest at G2/M phase and at S phase, respectively, and caused apoptosis in YES-2 cells *in vitro* [125]. *In vivo* results showed that the addition of Japanese apricot extract to 5-fluorouracil augmented the suppression of experimental metastasis of the peritoneum. Also a decrease in the number of peritoneal nodules in mice treated with 5-fluorouracil and Japanese apricot extracts was observed when compared with the individual treatment [125]. In other study the Japanese apricot extract inhibited A549 lung cancer cell proliferation at non-cytotoxic doses and suppressed NF-κB activation induced by tumor necrosis factor α (TNF-α) at a dose of 1 mg/mL and blocked proteasome activities in a dose-dependent manner at concentrations of 0.67 and 1 mg/mL [126].

Stone fruits anticancer properties can be also associated to their antioxidant and antiinflammatory effects. At physiological levels antioxidants may protect important cells macromolecules (proteins, DNA and lipids) from oxidative injury, preventing cell death and neoplasia, which are probably in the origin of many human diseases [6]. Moreover, by preventing inflammation, which is an adaptive immune response to tissue injury or infection [127], stone fruits can have a positive effect on chronic diseases such as cancer, Alzheimer's disease, diabetes among others that are related with inflammatory process [128 - 131]. In fact, peaches are responsible for attenuation on the oxidative stress and the inflammation *in vitro*, *ex vivo* and *in vivo* [132, 133]. The mechanism of action was found to be through its protective effect against lipids and proteins damage, increase of antioxidant enzymatic activities and blocking the induction of inflammatory mediators [133].

Several mechanisms of action can be used by different bioactive compounds from stone fruits. Catechin can induced apoptosis and inhibited G2/M phase in cell cycle in HepG2 cells [134]. In addition, catechin and epicatechin, present in all stone fruits, were reported as antioxidants, a possible anticancer mechanism. Inhibition of radicals [135] and prevention of the oxidative injure, induced by palmitic acid in rats, protecting the mitochondrial membrane from collapse and cell death [136]. Oxidative protection was also observed in rats pretreated with this compound [134, 137] that was capable to restore their antioxidant parameters (glutathione, SOD, CAT and lipid peroxidation).

Anthocyanins, present in all stone fruits, can suppress lipid peroxidation in Caco-2 cells [138] and attenuate ethanol-induced migration/invasion of breast cancer cells by blocking ethanol-induced activation of the ErbB2/cSrc/FAK pathway, which is necessary for cell migration/invasion [139]. Moreover, they also suppresses Benzo[a]pyrene-7,8-diol-9,10-epoxide-induced cyclooxygenase-2 (COX-2) expression mainly by blocking the activation of the Fyn (a proto-oncogene tyrosine-protein kinase) signaling pathway [140], inhibit nitric oxide synthase and COX-2 and decrease in nitric oxide and prostaglandin E2 production (PGE2) [141], and inhibit IκBα phosphorylation, thereby suppressing NF-κB activity in cell models [142] and *in vivo* models [143, 144], which may contribute to its chemopreventive potential. In addition, protective effect on DNA cleavage, free radical scavenging activity [145], reduction of reactive oxygen species generation [146], protection against oxidative injure by decreasing mitochondrial reactive oxygen species production and cell necrosis [78] and decreasing lipid peroxidation by increasing cell enzymatic levels [147 - 149] are other possible anticancer mechanisms.

The Flavanone, naringenin, and the flavonol, myricetin, exert a cytostatic effect by the impairment of cell cycle progression and inhibition of the cell migration [150]. Quercetin exerts an apoptotic activity in cancer cell lines mediated by the dissociation of Bax from Bcl-xL and the activation of caspase families [151]; and can also inhibits precursors of cellular differentiation; stimulates phagocytosis [152], antimitotic activity [153], suppression of COX-2 expression by inhibiting the p300 signaling and blocking the binding of multiple transactivators to COX-2 promoter [154], and interferes with cell cycle progression [155]. Moreover, flavonones were reported to have *in vitro* and *in vivo* antioxidant capacity [111, 112, 156, 157]. Reduction of ROS generation and induction of the cellular antioxidant enzymatic system are the *in vitro* and *in vivo* mechanisms described for these compounds [87, 158]. They can reduce nitric oxide production and prevent peroxinitrite formation in LPS-stimulated RAW 264.7 cells by strongly suppressing the phagocytic activity of activated macrophages, thus reducing the expression of mRNA and the secretion of pro-inflammatory cytokines by blockage of nuclear factor kappa-light-chain enhancer of activated B cells (NF-κB) activation and phosphorylation of p38 mitogen-activated protein kinase, ERK1/2 and c-Jun N terminal kinase [159].

The flavonol, quercetin-3-*O*-rutinoside, isolated from the ethyl acetate fraction of *P. domestica*, showed antiproliferative activity in MCF-7 and MDA-MB-468 at 80×10^{-3} mg/mL with inhibitions of 22 and 32%, respectively [117]. One possible mechanism of action of quercetin-3-*O*-rutinoside can be due to its antioxidant activity which has been reported in different *in vitro* and *in vivo* tests. This compound has scavenging activity against peroxyl radicals *in vitro* [160] and *in vivo* [88, 161, 162], recover levels of glutathione [160, 161] and antioxidant enzymes SOD, NO and CAT [88]. Quercetin, is also able to scavenge radicals [163] and have *in vitro* and *in vivo* ability to decrease the tumor necrosis factor-α production [164, 165] and increase total plasma antioxidant status [165] was reported for quercetin and quercetin-3-*O*-galactoside. Quercetin and quercetin-3-*O*-glucoside protect Caco-2 cells [166] and PC12 cells [167] against oxidative stress caused by different agents probably by their free radicals inhibitory effects [157]. In addition, they possess anti-inflammatory action by inhibition of 5-lipoxygenase [157], raft disrupting and anti-oxidant effects [105, 168].

Protocatechuic, 3-*O*-caffeoyl quinic and 5-*O*-caffeoyl quinic acids, isolated from *P. domestica and P. persica*, have been identified as potential chemopreventive dietary compounds due to their antiproliferative effect on the human breast cancer cell line MCF-7 and the estrogen-independent MDA-MB-435 breast cancer cell line [19, 117]. 3-*O*-Caffeoyl quinic and 5-*O*-caffeoyl quinic acids also presents low toxicity on normal cells [19]. Gallic acid also showed *in vitro* (DU145 cells) and *in vivo* (DU145 and 22Rv1 xenograft growth in nude mice) inhibitory and apoptotic activities against prostate cancer [169, 170]. These effects were also observed in other cancers models [171, 172]. Protocatechuic, vanilic, *p*-coumaric, ferulic and 3-*O*-caffeoylquinic (isolated from *P. domestica*), 2,5-dihydroxy benzoic and caffeic acids were reported with antioxidant activity in different models [117, 173 - 175]. *In vivo* tests indicate that hydroxycinnamic acids increases the antioxidants enzymes activities (*i.e.* SOD, CAT and glutathione peroxidase) and decreases levels of liver lipid peroxidation [95, 176 - 178]. Concerning their anti inflammatory properties, *p*-coumaric and ferulic acids showed inhibition of nitric oxide production and inducible nitric oxide synthase (iNOS) expression [179].

Epidemiological Studies

Diabetes and Obesity

Consumption of fruits and their role in the treatment of diabetes and obesity have been investigated in humans [16]. In respect to stone fruits, a study performed on overweight and obese individuals (n=123) fed with hypocaloric almond-enriched diet (56 g/day) resulted in a significant loss of weight when compared with individuals under a hypocaloric nut-free diet after 6 months with no significant weight gain after 18 months. In addition, a reduction on total cholesterol (4%), total:high density lipoprotein (HDL) cholesterol (5%) and triglycerides (12%), and an improvement of the lipid profile was also observed in the almond-enriched diet [11]. The same trend was observed in other study with prediabetes individuals subjected to an 20% almond-enriched diet (approximately 62 g/day) [10]. A clinically significant decline in human low density lipoprotein-cholesterol (LDL-C) was found in the almond-enriched intervention group (-12.4 mg/dl *vs.* -0.4 mg/dl) as compared with the nut-free control group after sixteen weeks of dietary modification. Besides, the almond-enriched intervention group showed improved markers of insulin sensitivity such as reductions in insulin, homeostasis model analysis for insulin resistance, and homeostasis model analysis for beta-cell function compared with the nut-free control group [10].

Cardiovascular Disease

The antithrombotic properties of anthocyanin-rich Queen Garnet plum juice (QGPJ) supplementation with and without exercise-induced oxidative stress was investigated in thirteen healthy participants in a randomized, double-blind, placebo-controlled, cross-over trial. The experiment was carried out during 28 days where participants consumed 200 mL/day of QGPJ and placebo juice, with treatments separated by a two-week wash-out period. QGPJ supplementation inhibited adenosine diphosphate-induced platelet aggregation in both groups (10.7 and 12.7%), reduced platelet activation-dependent P-selectin expression (32.9 and 38.7%) and exhibited favorable effects on coagulation parameters. The arachidonic acid-induced aggregation was reduced under oxidative stress by 28.8% [12]. Daily intake of a single dose of plums was reported to significantly reduce blood pressure in a placebo controlled clinical trial with 259 pre-hypertensive volunteers [13]. Consumption of plum as juice, whole fruit (dried plums), or 3 (about 11.5 g) or 6 plums soaked overnight in a glass of water, with control group (glass of plain water in the morning on empty stomach) also showed significantly reduction on serum cholesterol and LDL [13].

The cardiovascular risk of 108 overweight and obese women, under an almond-enriched diet during 3 months has been studied in a clinical study. Body mass index, waist circumference, waist to hip circumference ratio, total cholesterol, and triglyceride, total: High density lipoprotein-cholesterol, fasting blood sugar and diastolic blood pressure were significantly decreased in almond-enriched diet (50 g/day) group compared to the nut-free group [14]. In a study performed by Choudhury and co-workers [15], healthy young and middle-aged men subjected to a enriched almond diet (50 g/day) during 4 weeks showed an improvement on flow mediated dilatation and a reduction on systolic blood pressure for men with cardiovascular risk factors, but diastolic blood pressure was reduced only in healthy men [15].

Cancer

Epidemiological data reports that consumption of stone fruits has been inversely associated with estrogen receptor-negative (ER−) breast cancer [180, 181]. Breast cancer is a major concern worldwide and is responsible for one of the highest causes of death. ER-negative tumour growth is not estrogen-dependent; consequently, such tumors are not sensitive to hormones treatment that prevents the estrogen binding. Jung *et al.* 2013 [180], evaluated the association between fruit and vegetable intake and risk of ER− breast cancer. Authors analyzed 20 cohort studies that followed 993,466 women by 11 to 20 years and documented 4821 ER− breast cancers. They found that, for vegetable consumption, there is an inverse association with risk of ER− breast cancer, while for fruits only specific ones shows the same behaviour as the case of apples/pears, peaches/ nectarines/apricots, and strawberries [180]. In other study [181], the association of intake of specific fruits and vegetables with risk of ER- postmenopausal breast cancer has been checked. A total of 75,929 women aged 38-63 years at baseline where followed for up to 24 years and 792 incident cases of ER- postmenopausal breast cancer were found. The multivariate relative risk (RR) of ER− breast cancer was 0.82 for every 2 servings of berries/week, 0.69 for at least one serving of blueberries/week and 0.59 for 2 servings of peaches/nectarines per week compared with non-consumers [181].

The association of fruit and vegetable intake and risk of esophageal squamous cell carcinoma and esophageal adenocarcinoma was evaluated by Freedman 2007 [20]. A total of 490,802 participants of the National Institutes of Health - American Association of Retired Persons Diet and Health Study were followed 103 participants were diagnosed with esophageal squamous cell carcinoma and

213 with esophageal adenocarcinoma. A significant inverse association between total fruit and vegetable intake and esophageal squamous cell carcinoma risk was found. Protective effects were stronger for fruits being the Rosacea (apples, peaches, nectarines, plums, pears and strawberries) and Rutaceae (citrus fruits) those with higher action [20].

CONCLUDING REMARKS

The evidences found in the epidemiological studies, which relate the consumption of stone fruits and effects on chronic diseases, corroborate with the reported results for *in vitro* and *in vivo* tests with these fruits and their products (juices and extracts). Moreover, these effects can be attributed to the phenolic compounds present in stone fruits since the individual compounds exert similar effects at low concentrations. In conclusion, stone fruits are a good source of bioactive phenolic compounds and its consumption may have remarkable beneficial effects on human health due to their positive results in the treatment of diabetes and obesity, cardiovascular diseases and cancer.

CONFLICT OF INTEREST

The author confirms that author has no conflict of interest to declare for this publication.

ACKNOWLEDGEMENTS

Juliana Vinholes thanks the Science without Borders Program (CNPq) for the Young Talent attraction fellowship.

REFERENCES

[1]   Loef M, Walach H. Fruit, vegetables and prevention of cognitive decline or dementia: a systematic review of cohort studies. J Nutr Health Aging 2012; 16(7): 626-30.
[http://dx.doi.org/10.1007/s12603-012-0097-x] [PMID: 22836704]

[2]   Woodside JV, Young IS, McKinley MC. Fruits and vegetables: measuring intake and encouraging increased consumption. Proc Nutr Soc 2013; 72(2): 236-45.
[http://dx.doi.org/10.1017/S0029665112003059] [PMID: 23324158]

[3]   Barnes DE. The mediterranean diet: good for the heart = good for the brain? Ann Neurol 2011; 69(2): 226-8.
[http://dx.doi.org/10.1002/ana.22376] [PMID: 21387364]

[4]   Smith-Warner SA, Spiegelman D, Yaun SS, *et al.* Intake of fruits and vegetables and risk of breast cancer: a pooled analysis of cohort studies. JAMA 2001; 285(6): 769-76.
[http://dx.doi.org/10.1001/jama.285.6.769] [PMID: 11176915]

[5]   Briskin DP. Medicinal plants and phytomedicines. Linking plant biochemistry and physiology to human health. Plant Physiol 2000; 124(2): 507-14.
[http://dx.doi.org/10.1104/pp.124.2.507] [PMID: 11027701]

[6]   Praticò D, Delanty N. Oxidative injury in diseases of the central nervous system: focus on Alzheimer's disease. Am J Med 2000; 109(7): 577-85.
[http://dx.doi.org/10.1016/S0002-9343(00)00547-7] [PMID: 11063960]

[7]   Tapas AR, Sakarkar DM, Kakde RB. Flavonoids as nutraceuticals: A Review. Trop J Pharm Res 2008; 7(3): 1089-99.
[http://dx.doi.org/10.4314/tjpr.v7i3.14693]

[8]   Vicente AR, Manganaris GA, Cisneros-Zevallos L, Crisosto CH. Health-promoting Properties of Fruits and Vegetables In: Terry LA, Ed. Prunus. CABI 2011; pp. 59-238.

[9]   Abidi W, Jiménez S, Moreno MÁ, Gogorcena Y. Evaluation of antioxidant compounds and total sugar content in a nectarine [*Prunus persica* (L.) Batsch] progeny. Int J Mol Sci 2011; 12(10): 6919-35.
[http://dx.doi.org/10.3390/ijms12106919] [PMID: 22072927]

[10]   Wien M, Bleich D, Raghuwanshi M, *et al.* Almond consumption and cardiovascular risk factors in adults with prediabetes. J Am Coll Nutr 2010; 29(3): 189-97.
[http://dx.doi.org/10.1080/07315724.2010.10719833] [PMID: 20833991]

[11]   Foster GD, Shantz KL, Vander Veur SS, *et al.* A randomized trial of the effects of an almond-enriched, hypocaloric diet in the treatment of obesity. Am J Clin Nutr 2012; 96(2): 249-54.
[http://dx.doi.org/10.3945/ajcn.112.037895] [PMID: 22743313]

[12]   Santhakumar AB, Kundur AR, Sabapathy S, Stanley R, Singh I. The potential of anthocyanin-rich Queen Garnet plum juice supplementation in alleviating thrombotic risk under induced oxidative stress conditions. J Funct Foods 2015; 14: 747-57.
[http://dx.doi.org/10.1016/j.jff.2015.03.003]

[13]   Ahmed T, Sadia H, Batool S, Janjua A, Shuja F. Use of prunes as a control of hypertension. J Ayub Med Coll Abbottabad 2010; 22(1): 28-31.
[PMID: 21409897]

[14]   Abazarfard Z, Salehi M, Keshavarzi S. The effect of almonds on anthropometric measurements and lipid profile in overweight and obese females in a weight reduction program: A randomized controlled clinical trial. J Res Med Sci 2014; 19(5): 457-64.
[PMID: 25097630]

[15]   Choudhury K, Clark J, Griffiths HR. An almond-enriched diet increases plasma α-tocopherol and improves vascular function but does not affect oxidative stress markers or lipid levels. Free Radic Res 2014; 48(5): 599-606.
[http://dx.doi.org/10.3109/10715762.2014.896458] [PMID: 24555818]

[16]   Savory LA, Griffin SJ, Williams KM, *et al.* Changes in diet, cardiovascular risk factors and modelled cardiovascular risk following diagnosis of diabetes: 1-year results from the ADDITION-Cambridge

trial cohort. Diabet Med 2014; 31(2): 148-55.
[http://dx.doi.org/10.1111/dme.12316] [PMID: 24102972]

[17]   WHO.   Cardiovascular   diseases.   World   Health   Organization   Available   from:
       http://www.who.int/mediacentre/factsheets/fs317/en/ , 2015 [[Accessed on 14/01/2015]];

[18]   Vizzotto M, Porter W, Byrne D, Cisneros-Zevallos L. Polyphenols of selected peach and plum
       genotypes reduce cell viability and inhibit proliferation of breast cancer cells while not affecting
       normal cells. Food Chem 2014; 164: 363-70.
       [http://dx.doi.org/10.1016/j.foodchem.2014.05.060] [PMID: 24996346]

[19]   Noratto G, Porter W, Byrne D, Cisneros-Zevallos L. Identifying peach and plum polyphenols with
       chemopreventive potential against estrogen-independent breast cancer cells. J Agric Food Chem 2009;
       57(12): 5219-26.
       [http://dx.doi.org/10.1021/jf900259m] [PMID: 19530711]

[20]   Freedman ND, Park Y, Subar AF, *et al.* Fruit and vegetable intake and esophageal cancer in a large
       prospective cohort study. Int J Cancer 2007; 121(12): 2753-60.
       [http://dx.doi.org/10.1002/ijc.22993] [PMID: 17691111]

[21]   Campbell OE, Padilla-Zakour OI. Phenolic and carotenoid composition of canned peaches (*Prunus
       persica*) and apricots (*Prunus armeniaca*) as affected by variety and peeling. Food Res Int 2013;
       54(1): 448-55.
       [http://dx.doi.org/10.1016/j.foodres.2013.07.016]

[22]   Oliveira A, Pintado M, Almeida DP. Phytochemical composition and antioxidant activity of peach as
       affected by pasteurization and storage duration. LWT - Food. Sci Tech (Paris) 2012; 49(2): 202-7.

[23]   Bohoyo-Gil D, Dominguez-Valhondo D, García-Parra JJ, González-Gómez D. UHPLC as a suitable
       methodology for the analysis of carotenoids in food matrix. Eur Food Res Technol 2012; 235(6):
       1055-61.
       [http://dx.doi.org/10.1007/s00217-012-1838-0]

[24]   Bobrich A, Fanning KJ, Rychlik M, Russell D, Topp B, Netzel M. Phytochemicals in Japanese plums:
       impact of maturity and bioaccessibility 2014.
       [http://dx.doi.org/10.1016/j.foodres.2014.06.030]

[25]   Gil MI, Tomás-Barberán FA, Hess-Pierce B, Kader AA. Antioxidant capacities, phenolic compounds,
       carotenoids, and vitamin C contents of nectarine, peach, and plum cultivars from California. J Agric
       Food Chem 2002; 50(17): 4976-82.
       [http://dx.doi.org/10.1021/jf020136b] [PMID: 12166993]

[26]   Ruiz D, Egea J, Tomás-Barberán FA, Gil MI. Carotenoids from new apricot (*Prunus armeniaca* L.)
       varieties and their relationship with flesh and skin color. J Agric Food Chem 2005; 53(16): 6368-74.
       [http://dx.doi.org/10.1021/jf0480703] [PMID: 16076120]

[27]   Valero D, Díaz-Mula HM, Zapata PJ, *et al.* Postharvest treatments with salicylic acid, acetylsalicylic
       acid or oxalic acid delayed ripening and enhanced bioactive compounds and antioxidant capacity in
       sweet cherry. J Agric Food Chem 2011; 59(10): 5483-9.
       [http://dx.doi.org/10.1021/jf200873j] [PMID: 21506518]

[28]   von Lintig J. Metabolism of carotenoids and retinoids related to vision. J Biol Chem 2012; 287(3):

1627-34.
[http://dx.doi.org/10.1074/jbc.R111.303990] [PMID: 22074927]

[29]   Stahl W, Sies H. Antioxidant activity of carotenoids. Mol Aspects Med 2003; 24(6): 345-51.
        [http://dx.doi.org/10.1016/S0098-2997(03)00030-X] [PMID: 14585305]

[30]   Fiedor J, Burda K. Potential role of carotenoids as antioxidants in human health and disease. Nutrients
        2014; 6(2): 466-88.
        [http://dx.doi.org/10.3390/nu6020466] [PMID: 24473231]

[31]   Traber MG, Atkinson J. Vitamin E, antioxidant and nothing more. Free Radic Biol Med 2007; 43(1):
        4-15.
        [http://dx.doi.org/10.1016/j.freeradbiomed.2007.03.024] [PMID: 17561088]

[32]   Engin KN. Alpha-tocopherol: looking beyond an antioxidant. Mol Vis 2009; 15: 855-60.
        [PMID: 19390643]

[33]   Hollman PC. Evidence for health benefits of plant phenols: local or systemic effects? J Sci Food Agric
        2001; 81(9): 842-52.
        [http://dx.doi.org/10.1002/jsfa.900]

[34]   Sánchez-Moreno C. Compuestos polifenólicos: estructura y classificación: presencia en alimentos y
        consumo: biodisponibilidad y metabolismo. Alimentaria 2002; 329: 19-28.

[35]   Robards K, Prenzler PD, Tucker G, Swatsitang P, Glover W. Phenolic compounds and their role in
        oxidative processes in fruits. Food Chem 1999; 66(4): 401-36.
        [http://dx.doi.org/10.1016/S0308-8146(99)00093-X]

[36]   Strack D, Wray V. The Flavonoids: Advances in Research Since 1986. London: Chapman and Hall
        1992.

[37]   Tapiero H, Tew KD, Ba GN, Mathé G. Polyphenols: do they play a role in the prevention of human
        pathologies? Biomed Pharmacother 2002; 56(4): 200-7.
        [http://dx.doi.org/10.1016/S0753-3322(02)00178-6] [PMID: 12109813]

[38]   Durst RW, Weaver GW. Nutritional content of fresh and canned peaches. J Sci Food Agric 2013;
        93(3): 593-603.
        [http://dx.doi.org/10.1002/jsfa.5849] [PMID: 22968977]

[39]   Abidi W, Cantin CM, Buhner T, Gonzalo MJ, Moreno MÁ, Gogorcena Y. Genetic control and
        location of QTLs involved in antioxidant capacity and fruit quality traits in peach. Acta Hortic 2012;
        962: 129-34. [*Prunus Persica* (L.) BATSCH].
        [http://dx.doi.org/10.17660/ActaHortic.2012.962.17]

[40]   Cevallos-Casals BA, Byrne D, Okie WR, Cisneros-Zevallos L. Selecting new peach and plum
        genotypes rich in phenolic compounds and enhanced functional properties. Food Chem 2006; 96(2):
        273-80.
        [http://dx.doi.org/10.1016/j.foodchem.2005.02.032]

[41]   Vizzotto M, Cisneros-Zevallos L, Byrne DH, Ramming DW, Okie WR. Large variation found in the
        phytochemical and antioxidant activity of peach and plum germplasm. J Am Soc Hortic Sci 2007;
        132(3): 334-40.

[42]   Neveu V, Perez-Jiménez J, Vos F, *et al.* Phenol-Explorer: an online comprehensive database on polyphenol contents in foods 2010.
[http://dx.doi.org/10.1093/database/bap024]

[43]   Kim D-O, Chun OK, Kim YJ, Moon H-Y, Lee CY. Quantification of polyphenolics and their antioxidant capacity in fresh plums. J Agric Food Chem 2003; 51(22): 6509-15.
[http://dx.doi.org/10.1021/jf0343074] [PMID: 14558771]

[44]   Cosmulescu S, Trandafir I, Nour V, Botu M. Total phenolic, flavonoid distribution and antioxidant capacity in skin, pulp and fruit extracts of plum cultivars. J Food Biochem 2015; 39(1): 64-9.
[http://dx.doi.org/10.1111/jfbc.12112]

[45]   Jaiswal R, Karaköse H, Rühmann S, *et al.* Identification of phenolic compounds in plum fruits (*Prunus salicina* L. and *Prunus domestica* L.) by high-performance liquid chromatography/tandem mass spectrometry and characterization of varieties by quantitative phenolic fingerprints. J Agric Food Chem 2013; 61(49): 12020-31.
[http://dx.doi.org/10.1021/jf402288j] [PMID: 24152059]

[46]   Kamiloglu S, Pasli AA, Ozcelik B, Capanoglu E. Evaluating the *in vitro* bioaccessibility of phenolics and antioxidant activity during consumption of dried fruits with nuts. LWT - Food. Sci Tech (Paris) 2014; 56(2): 284-9.

[47]   Kiat VV, Siang WK, Madhavan P, Chin JH, Ahmad M, Akowuah GA. FT-IR Profile and antiradical activity of dehulled kernels of apricot, almond and pumpkin. Res J Pharm Biol Chem Sci 2014; 5(2): 112-20.

[48]   Milbury PE, Chen C-Y, Dolnikowski GG, Blumberg JB. Determination of flavonoids and phenolics and their distribution in almonds. J Agric Food Chem 2006; 54(14): 5027-33.
[http://dx.doi.org/10.1021/jf0603937] [PMID: 16819912]

[49]   Roussos PA, Sefferou V, Denaxa N-K, Tsantili E, Stathis V. Apricot (*Prunus armeniaca* L.) fruit quality attributes and phytochemicals under different crop load. Sci Hortic (Amsterdam) 2011; 129(3): 472-8.
[http://dx.doi.org/10.1016/j.scienta.2011.04.021]

[50]   Ruiz D, Egea J, Gil MI, Tomás-Barberán FA. Characterization and quantitation of phenolic compounds in new apricot (*Prunus armeniaca* L.) varieties. J Agric Food Chem 2005; 53(24): 9544-52.
[http://dx.doi.org/10.1021/jf051539p] [PMID: 16302775]

[51]   Kim D-O, Heo HJ, Kim YJ, Yang HS, Lee CY. Sweet and sour cherry phenolics and their protective effects on neuronal cells. J Agric Food Chem 2005; 53(26): 9921-7.
[http://dx.doi.org/10.1021/jf0518599] [PMID: 16366675]

[52]   Melicháčová S, Timoracká M, Bystrická J, Vollmannová A, Čéry J. Relation of total antiradical activity and total polyphenol content of sweet cherries (*Prunus avium* L.) and tart cherries (*Prunus cerasus* L.). Acta Agric Slov 2010; 95(1): 21-8.
[http://dx.doi.org/10.2478/v10014-010-0003-3]

[53]   Pérez-Jiménez J, Fezeu L, Touvier M, *et al.* Dietary intake of 337 polyphenols in French adults. Am J Clin Nutr 2011; 93(6): 1220-8.

[http://dx.doi.org/10.3945/ajcn.110.007096] [PMID: 21490142]

[54]   Mullie P, Clarys P, Deriemaeker P, Hebbelinck M. Estimation of daily human intake of food flavonoids. Int J Food Sci Nutr 2008; 59(4): 291-8.
[http://dx.doi.org/10.1080/09687630701539293] [PMID: 17852474]

[55]   Chun OK, Chung SJ, Song WO. Estimated dietary flavonoid intake and major food sources of U.S. adults. J Nutr 2007; 137(5): 1244-52.
[PMID: 17449588]

[56]   Andreotti C, Ravaglia D, Ragaini A, Costa G. Phenolic compounds in peach (*Prunus persica*) cultivars at harvest and during fruit maturation. Ann Appl Biol 2008; 153(1): 11-23.
[http://dx.doi.org/10.1111/j.1744-7348.2008.00234.x]

[57]   de Pascual-Teresa S, Santos-Buelga C, Rivas-Gonzalo JC. Quantitative analysis of flavan-3-ols in Spanish foodstuffs and beverages. J Agric Food Chem 2000; 48(11): 5331-7.
[http://dx.doi.org/10.1021/jf000549h] [PMID: 11087482]

[58]   Arts IC, van de Putte B, Hollman PC. Catechin contents of foods commonly consumed in The Netherlands. 1. Fruits, vegetables, staple foods, and processed foods. J Agric Food Chem 2000; 48(5): 1746-51.
[http://dx.doi.org/10.1021/jf000025h] [PMID: 10820089]

[59]   Ozturk B, Yıldız K, Kucuker E. Effect of pre-harvest methyl jasmonate treatments on ethylene production, water-soluble phenolic compounds and fruit quality of Japanese plums. J Sci Food Agric 2015; 95(3): 583-91.
[http://dx.doi.org/10.1002/jsfa.6787] [PMID: 24930710]

[60]   Ozturk B, Kucuker E, Karaman S, Yıldız K, Kılıc K. Effect of aminoethoxyvinylglycine and methyl jasmonate on individual phenolics and post-harvest fruit quality of three different Japanese plums (*Prunus salicina* Lindell). Int J Food Eng 2013; 9(4): 421.
[http://dx.doi.org/10.1515/ijfe-2012-0257]

[61]   Venter A, Joubert E, de Beer D. Characterisation of phenolic compounds in South African plum fruits (*Prunus salicina* Lindl.) using HPLC coupled with diode-array, fluorescence, mass spectrometry and on-line antioxidant detection. Molecules 2013; 18(5): 5072-90.
[http://dx.doi.org/10.3390/molecules18055072] [PMID: 23644975]

[62]   Venter A, Joubert E, de Beer D. Nutraceutical value of yellow- and red-fleshed South African plums (*Prunus salicina* Lindl.): evaluation of total antioxidant capacity and phenolic composition. Molecules 2014; 19(3): 3084-109.
[http://dx.doi.org/10.3390/molecules19033084] [PMID: 24619353]

[63]   Bolling BW, Dolnikowski G, Blumberg JB, Chen CO. Polyphenol content and antioxidant activity of California almonds depend on cultivar and harvest year. Food Chem 2010; 122(3): 819-25.
[http://dx.doi.org/10.1016/j.foodchem.2010.03.068] [PMID: 25544797]

[64]   Yıldırım AN, San B, Koyuncu F, Yıldırım F. Variability of phenolics, α-tocopherol and amygdalin contents of selected almond (*Prunus amygdalus* Batsch.) genotypes. J Food Agric Environ 2010; 8(1): 76-9.

[65]   Čanadanović-Brunet JM, Vulic J, Cetkovic G, Djilas S, Tumbas-Saponjac V. Bioactive compounds

and antioxidant properties of dried apricot. Acta Per Technol 2013; 44: 193-205.
[http://dx.doi.org/10.2298/APT1344193C]

[66] Campbell OE, Merwin IA, Padilla-Zakour OI. Characterization and the effect of maturity at harvest on the phenolic and carotenoid content of Northeast USA Apricot (*Prunus armeniaca*) varieties. J Agric Food Chem 2013; 61(51): 12700-10.
[http://dx.doi.org/10.1021/jf403644r] [PMID: 24328399]

[67] Madrau MA, Piscopo A, Sanguinetti A, *et al*. Effect of drying temperature on polyphenolic content and antioxidant activity of apricots. Eur Food Res Technol 2009; 228(3): 441-8.
[http://dx.doi.org/10.1007/s00217-008-0951-6]

[68] Sochor J, Zitka O, Skutkova H, *et al*. Content of phenolic compounds and antioxidant capacity in fruits of apricot genotypes. Molecules 2010; 15(9): 6285-305.
[http://dx.doi.org/10.3390/molecules15096285] [PMID: 20877223]

[69] Mozetic B, Trebse P, Hribar J. Determination and quantitation of anthocyanins and hydroxycinnamic acids in different cultivars of sweet cherries (*Prunus avium* L.) from Nova Gorica Region (Slovenia). Food Technol Biotechnol 2002; 40(3): 207-12.

[70] Macheix JJ, Fleuriet A. Fruit Phenolics. Taylor & Francis 1990.

[71] Zamora-Ros R, Rothwell JA, Scalbert A, *et al*. Dietary intakes and food sources of phenolic acids in the European Prospective Investigation into Cancer and Nutrition (EPIC) study. Br J Nutr 2013; 110(8): 1500-11.
[http://dx.doi.org/10.1017/S0007114513000688] [PMID: 23507418]

[72] Grosso G, Stepaniak U, Topor-Mądry R, Szafraniec K, Pająk A. Estimated dietary intake and major food sources of polyphenols in the Polish arm of the HAPIEE study. Nutrition 2014; 30(11-12): 1398-403.
[http://dx.doi.org/10.1016/j.nut.2014.04.012] [PMID: 25280419]

[73] Brandi F, Liverani A, Giovannini D, *et al*. Molecular and biochemical studies on phytonutrient accumulation in peach fruit. Acta Hortic 2013; 976: 389-95.
[http://dx.doi.org/10.17660/ActaHortic.2013.976.53]

[74] Veberic R, Stampar F. Selected polyphenols in fruits of different cultivars of genus Prunus. Phyton - Ann Rei Bot 2005; 45(3): 83-375.

[75] Noratto G, Martino HS, Simbo S, Byrne D, Mertens-Talcott SU. Consumption of polyphenol-rich peach and plum juice prevents risk factors for obesity-related metabolic disorders and cardiovascular disease in Zucker rats. J Nutr Biochem 2015; 26(6): 633-41.
[http://dx.doi.org/10.1016/j.jnutbio.2014.12.014] [PMID: 25801980]

[76] Després J-P, Lemieux I. Abdominal obesity and metabolic syndrome. Nature 2006; 444(7121): 881-7.
[http://dx.doi.org/10.1038/nature05488] [PMID: 17167477]

[77] Utsunomiya H, Yamakawa T, Kamei J, Kadonosono K, Tanaka S. Anti-hyperglycemic effects of plum in a rat model of obesity and type 2 diabetes, Wistar fatty rat. Biomed Res 2005; 26(5): 193-200.
[http://dx.doi.org/10.2220/biomedres.26.193] [PMID: 16295695]

[78] Sun C-D, Zhang B, Zhang J-K, *et al*. Cyanidin-3-glucoside-rich extract from Chinese bayberry fruit protects pancreatic β cells and ameliorates hyperglycemia in streptozotocin-induced diabetic mice. J

Med Food 2012; 15(3): 288-98.
[http://dx.doi.org/10.1089/jmf.2011.1806] [PMID: 22181073]

[79]   Guo H, Xia M, Zou T, Ling W, Zhong R, Zhang W. Cyanidin 3-glucoside attenuates obesity-associated insulin resistance and hepatic steatosis in high-fat diet-fed and db/db mice *via* the transcription factor FoxO1. J Nutr Biochem 2012; 23(4): 349-60.
[http://dx.doi.org/10.1016/j.jnutbio.2010.12.013] [PMID: 21543211]

[80]   Guo H, Ling W, Wang Q, Liu C, Hu Y, Xia M. Cyanidin 3-glucoside protects 3T3-L1 adipocytes against H_2O_2- or TNF-α-induced insulin resistance by inhibiting c-Jun NH_2-terminal kinase activation. Biochem Pharmacol 2008; 75(6): 1393-401.
[http://dx.doi.org/10.1016/j.bcp.2007.11.016] [PMID: 18179781]

[81]   Scazzocchio B, Varì R, Filesi C, *et al.* Cyanidin-3-*O*-β-glucoside and protocatechuic acid exert insulin-like effects by upregulating PPARγ activity in human omental adipocytes. Diabetes 2011; 60(9): 2234-44.
[http://dx.doi.org/10.2337/db10-1461] [PMID: 21788573]

[82]   Guo H, Guo J, Jiang X, Li Z, Ling W. Cyanidin-3-*O*-β-glucoside, a typical anthocyanin, exhibits antilipolytic effects in 3T3-L1 adipocytes during hyperglycemia: involvement of FoxO1-mediated transcription of adipose triglyceride lipase. Food Chem Toxicol 2012; 50(9): 3040-7.
[http://dx.doi.org/10.1016/j.fct.2012.06.015] [PMID: 22721980]

[83]   Adisakwattana S, Yibchok-Anun S, Charoenlertkul P, Wongsasiripat N. Cyanidin-3-rutinoside alleviates postprandial hyperglycemia and its synergism with acarbose by inhibition of intestinal α-glucosidase. J Clin Biochem Nutr 2011; 49(1): 36-41.
[http://dx.doi.org/10.3164/jcbn.10-116] [PMID: 21765605]

[84]   Akkarachiyasit S, Charoenlertkul P, Yibchok-Anun S, Adisakwattana S. Inhibitory activities of cyanidin and its glycosides and synergistic effect with acarbose against intestinal α-glucosidase and pancreatic α-amylase. Int J Mol Sci 2010; 11(9): 3387-96.
[http://dx.doi.org/10.3390/ijms11093387] [PMID: 20957102]

[85]   Akkarachiyasit S, Yibchok-Anun S, Wacharasindhu S, Adisakwattana S. *In vitro* inhibitory effects of cyandin-3-rutinoside on pancreatic α-amylase and its combined effect with acarbose. Molecules 2011; 16(3): 2075-83.
[http://dx.doi.org/10.3390/molecules16032075] [PMID: 21368719]

[86]   Zhang R, Yao Y, Wang Y, Ren G. Antidiabetic activity of isoquercetin in diabetic KK -A y mice. Nutr Metab 2011; 8(1): 1.

[87]   Annadurai T, Muralidharan AR, Joseph T, Hsu MJ, Thomas PA, Geraldine P. Antihyperglycemic and antioxidant effects of a flavanone, naringenin, in streptozotocin-nicotinamide-induced experimental diabetic rats. J Physiol Biochem 2012; 68(3): 307-18.
[http://dx.doi.org/10.1007/s13105-011-0142-y] [PMID: 22234849]

[88]   Niture NT, Patil DG, Somani RS, Sahane RS. Effect of rutin on early diabetic neuropathy in experimental animals J Nat Prod Plant Resour 2014.

[89]   Torres-Piedra M, Ortiz-Andrade R, Villalobos-Molina R, *et al.* A comparative study of flavonoid analogues on streptozotocin-nicotinamide induced diabetic rats: quercetin as a potential antidiabetic agent acting *via* 11β-hydroxysteroid dehydrogenase type 1 inhibition. Eur J Med Chem 2010; 45(6):

2606-12.
[http://dx.doi.org/10.1016/j.ejmech.2010.02.049] [PMID: 20346546]

[90] Kandasamy N, Ashokkumar N. Myricetin, a natural flavonoid, normalizes hyperglycemia in streptozotocin-cadmium-induced experimental diabetic nephrotoxic rats. Biomed Prev Nutr 2012; 2(4): 246-51.
[http://dx.doi.org/10.1016/j.bionut.2012.04.003]

[91] Kandasamy N, Ashokkumar N. Renoprotective effect of myricetin restrains dyslipidemia and renal mesangial cell proliferation by the suppression of sterol regulatory element binding proteins in an experimental model of diabetic nephropathy. Eur J Pharmacol 2014; 743(0): 53-62.
[http://dx.doi.org/10.1016/j.ejphar.2014.09.014] [PMID: 25240712]

[92] Kandasamy N, Ashokkumar N. Protective effect of bioflavonoid myricetin enhances carbohydrate metabolic enzymes and insulin signaling molecules in streptozotocin-cadmium induced diabetic nephrotoxic rats. Toxicol Appl Pharmacol 2014; 279(2): 173-85.
[http://dx.doi.org/10.1016/j.taap.2014.05.014] [PMID: 24923654]

[93] Harini R, Pugalendi KV. Antihyperglycemic effect of protocatechuic acid on streptozotocin-diabetic rats. J Basic Clin Physiol Pharmacol 2010; 21(1): 79-91.
[http://dx.doi.org/10.1515/JBCPP.2010.21.1.79] [PMID: 20506690]

[94] Azay-Milhau J, Ferrare K, Leroy J, *et al.* Antihyperglycemic effect of a natural chicoric acid extract of chicory (*Cichorium intybus* L.): a comparative *in vitro* study with the effects of caffeic and ferulic acids. J Ethnopharmacol 2013; 150(2): 755-60.
[http://dx.doi.org/10.1016/j.jep.2013.09.046] [PMID: 24126061]

[95] Jung UJ, Lee M-K, Park YB, Jeon S-M, Choi M-S. Antihyperglycemic and antioxidant properties of caffeic acid in db/db mice. J Pharmacol Exp Ther 2006; 318(2): 476-83.
[http://dx.doi.org/10.1124/jpet.106.105163] [PMID: 16644902]

[96] Muthusamy VS, Saravanababu C, Ramanathan M, *et al.* Inhibition of protein tyrosine phosphatase 1B and regulation of insulin signalling markers by caffeoyl derivatives of chicory (*Cichorium intybus*) salad leaves. Br J Nutr 2010; 104(6): 813-23.
[http://dx.doi.org/10.1017/S0007114510001480] [PMID: 20444318]

[97] Muramatsu K, Fukuyo M, Hara Y. Effect of green tea catechins on plasma cholesterol level in cholesterol-fed rats. J Nutr Sci Vitaminol (Tokyo) 1986; 32(6): 613-22.
[http://dx.doi.org/10.3177/jnsv.32.613] [PMID: 3585557]

[98] Mangiapane H, Thomson J, Salter A, Brown S, Bell GD, White DA. The inhibition of the oxidation of low density lipoprotein by (+)-catechin, a naturally occurring flavonoid. Biochem Pharmacol 1992; 43(3): 445-50.
[http://dx.doi.org/10.1016/0006-2952(92)90562-W] [PMID: 1540202]

[99] Stevens JF, Miranda CL, Wolthers KR, Schimerlik M, Deinzer ML, Buhler DR. Identification and *in vitro* biological activities of hop proanthocyanidins: inhibition of nNOS activity and scavenging of reactive nitrogen species. J Agric Food Chem 2002; 50(12): 3435-43.
[http://dx.doi.org/10.1021/jf0116202] [PMID: 12033808]

[100] Cheng J-C, Dai F, Zhou B, Yang L, Liu Z-L. Antioxidant activity of hydroxycinnamic acid derivatives in human low density lipoprotein: Mechanism and structure–activity relationship. Food Chem 2007;

104(1): 132-9.
[http://dx.doi.org/10.1016/j.foodchem.2006.11.012]

[101] Ikeda I, Imasato Y, Sasaki E, *et al.* Tea catechins decrease micellar solubility and intestinal absorption of cholesterol in rats 1992.
[http://dx.doi.org/10.1016/0005-2760(92)90269-2]

[102] Wei X, Wang D, Yang Y, *et al.* Cyanidin-3-*O*-β-glucoside improves obesity and triglyceride metabolism in KK-Ay mice by regulating lipoprotein lipase activity. J Sci Food Agric 2011; 91(6): 1006-13.
[http://dx.doi.org/10.1002/jsfa.4275] [PMID: 21360538]

[103] Wang D, Xia M, Gao S, *et al.* Cyanidin-3-*O*-β-glucoside upregulates hepatic cholesterol 7α-hydroxylase expression and reduces hypercholesterolemia in mice. Mol Nutr Food Res 2012; 56(4): 610-21.
[http://dx.doi.org/10.1002/mnfr.201100659] [PMID: 22495986]

[104] Gong M, Garige M, Varatharajalu R, *et al.* Quercetin up-regulates paraoxonase 1 gene expression with concomitant protection against LDL oxidation. Biochem Biophys Res Commun 2009; 379(4): 1001-4.
[http://dx.doi.org/10.1016/j.bbrc.2009.01.015] [PMID: 19141295]

[105] Fang X-K, Gao J, Zhu D-N. Kaempferol and quercetin isolated from *Euonymus alatus* improve glucose uptake of 3T3-L1 cells without adipogenesis activity. Life Sci 2008; 82(11-12): 615-22.
[http://dx.doi.org/10.1016/j.lfs.2007.12.021] [PMID: 18262572]

[106] Haas MJ, Onstead-Haas LM, Szafran-Swietlik A, *et al.* Induction of hepatic apolipoprotein A-I gene expression by the isoflavones quercetin and isoquercetin. Life Sci 2014; 110(1): 8-14.
[http://dx.doi.org/10.1016/j.lfs.2014.06.014] [PMID: 24963805]

[107] WHO. Cardiovascular diseases. World Health Organization Available from: http://www.who.int/mediacentre/factsheets/fs317/en/ , 2015 [[Accessed on 2015 14/01/2015]];

[108] Parlakpinar H, Olmez E, Acet A, *et al.* Beneficial effects of apricot-feeding on myocardial ischemia-reperfusion injury in rats. Food Chem Toxicol 2009; 47(4): 802-8.
[http://dx.doi.org/10.1016/j.fct.2009.01.014] [PMID: 19271314]

[109] Auclair S, Milenkovic D, Besson C, *et al.* Catechin reduces atherosclerotic lesion development in apo E-deficient mice: a transcriptomic study. Atherosclerosis 2009; 204(2): e21-7.
[http://dx.doi.org/10.1016/j.atherosclerosis.2008.12.007] [PMID: 19152914]

[110] Xie X, Zhao R, Shen GX. Impact of cyanidin-3-glucoside on glycated LDL-induced NADPH oxidase activation, mitochondrial dysfunction and cell viability in cultured vascular endothelial cells. Int J Mol Sci 2012; 13(12): 15867-80.
[http://dx.doi.org/10.3390/ijms131215867] [PMID: 23443099]

[111] Han X, Pan J, Ren D, Cheng Y, Fan P, Lou H. Naringenin-7-O-glucoside protects against doxorubicin-induced toxicity in H9c2 cardiomyocytes by induction of endogenous antioxidant enzymes. Food Chem Toxicol 2008; 46(9): 3140-6.
[http://dx.doi.org/10.1016/j.fct.2008.06.086] [PMID: 18652870]

[112] Han XZ, Gao S, Cheng YN, *et al.* Protective effect of naringenin-7-*O*-glucoside against oxidative stress induced by doxorubicin in H9c2 cardiomyocytes. Biosci Trends 2012; 6(1): 19-25.

[PMID: 22426099]

[113] Perez-Vizcaino F, Duarte J, Jimenez R, Santos-Buelga C, Osuna A. Antihypertensive effects of the flavonoid quercetin. Pharmacol Rep 2009; 61(1): 67-75.
[http://dx.doi.org/10.1016/S1734-1140(09)70008-8] [PMID: 19307694]

[114] Shen Y, Croft KD, Hodgson JM, *et al.* Quercetin and its metabolites improve vessel function by inducing eNOS activity *via* phosphorylation of AMPK. Biochem Pharmacol 2012; 84(8): 1036-44.
[http://dx.doi.org/10.1016/j.bcp.2012.07.016] [PMID: 22846602]

[115] Orbán-Gyapai O, Raghavan A, Vasas A, Forgo P, Hohmann J, Shah ZA. Flavonoids isolated from Rumex aquaticus exhibit neuroprotective and neurorestorative properties by enhancing neurite outgrowth and synaptophysin. CNS Neurol Disord Drug Targets 2014; 13(8): 1458-64.
[http://dx.doi.org/10.2174/1871527313666141023154446] [PMID: 25345505]

[116] Byrne DH, Noratto G, Cisneros-Zevallos L, Porter W, Vizzotto M. Health benefits of peach, nectarine and plums. Acta Hortic 2009; 841: 267-73.
[http://dx.doi.org/10.17660/ActaHortic.2009.841.32]

[117] Dhingra N, Sharma R, Kar A. Antioxidative and antiproliferative activities of isolated compounds from *Prunus domestica*: an *in vitro* study. Int J Phytomed 2013; 5(3): 6.

[118] Noratto G, Porter W, Byrne D, Cisneros-Zevallos L. Polyphenolics from peach (*Prunus persica* var. Rich Lady) inhibit tumor growth and metastasis of MDA-MB-435 breast cancer cells *in vivo*. J Nutr Biochem 2014; 25(7): 796-800.
[http://dx.doi.org/10.1016/j.jnutbio.2014.03.001] [PMID: 24745759]

[119] Ogur R, Istanbulluoglu H, Korkmaz A, Barla A, Tekbas OF, Oztas E. Report: investigation of anti-cancer effects of cherry in vitro. Pak J Pharm Sci 2014; 27(3): 587-92.
[PMID: 24811821]

[120] Lea MA, Ibeh C, desBordes C, *et al.* Inhibition of growth and induction of differentiation of colon cancer cells by peach and plum phenolic compounds. Anticancer Res 2008; 28(4B): 2067-76.
[PMID: 18751377]

[121] Yang Y, Gallaher DD. Effect of dried plums on colon cancer risk factors in rats. Nutr Cancer 2005; 53(1): 117-25.
[http://dx.doi.org/10.1207/s15327914nc5301_14] [PMID: 16351514]

[122] Tada K, Kawahara K, Matsushita S, Hashiguchi T, Maruyama I, Kanekura T. MK615, a *Prunus mume* Steb. Et Zucc ('Ume') extract, attenuates the growth of A375 melanoma cells by inhibiting the ERK1/2-Id-1 pathway. Phytother Res 2012; 26(6): 833-8.
[http://dx.doi.org/10.1002/ptr.3645] [PMID: 22076920]

[123] Hattori M, Kawakami K, Akimoto M, Takenaga K, Suzumiya J, Honma Y. Antitumor effect of Japanese apricot extract (MK615) on human cancer cells *in vitro* and *in vivo* through a reactive oxygen species-dependent mechanism. Tumori 2013; 99(2): 239-48.
[PMID: 23748821]

[124] Matsushita S, Tada KI, Kawahara KI, *et al.* Advanced malignant melanoma responds to *Prunus mume* Sieb. Et Zucc (Ume) extract: Case report and *in vitro* study. Exp Ther Med 2010; 1(4): 569-74.
[PMID: 22993577]

[125] Yamai H, Sawada N, Yoshida T, *et al.* Triterpenes augment the inhibitory effects of anticancer drugs on growth of human esophageal carcinoma cells *in vitro* and suppress experimental metastasis *in vivo*. Int J Cancer 2009; 125(4): 952-60.
[http://dx.doi.org/10.1002/ijc.24433] [PMID: 19462449]

[126] Yoon H. Japanese apricot extract attenuates nuclear factor-κB activation and proteasomal activity in A549 lung cancer cells. Food Sci Biotechnol 2012; 21(5): 1507-10.
[http://dx.doi.org/10.1007/s10068-012-0200-4]

[127] Medzhitov R. Origin and physiological roles of inflammation. Nature 2008; 454(7203): 428-35.
[http://dx.doi.org/10.1038/nature07201] [PMID: 18650913]

[128] Bastard JP, Maachi M, Lagathu C, *et al.* Recent advances in the relationship between obesity, inflammation, and insulin resistance. Eur Cytokine Netw 2006; 17(1): 4-12.
[PMID: 16613757]

[129] Pearson TA, Mensah GA, Alexander RW, *et al.* Markers of inflammation and cardiovascular disease: application to clinical and public health practice: A statement for healthcare professionals from the Centers for Disease Control and Prevention and the American Heart Association. Circulation 2003; 107(3): 499-511.
[http://dx.doi.org/10.1161/01.CIR.0000052939.59093.45] [PMID: 12551878]

[130] Coussens LM, Werb Z. Inflammation and cancer. Nature 2002; 420(6917): 860-7.
[http://dx.doi.org/10.1038/nature01322] [PMID: 12490959]

[131] Rakoff-Nahoum S. Why cancer and inflammation? Yale J Biol Med 2006; 79(3-4): 123-30.
[PMID: 17940622]

[132] Gasparotto J, Somensi N, Bortolin RC, *et al.* Effects of different products of peach (*Prunus persica* L. Batsch) from a variety developed in southern Brazil on oxidative stress and inflammatory parameters *in vitro* and *ex vivo*. J Clin Biochem Nutr 2014; 55(2): 110-9.
[http://dx.doi.org/10.3164/jcbn.13-97] [PMID: 25320458]

[133] Gasparotto J, Somensi N, Bortolin RC, *et al.* Preventive supplementation with fresh and preserved peach attenuates CCl4-induced oxidative stress, inflammation and tissue damage. J Nutr Biochem 2014; 25(12): 1282-95.
[http://dx.doi.org/10.1016/j.jnutbio.2014.07.004] [PMID: 25287815]

[134] Jain P, Kumar N, Josyula VR, *et al.* A study on the role of (+)-catechin in suppression of HepG2 proliferation *via* caspase dependent pathway and enhancement of its *in vitro* and *in vivo* cytotoxic potential through liposomal formulation. Eur J Pharm Sci 2013; 50(3-4): 353-65.
[http://dx.doi.org/10.1016/j.ejps.2013.08.005] [PMID: 23954456]

[135] Iacopini P, Baldi M, Storchi P, Sebastiani L. Catechin, epicatechin, quercetin, rutin and resveratrol in red grape: Content, *in vitro* antioxidant activity and interactions. J Food Compos Anal 2008; 21(8): 589-98.
[http://dx.doi.org/10.1016/j.jfca.2008.03.011]

[136] Wong K-L, Wu Y-R, Cheng K-S, *et al.* Palmitic acid-induced lipotoxicity and protection by (+)-catechin in rat cortical astrocytes. Pharmacol Rep 2014; 66(6): 1106-13.
[http://dx.doi.org/10.1016/j.pharep.2014.07.009] [PMID: 25443742]

[137] Teixeira MD, Souza CM, Menezes AP, *et al.* Catechin attenuates behavioral neurotoxicity induced by 6-OHDA in rats. Pharmacol Biochem Behav 2013; 110: 1-7.
[http://dx.doi.org/10.1016/j.pbb.2013.05.012] [PMID: 23714698]

[138] Elisia I, Kitts DD. Anthocyanins inhibit peroxyl radical-induced apoptosis in Caco-2 cells. Mol Cell Biochem 2008; 312(1-2): 139-45.
[http://dx.doi.org/10.1007/s11010-008-9729-1] [PMID: 18327700]

[139] Xu M, Bower KA, Wang S, *et al.* Cyanidin-3-glucoside inhibits ethanol-induced invasion of breast cancer cells overexpressing ErbB2. Mol Cancer 2010; 9(1): 285.
[http://dx.doi.org/10.1186/1476-4598-9-285] [PMID: 21034468]

[140] Lim T-G, Kwon JY, Kim J, *et al.* Cyanidin-3-glucoside suppresses B[a]PDE-induced cyclooxygenase-2 expression by directly inhibiting Fyn kinase activity. Biochem Pharmacol 2011; 82(2): 167-74.
[http://dx.doi.org/10.1016/j.bcp.2011.03.032] [PMID: 21501596]

[141] Wang Q, Xia M, Liu C, *et al.* Cyanidin-3-*O*-β-glucoside inhibits iNOS and COX-2 expression by inducing liver X receptor alpha activation in THP-1 macrophages. Life Sci 2008; 83(5-6): 176-84.
[http://dx.doi.org/10.1016/j.lfs.2008.05.017] [PMID: 18619979]

[142] Zhang Y, Lian F, Zhu Y, *et al.* Cyanidin-3-O-β-glucoside inhibits LPS-induced expression of inflammatory mediators through decreasing IkappaBalpha phosphorylation in THP-1 cells. Inflamm Res 2010; 59(9): 723-30.
[http://dx.doi.org/10.1007/s00011-010-0183-7] [PMID: 20309718]

[143] Min S-W, Ryu S-N, Kim D-H. Anti-inflammatory effects of black rice, cyanidin-3-*O*-β-D-glycoside, and its metabolites, cyanidin and protocatechuic acid. Int Immunopharmacol 2010; 10(8): 959-66.
[http://dx.doi.org/10.1016/j.intimp.2010.05.009] [PMID: 20669401]

[144] Hassimotto NMA, Moreira V. Nascimento NGd, Souto PCMdC, Teixeira C, Lajolo FM. Inhibition of carrageenan-induced acute inflammation in mice by oral administration of anthocyanin mixture from wild mulberry and cyanidin-3-glucoside. Bio Med Res Int 2013; 2013: 10.

[145] Acquaviva R, Russo A, Galvano F, *et al.* Cyanidin and cyanidin 3-*O*-beta-D -glucoside as DNA cleavage protectors and antioxidants. Cell Biol Toxicol 2003; 19(4): 243-52.
[http://dx.doi.org/10.1023/B:CBTO.0000003974.27349.4e] [PMID: 14686616]

[146] Song J, Zhao M, Liu X, Zhu Y, Hu X, Chen F. Protection of cyanidin-3-glucoside against oxidative stress induced by acrylamide in human MDA-MB-231 cells. Food Chem Toxicol 2013; 58(0): 306-10.
[http://dx.doi.org/10.1016/j.fct.2013.05.003] [PMID: 23685245]

[147] Li C-Y, Xu H-D, Zhao B-T, Chang H-I, Rhee H-I. Gastroprotective effect of cyanidin 3-glucoside on ethanol-induced gastric lesions in rats. Alcohol 2008; 42(8): 683-7.
[http://dx.doi.org/10.1016/j.alcohol.2008.08.009] [PMID: 19038699]

[148] Nasri S, Roghani M, Baluchnejadmojarad T, Rabani T, Balvardi M. Vascular mechanisms of cyanidin-3-glucoside response in streptozotocin-diabetic rats. Pathophysiology 2011; 18(4): 273-8.
[http://dx.doi.org/10.1016/j.pathophys.2011.03.001] [PMID: 21546226]

[149] Nasri S, Roghani M, Baluchnejadmojarad T, Balvardi M, Rabani T. Chronic cyanidin-3-glucoside administration improves short-term spatial recognition memory but not passive avoidance learning and memory in streptozotocin-diabetic rats. Phytother Res 2012; 26(8): 1205-10.

[http://dx.doi.org/10.1002/ptr.3702] [PMID: 22228592]

[150] Maggioni D, Nicolini G, Rigolio R, *et al.* Myricetin and naringenin inhibit human squamous cell carcinoma proliferation and migration *in vitro*. Nutr Cancer 2014; 66(7): 1257-67.
[http://dx.doi.org/10.1080/01635581.2014.951732] [PMID: 25256786]

[151] Lee D-H, Szczepanski M, Lee YJ. Role of Bax in quercetin-induced apoptosis in human prostate cancer cells. Biochem Pharmacol 2008; 75(12): 2345-55.
[http://dx.doi.org/10.1016/j.bcp.2008.03.013] [PMID: 18455702]

[152] Yu C-S, Lai K-C, Yang J-S, *et al.* Quercetin inhibited murine leukemia WEHI-3 cells *in vivo* and promoted immune response. Phytother Res 2010; 24(2): 163-8.
[PMID: 19449452]

[153] Boly R, Gras T, Lamkami T, *et al.* Quercetin inhibits a large panel of kinases implicated in cancer cell biology. Int J Oncol 2011; 38(3): 833-42.
[PMID: 21206969]

[154] Xiao X, Shi D, Liu L, *et al.* Quercetin suppresses cyclooxygenase-2 expression and angiogenesis through inactivation of P300 signaling. PLoS One 2011; 6(8): e22934.
[http://dx.doi.org/10.1371/journal.pone.0022934] [PMID: 21857970]

[155] Choi EJ, Bae SM, Ahn WS. Antiproliferative effects of quercetin through cell cycle arrest and apoptosis in human breast cancer MDA-MB-453 cells. Arch Pharm Res 2008; 31(10): 1281-5.
[http://dx.doi.org/10.1007/s12272-001-2107-0] [PMID: 18958418]

[156] Cavia-Saiz M, Busto MD, Pilar-Izquierdo MC, Ortega N, Perez-Mateos M, Muñiz P. Antioxidant properties, radical scavenging activity and biomolecule protection capacity of flavonoid naringenin and its glycoside naringin: a comparative study. J Sci Food Agric 2010; 90(7): 1238-44.
[http://dx.doi.org/10.1002/jsfa.3959] [PMID: 20394007]

[157] Wijaya S, Jin KT, Nee TK, Wiart C. *In vitro* 5-LOX inhibitory and antioxidant activities of extracts and compounds from the aerial parts of *Lopholaena coriifolia* (Sond.) E. Phillips & C.A. Sm. J Complement Integr Med 2012; 9(1): 11.
[http://dx.doi.org/10.1515/1553-3840.1615] [PMID: 22728459]

[158] Jayaraman J, Veerappan M, Namasivayam N. Potential beneficial effect of naringenin on lipid peroxidation and antioxidant status in rats with ethanol-induced hepatotoxicity. J Pharm Pharmacol 2009; 61(10): 1383-90.
[http://dx.doi.org/10.1211/jpp.61.10.0016] [PMID: 19814872]

[159] Lee JK. Anti-inflammatory effects of eriodictyol in lipopolysaccharide-stimulated raw 264.7 murine macrophages. Arch Pharm Res 2011; 34(4): 671-9.
[http://dx.doi.org/10.1007/s12272-011-0418-3] [PMID: 21544733]

[160] Dhanya R, Arun KB, Syama HP, *et al.* Rutin and quercetin enhance glucose uptake in L6 myotubes under oxidative stress induced by tertiary butyl hydrogen peroxide. Food Chem 2014; 158: 546-54.
[http://dx.doi.org/10.1016/j.foodchem.2014.02.151] [PMID: 24731381]

[161] Mahmoud M, Hamdan D, Wink M, El-Shazly A. Naringin and rutin prevent d-galactosamine-induced hepatic injury in rats via attenuation of the inflammatory cascade and oxidative stress. Eur Sci J 2013; 9(30): 55-141.

[162] Dixit S. Anticancer effect of rutin isolated from the methanolic extract of *Triticum aestivum* straw in mice. Med Sci 2014; 2(4): 153-60.

[163] Owolabi MA, Coker HA, Jaja SI. Bioactivity of the phytoconstituents of the leaves of *Persea americana.* J Med Plants Res 2010; 4(12): 1130-5.

[164] Kim SJ, Um JY, Lee JY, Lee JY. Anti-inflammatory activity of hyperoside through the suppression of nuclear factor-κB activation in mouse peritoneal macrophages. Am J Chin Med 2011; 39(1): 171-81. [http://dx.doi.org/10.1142/S0192415X11008737] [PMID: 21213407]

[165] Boots AW, Wilms LC, Swennen EL, Kleinjans JC, Bast A, Haenen GR. *In vitro* and *ex vivo* anti-inflammatory activity of quercetin in healthy volunteers. Nutrition 2008; 24(7-8): 703-10. [http://dx.doi.org/10.1016/j.nut.2008.03.023] [PMID: 18549926]

[166] Carrasco-Pozo C, Mizgier ML, Speisky H, Gotteland M. Differential protective effects of quercetin, resveratrol, rutin and epigallocatechin gallate against mitochondrial dysfunction induced by indomethacin in Caco-2 cells. Chem Biol Interact 2012; 195(3): 199-205. [http://dx.doi.org/10.1016/j.cbi.2011.12.007] [PMID: 22214982]

[167] Liu Z, Tao X, Zhang C, Lu Y, Wei D. Protective effects of hyperoside (quercetin-3-*o*-galactoside) to PC12 cells against cytotoxicity induced by hydrogen peroxide and *tert*-butyl hydroperoxide. Biomed Pharmacother 2005; 59(9): 481-90. [http://dx.doi.org/10.1016/j.biopha.2005.06.009] [PMID: 16271843]

[168] Kao T-K, Ou Y-C, Raung S-L, Lai C-Y, Liao S-L, Chen C-J. Inhibition of nitric oxide production by quercetin in endotoxin/cytokine-stimulated microglia. Life Sci 2010; 86(9-10): 315-21. [http://dx.doi.org/10.1016/j.lfs.2009.12.014] [PMID: 20060843]

[169] Veluri R, Singh RP, Liu Z, Thompson JA, Agarwal R, Agarwal C. Fractionation of grape seed extract and identification of gallic acid as one of the major active constituents causing growth inhibition and apoptotic death of DU145 human prostate carcinoma cells. Carcinogenesis 2006; 27(7): 1445-53. [http://dx.doi.org/10.1093/carcin/bgi347] [PMID: 16474170]

[170] Kaur M, Velmurugan B, Rajamanickam S, Agarwal R, Agarwal C. Gallic acid, an active constituent of grape seed extract, exhibits anti-proliferative, pro-apoptotic and anti-tumorigenic effects against prostate carcinoma xenograft growth in nude mice. Pharm Res 2009; 26(9): 2133-40. [http://dx.doi.org/10.1007/s11095-009-9926-y] [PMID: 19543955]

[171] Lu Y, Jiang F, Jiang H, *et al.* Gallic acid suppresses cell viability, proliferation, invasion and angiogenesis in human glioma cells. Eur J Pharmacol 2010; 641(2-3): 102-7. [http://dx.doi.org/10.1016/j.ejphar.2010.05.043] [PMID: 20553913]

[172] You BR, Moon HJ, Han YH, Park WH. Gallic acid inhibits the growth of HeLa cervical cancer cells *via* apoptosis and/or necrosis. Food Chem Toxicol 2010; 48(5): 1334-40. [http://dx.doi.org/10.1016/j.fct.2010.02.034] [PMID: 20197077]

[173] Mori H, Iwahashi H. Antioxidant activity of caffeic acid through a novel mechanism under UVA irradiation. J Clin Biochem Nutr 2009; 45(1): 49-55. [http://dx.doi.org/10.3164/jcbn.08-258] [PMID: 19590707]

[174] Ashidate K, Kawamura M, Mimura D, *et al.* Gentisic acid, an aspirin metabolite, inhibits oxidation of low-density lipoprotein and the formation of cholesterol ester hydroperoxides in human plasma. Eur J

Pharmacol 2005; 513(3): 173-9.
[http://dx.doi.org/10.1016/j.ejphar.2005.03.012] [PMID: 15862799]

[175] Gülçin I. Antioxidant activity of caffeic acid (3,4-dihydroxycinnamic acid). Toxicology 2006; 217(2-3): 213-20.
[http://dx.doi.org/10.1016/j.tox.2005.09.011] [PMID: 16243424]

[176] Yeh Y-H, Lee Y-T, Hsieh H-S, Hwang D-F. Dietary caffeic acid, ferulic acid and coumaric acid supplements on cholesterol metabolism and antioxidant activity in rats. J Food Drug Anal 2009; 17(2): 123-32.

[177] Koriem KM, Abdelhamid AZ, Younes HF. Caffeic acid protects tissue antioxidants and DNA content in methamphetamine induced tissue toxicity in Sprague Dawley rats. Toxicol Mech Methods 2013; 23(2): 134-43.
[http://dx.doi.org/10.3109/15376516.2012.730561] [PMID: 22992185]

[178] Koriem KMM, Soliman RE. Chlorogenic and Caftaric acids in liver toxicity and oxidative stress induced by methamphetamine. J Toxicol 2014; 2014: 10.
[http://dx.doi.org/10.1155/2014/583494]

[179] Kim EO, Min KJ, Kwon TK, Um BH, Moreau RA, Choi SW. Anti-inflammatory activity of hydroxycinnamic acid derivatives isolated from corn bran in lipopolysaccharide-stimulated Raw 264.7 macrophages. Food Chem Toxicol 2012; 50(5): 1309-16.
[http://dx.doi.org/10.1016/j.fct.2012.02.011] [PMID: 22366099]

[180] Jung S, Spiegelman D, Baglietto L, *et al.* Fruit and vegetable intake and risk of breast cancer by hormone receptor status. J Natl Cancer Inst 2013; 105(3): 219-36.
[http://dx.doi.org/10.1093/jnci/djs635] [PMID: 23349252]

[181] Fung TT, Chiuve SE, Willett WC, Hankinson SE, Hu FB, Holmes MD. Intake of specific fruits and vegetables in relation to risk of estrogen receptor-negative breast cancer among postmenopausal women. Breast Cancer Res Treat 2013; 138(3): 925-30.
[http://dx.doi.org/10.1007/s10549-013-2484-3] [PMID: 23532538]

CHAPTER 5

Pomegranate (*Punica granatum*): A Natural Approach to Combat Oxidative Stress-Related Diseases

Ana Paula Duarte[*], Ângelo Luís , Fernanda C. Domingues

CICS-UBI – Health Sciences Research Centre, Faculty of Health Sciences, University of Beira Interior, Covilhã, Portugal

Abstract: Pomegranate fruit *(Punica granatum)* has been widely studied as a dietary component that is an important source of biologically active compounds. Mainly in the over the last ten years, much research has been done into what concerns the health benefits of this fruit (peel, seeds and juice) producing encouraging results to prevent and treat specific diseases such as obesity, diabetes, neurodegeneration and cancer. Additionally, the use of pomegranate by traditional medicine as antidiarrheal, anti-helminthic, diuretic and digestive, has been also reported in ethnobotanical studies. Pomegranate is described as containing a very high content of polyphenolic compounds when compared to other fruits or vegetables, namely ellagitannins and hydrolysable tannins and flavonoids (anthocyanins, flavones and isoflavones). These phytochemicals have been identified for their many health benefits including its very high antioxidant activity and anti-inflammatory, antitumor, antimicrobial activities, among others. Considering that polyphenols have been widely reported to counteract the oxidative stress effects in different model systems, this chapter will highlight the phytomedicinal potential of pomegranate by describing its health impact and mechanisms of action in oxidative stress-related conditions. In addition, its origin, production, consumption and traditional uses will be briefly discussed and those related to its phytochemical composition and health promoting properties, namely those related to its antioxidant activity will be detailed.

---

[*] **Corresponding author Ana Paula Duarte:** Health Sciences Research Centre, University of Beira Interior, Avenida Infante D. Henrique, 6200-506 Covilhã, Portugal Te: +351275329020; Fax: +351275329099 Email: apcd@ubi.pt

The main focus will be the action of pomegranate in oxidative stress-related diseases, including inflammatory processes, cancer, diabetes, cardiovascular diseases and neurodegenerative disorders. Bioavailability aspects will also be presented.

Keywords: Cancer, Cardiovascular diseases, Diabetes, Inflammatory processes, Natural antioxidants, Neurodegenerative disorders, Polyphenols, Pomegranate, Heart-disease, *Punica granatum.*

INTRODUCTION

Oxygen is of vital importance in providing the necessary energy for the survival and normal functioning of most eukaryotic organisms [1, 2]. Through the respiratory chain, oxygen is partly reduced at a low ratio into superoxide. As this is a basic free radical, it can be transformed into other forms of reactive oxygen species (ROS) [1, 2]. Other free radicals can also be generated from nitrogen and therefore classified into the family of reactive nitrogen species (RNS) [1, 2]. At a physiological level, ROS and RNS have been recently shown to mediate a variety of normal functions including the regulation of signal transduction, the induction of mitogenic response and are involved in the defense against infectious agents [1, 2].

ROS are balanced with antioxidant systems to keep their levels constant in living organisms [1, 2]. These antioxidant systems are both enzymatic and non-enzymatic.The imbalance between ROS and antioxidant systems is termed oxidative stress, and can be caused by the over production of ROS and/or the reduction of antioxidants; both situations can be harmful [1, 2]. During oxidative stress, excessive free radicals may freely pass through the plasma membrane, promoting damage in the cell membrane *via* lipid peroxidation. This process modifies the signal and structural proteins leading to misfolding and aggregation as well as the oxidation of RNA/DNA. Due to this, transcription is interrupted resulting in gene mutation [1, 2].

It is commonly accepted that in a situation of oxidative stress, ROS such as superoxide ($O2^-$, OOH^-), hydroxyl (OH^-) and peroxyl (ROO^-) radicals are generated [3]. The ROS play an important role in the pathogenesis of various serious diseases, such as neurodegenerative disorders [4], cancer [5], cardiovascular diseases [6], atherosclerosis [7], cataracts [8], and inflammation [9].

The use of traditional medicine is widespread and plants still present a large source of natural antioxidants that might serve as leads for the development of novel drugs [3, 10]. Several ant-inflammatory, digestive, antinecrotic, neuro-protective, and hepatoprotective drugs have recently been shown to have an antioxidant and/or antiradical scavenging mechanism as part of their activity [3]. Natural antioxidant ingredients of fruits and vegetables could act as protective factors against oxidative damage [11]. According to the recent studies, the antioxidant features of many fruits and vegetables come from their flavonoid and related polyphenolic components [11 - 15]. In human, daily intake of these compounds is approximately between a few hundred milligrams and one gram [11]. Moreover, there are several studies which have demonstrated the bioavailability of the antioxidants in many fruits and plants [16 - 18].

Phenolics are compounds that have one or more aromatic rings with one or more hydroxyl groups [19]. They are common in the plant kingdom and are the most abundant secondary metabolite of plants, with more than 8.000 phenolic structures currently known. They range from simple molecules like phenolic acids up to highly polymerized substances such as tannins [19]. Plant phenolics are usually involved in the defense of the plant against ultraviolet radiation or aggression by pathogens, parasites and predators. They also contribute to the colors of plants. Since they can be found in virtually all the organs of the plants, they are therefore an integral part of the human diet [19]. Still, the health effects of dietary polyphenols have only come to the attention of nutritionists in the recent years [20]. The potent antioxidant properties of polyphenols, their abundance in the diet, and their credible effects in the prevention of various diseases associated with oxidative stresses have resulted in an increasing interest by researchers and food manufacturers [19]. The effects of these secondary plant metabolites, that can help to prevent cardiovascular diseases, neurodegenerative diseases and cancer result from epidemiologic data as well as *in vitro* and *in vivo* results, which have originated several nutritional recommendations [19, 21]. Polyphenols have also been found to modulate the activity of a wide range of enzyme and cell receptors. Thus, in addition to having relevant antioxidant properties, polyphenols have several other specific biological actions that can help to prevent and treat several diseases [19].

Fig. (1). Pomegranate tree and pomegranate fruit (Photos by A.P. Duarte, June 2014).

Pomegranate (*Punica granatum* L.) (Fig. **1**) belongs to the Punicacea family [22]. The pomegranate originated in the region of the Middle East and was subsequently introduced into the Mediterranean diet [23]. It grows and yields well in semi-arid, mild to subtropical climates, where the air not humid such as Afghanistan, Iran, India, China, Japan, The United States (California), Spain, Egypt, Turkey, Greece and Russia [24]. Tunisia is one of the countries of origin as well as one of the places where pomegranates continue to grow. Its area of cultivation covers more than 11.000 ha and has spread into all areas except in the areas of high altitude [24]. According to recent studies, the production of pomegranates is estimated to have increased from 67. 000 tons in 2009 to 71.597 tons in 2010 [24]. Pomegranate peels are rich in tannins. Their use is widespread essentially because of their medicinal properties; due to the fact that they contain anti-inflammatory, antioxidant, anticancer and antihelminthic properties; and also due to their other uses such as tanning, dyeing and heavy metal removal [24]. The various parts of a pomegranate are recognized as a reservoir of bioactive compounds with prospective biological activities [25]. For example, the dyslipidemia of obesity and cardiovascular risk factors have been decreased due to bioactive compounds found in the leaves of pomegranates [26]. In addition, the anti-parasitic, anti-microbial and antioxidant activities of pomegranate leaves extracts have also been reported [27]. Many papers have been published highlighting the ability of pomegranate leaf extracts to tackle obesity, cancer and other human diseases [28 - 30]. Phenolics in pomegranate leaves are considered to be primarily responsible for their health benefits [25].

Pomegranate has been used for centuries to treat number of inflammatory diseases [31]. Based on its usage in Ayurvedic and Unani medicine, dietary supplements containing pomegranate extract are becoming popular in the Western world for the treatment and prevention of arthritis and other inflammatory diseases [31]. More recently standardized extracts of pomegranate fruit have been shown to possess anti-inflammatory and cartilage sparing effects *in vitro* [31].

Considering that polyphenols have been widely reported to counteract the oxidative stress effects in different model systems, this chapter will highlight the phytomedicinal potential of pomegranate by describing its health effects and mechanisms of action in oxidative stress-related diseases. In addition, its origin, production, consumption and traditional uses will be briefly discussed and those related to its phytochemical composition and health promoting properties, namely those related to its antioxidant activity will be detailed.

The main focus will be the action of pomegranate in oxidative stress-related diseases, including inflammatory processes, cancer, diabetes, cardiovascular diseases and neurodegenerative disorders. Bioavailability aspects will also be presented.

METHODS

A literature search was conducted in ISI Web of Knowledge, Scopus, Pubmed and Science Direct to collect all articles, including original research articles and reviews from 2000 to 2015.

Key search words included ROS, oxidative stress, free radicals, natural antioxidants, antioxidant activity, polyphenols, flavonoids, plant polyphenols, pomegranate, Punica granatum, punicic acid, punicalagin, punicalin, tannins, ellagic acid, bioavailability, inflammation, anti-inflammatory activity, cancer, anti-tumoral activity, prostate cancer, diabetes, glycemia, cardiovascular diseases, atherosclerosis, hypertension, neurodegenerative diseases, Alzheimer's disease, Parkinson's disease, isolated or in different combinations.

Additional studies were identified from reference lists.

CHEMICAL COMPOSITION - POMEGRANATE POLYPHENOLS: THE STRATEGIC PLAYERS

Pomegranates are one of the oldest fruits in the promotion of health [24, 32 - 34]. It was initiated in the Middle East, and has spread throughout the Mediterranean, China as well as all of the different tropical and subtropical countries around the world [32 - 34]. One of the most important producers and exporters of pomegranates is Iran, as its total production in 2005 was 670.000 tons [35]. 1000 cultivars of pomegranate can be found in existence, classified by the appearance and characteristics of fruit, flower and tree [22, 24, 32]. In general, there are two main varieties, an ornamental and an edible one. The edible is classified into a sweet or sour type depending on the taste of the juice [23, 36].

Interest in pomegranates and derived products (juice, jelly, jam etc.) were pushed up by consumers and researchers mainly due to the benefits associated to its consumption [32, 34, 37], namely prevention and treatment of cancer [38], cardiovascular diseases [39], diabetes [40], Alzheimer's disease [41, 42], arthritis and colitis [31, 43].

Pomegranate fruits are widely consumed fresh or processed into juice, jams, syrup and sauce [35]. The edible portion (aril) of the fruit is about 55–60% of the total fruit weight and consists of about 75–85% juice and 15–25% seeds [35]. This fruit can be divided into several anatomical compartments: outside peel, inside peel (pellicles), and arils (pulp and seeds) [23]. Pomegranate is known to contain considerable amounts of phenolic compounds, including anthocyanins (3-glucosides and 3,5-diglucosides of delphinidin, cyanidin, and pelargonidin), punicalin, punicalagin, pedunculagin and different flavanols [35, 44, 45]. This fruit also contain hydrolysable tannins, as major active chemical constituents and phytoconstituents, namely, corilagin, ellagic acid, kaempferol, luteolin, myricetin, quercetin, quercimetrine, and quercetin-3-*O*-rutinoside [37].

About 18% of dried and cleaned white pomegranate seeds are made of oil. This oil is rich in cis-9,trans-11,cis-13 octadecatrienoic acid, also known as punicic acid (65-80%) (Table **1**), which is a triple conjugated 18-carbon fatty acid [43], [46]. Punicic acid is metabolized to cis-9,trans-11 conjugated linoleic acid *via* a saturation mechanism [46]. Punicic acid has anti-inflammatory properties since it can inhibit the synthesis of prostaglandins, tumor necrosis factor (TNF-α) and neutrophils myeloprexidase activity [47, 48]. Studies have also shown that this acid has anti-cancer activity due to an effect on the tumor suppressor agent, protein tyrosine phosphatase receptor type-γ [47, 48]. Punicic acid has also the capacity to induce vasorelaxation *via* nitric oxid-guanylyl cyclase pathway [49]. There are some phytoestrogen compounds in pomegranate seeds that have sex steroid hormones similar to those in humans [43]. The 17-alpha-estradiol is a mirror-image version of estrogen [43]. Due to existence of these phytoestrogenic compounds, pomegranate has good potential for prophylaxis and treating menopausal complains and osteoporosis [47].

Pomegranate peel extracts were reported to possess remarkable antioxidant, antibacterial, anti-inflammatory and hypolipidemic bioactivities [36]. These potential health benefits are attributed to the polyphenolic compounds that the pomegranate peel extracts contain, including punicalagin (Table **1**), gallic acid, ellagic acid, chlorogenic acid, caffeic acid, catechin, epicatechin, rutin, quercetin, and galangal [36]. Pomegranate peels are also rich in tannins [24] and have been used traditionally for their medicinal properties as anticancer, anti-inflammatory, antioxidant and antihelminthic [24] and for other purposes such as tanning, dyeing and heavy metal removal [24]. Punicalagin is known to be a potent antioxidant, behaving as free radical scavenger, having superoxide anion, singlet oxygen and hydroxyl radical scavenging activities along with lipid peroxidation inhibitory activity [7, 50]; and has been reported to have a wide range of other biological effects including hepatoprotective, antimicrobial, antiproliferative, apoptotic, chemopreventive and immune-suppressive activities [51, 52]. More importantly, it has also been shown to exhibit antiviral effects on human immunodeficiency virus, influenza virus, and herpes simplex virus [51, 52].

Pomegranate arils are usually freshly consumed or used for the production of juice, jams and jellies production, in addition also to develop extracts to be used as ingredients in medicinal herb preparations and dietary supplements [23]. Arils are rich in sugars, organic acids, vitamins, minerals and polyphenols, which are the major antioxidant and health functional factors, and they mainly consist of ellagitannin (punicalagin), gallic acid, ellagic acid, anthocyanins, catechins, caffeic acid, and quercetin [32]. The arils also contain proteins, crude fibers, vitamins, minerals, pectin and sugars. The arils are used to elaborate increasingly popular pomegranate products such as canned beverages, paste and, especially, fresh juice which can be obtained from the arils or the whole fruit [34].

Most pomegranate fruits are freshly eaten, and some are processed to juice and wine [33, 36]. Pomegranate juice is a good source of fructose, sucrose, and glucose. It also has some simple organic acids such as ascorbic acid, citric acid, fumaric acid, and malic acid [43]. In addition, it contains small amounts of all amino acids, specifically proline, methionine, and valine [43]. Pomegranate juice exceeds the *in vitro* antioxidant potency of other common commercial fruit juices [53]. The most abundant type of polyphenols in pomegranate juice are ellagitannins [53]. The major anthocyanins in pomegranate juice across several cultivars were delphinidin 3,5-diglucoside, cyanidin 3,5-diglucoside, pelargonidin 3,5-diglucoside, delphinidin 3-glucoside, cyanidin 3-glucoside, and pelargonidin 3-glucoside [28]. Punicalagin and punicalin (Table 1) are unique to pomegranate and are part of a family of ellagitannins [53]. The ellagitannins are hydrolyzed to ellagic acid in the gut, and further metabolized by the colon microbiota to form urolithin [53].

Punicalin, as a byproduct of the hydrolyzed pomegranate husk industry, has generated particular interest with regard to human health effects including antioxidant activities, multitargeted therapy of cancer, inhibitory activity on HIV-1 reverse transcriptase, anti-hepatotoxic and antibacterial properties, among others [54, 55].

Several classes of pomegranate flavonoids include anthocyanins, flavan-3-ols, and flavonols [53]. Pomegranate juice and peel have catechins with a high antioxidant activity. They are essential compounds of anthocyanin's production with antioxidant and inflammatory role [43, 53]. Anthocyanins cause the red color of juice, which is not found in the peel. All pomegranate flavonoids have shown antioxidant activity with indirect inhibition of inflammatory markers such as TNF-α [43, 53]. Pomegranate juice may provide protection against cardiovascular diseases and stroke, by acting as a potent antioxidant against LDL oxidation and inhibition of atherosclerosis development [56]. Pomegranate juice phytochemicals also show potential in chemoprevention of various types of cancers, by exerting antiproliferative effects on tumor cells [56].

Punicalagin and punicalin significantly reduce the production of nitric oxide and prostaglandin E2 (PGE2), so they act as an anti-inflammatory compounds [47]. Another effect of pomegranate polyphenols is the prevention of atherosclerosis development and cardiovascular disease [47]. The anti-cancer effect of pomegranate fruit's ellagitannins and its hydrolyzed products especially ellagic acid and punicalagin on colon cancer cells and prostate cancer cells have also been demonstrated [47].

The pomegranate husks, a food processing residual, appears to be exceptionally rich in polyphenols, particularly hydrolysable tannins [57]. This peculiar composition makes pomegranate husks unique when compared with other plant food wastes, which contain primarily flavonoids and other simpler phenolics [57]. Amyrgialaki *et al.*, 2014 [57] have conducted a study dealing with the phenolics extraction from pomegranate husks and have concluded that the major polyphenolic phytochemicals detected were identified as punicalin (Table **1**), and ellagic acid, using liquid chromatography–mass spectroscopy [57]. Moreover, they observed that the polyphenolic concentration in the extracts correlated significantly with antiradical activity, suggesting that the totality of polyphenols accounts for the antioxidant properties observed [57].

Pure punicalagin, total pomegranate tannin extract, or pomegranate juice have been shown to inhibit, in a dose-dependent manner, TNF-α-induced cyclo-oxygenase 2 (COX-2) expression in HT-29 human colon cancer cells [33]. In addition, a whole pomegranate methanol extract has been shown to inhibit TNF-α expression and release in microglial cells activated with lipopolysaccharides [33]. A recent study suggested that a pomegranate extract could be particularly promising in the dietary prevention of intestinal inflammation since it was able to inhibit NF-κB activity and to decrease inflammatory cytokine (IL-8) and PGE2 in human intestinal Caco-2 cells stimulated by cytokines [33].

The pomegranate tree's bark and roots are rich sources of chemicals called alkaloids (pelletierines and its derivative), which are carbon-based substances used to treat worms in the human gastrointestinal tract in traditional medicine [43, 47].

In sum, several phytochemicals can be found in pomegranate fruit; amongst these phytochemicals, high molecular weight polyphenols (*e.g.* ellagitannins and the pomegranate-peculiar punicalagin) are likely to mediate the protective effects against a wide range of oxidative and inflammatory disorders [26, 27, 58]. Other phenolic compounds (anthocyanins, gallotannins, hydroxycinnamic acids, hydroxybenzoic acids and hydrolysable tannins *i.e.* ellagitannins, and gallagyl esters) have been identified in peel and other anatomical parts of the fruit. The whole fruit is rich in large polyphenolic compounds such as punicalagin isomers, ellagic acid derivatives and anthocyanins [26, 27, 58].

Table 1. Principal compounds found in pomegranate, and its known bioactivities (Adapted from Lansky and Newman, 2007 [22]).

Chemical class	Compound name	Compound structure	Plant part	Bioactivities
Conjugated fatty acids	Punicic acid (*cis*-9, *trans*-11, *cis*-13 octadecatrienoic acid)		Seeds	Inhibits PC-3 invasion; May enhance B-cell function *in vivo*; Cytotoxic in leukemia cells *via* lipid peroxidation.

(Table 1) contd.....

Chemical class	Compound name	Compound structure	Plant part	Bioactivities
Ellagitannins	Punicalagin		Peel, leaves, bark of tree and bark of tree root	↓ UV-B mediated activation of NF-kB mitogen-activated protein kinases (MAPK); ↓ Inflammatory cell signaling in colon cancer; Antiproliferative, apoptotic and antioxidant activities.
	Punicalin			

BIOAVAILABILITY OF POMEGRANATE POLYPHENOLS

Several factors were proposed in an attempt to explain the differences observed between the positive effects of polyphenolic consumption reported in epidemiological studies and the often negative findings reported in intervention trials with supplements [16, 59]. These factors included: differing doses of administered compounds; additive/synergistic effects – such as those between polyphenols and other antioxidants – present in whole foods but not in supplements; differences in bioavailability due to varying matrix release kinetics; and compounds present in the matrix, enhancing or reducing polyphenolics bioavailability [16, 59, 60].

The absorption, distribution, metabolism, and excretion of polyphenols after dietary intake have been topics of increased research in recent years [61, 62]. After the acute ingestion, absorption of some, but by no means all components into the circulatory system, occurs in the small intestine [61, 63].

Phytochemicals ingested daily from vegetal sources or natural extracts which are used to develop new functional foods, are complex mixtures of molecules that have to be absorbed and metabolized if they are to finally reach the cells *in vivo* [16, 64]. Therefore, the primary aspects that influence polyphenols positive effects is their bioavailability and metabolic rate [16, 65]. Bioaccessibility is defined as the amount of a food constituent, present in the gut as a result of its release from the solid food matrix, which might be able to pass through the intestinal barrier [59, 60, 64, 66]. In addition, bioavailability is the proportion of the nutrient that is digested, absorbed, and metabolized through the normal pathways [64, 67]. In general, the bioavailability of dietary compounds; particularly phytochemicals; is dependent on their digestive stability, their release from the food matrix and the efficiency of their transepithelial passage [16, 64, 65]. Consequently, bioaccessibility has to be taken into consideration in studies concerning the bioavailability of polyphenols [21, 64].

A significant portion of the pomegranate juice contains the pomegranate polyphenols called ellagitannins and they often coexist with ellagic acid, the main product obtained through hydrolysis of this class of tannins [68, 69]. This acid is obtained through the metabolization of the ellagitannins by the intestinal bacteria and is found to be analogous to urolithins [69, 70]. The urolithins are reported to be systematically bioavailable and they accumulate in organs such as colon, prostate and intestine [69].

The bioavailability of ellagitannins has been studied both in animals and in humans [61]. Ellagitannins are large molecules and because of that, they are not easily absorbed by humans; however, in a single reported experiment with rats, that were fed a standard diet plus 6% punicalagin, the resulting ellagitannin punicalagin was identified in the serum (30 $\mu g.mL^{-1}$) and 3–6% of all the ingested punicalagin was excreted in urine and faeces [26, 71]. Ellagic acid is a hydrolysate of ellagitannins such as punicalagin and is naturally present in pomegranate. Several studies have considered it to be highly bioactive [26, 71]. Following the consumption of pomegranate juice or pomegranate extract, ellagic acid is released into the bloodstream. The hydrolysis of ellagitannins and the resulting release of ellagic acid are not caused by an enzymatic reaction but are pH-dependent [26, 71].

An extensive description of the bioavailability of pomegranate polyphenols, namely ellagitannins, was done and summarized by Johanningsmeier and Harris, 2011 [28]. In this work, the authors have described that ellagic acid and its metabolites were detected in the plasma of individuals that had consumed pomegranate juice [28]. Furthermore, the authors have also reported that there was no difference in bioavailability, as indicated by plasma ellagic acid or its metabolites, when individuals consumed either pomegranate juice, liquid extract, or powdered extract with similar levels of total polyphenols [28]. In addition, the authors have also identified urolithin metabolites, which are a result of the colonic microbial metabolism of the pomegranate juice polyphenols, based on the appearance time of those metabolites in plasma and urine samples. Those metabolites have been demonstrated to be readily absorbed in mouse models, with the highest levels accumulating in prostate, colon, and intestinal tissues [28]. Similarly, after three days of supplementation with pomegranate juice, urolithin A glucuronide, urolithin B glucuronide, and dimethylellagic acid were the only ellagic acid metabolites detected in human prostate tissues [28]. Those results appear to indicate that ellagitannins are hydrolyzed in the stomach where some portion of ellagic acid is absorbed into circulation. Gut microflora metabolizes then, the remaining ellagic acid into urolithin derivatives [28]. The less polar of these urolithin derivatives (A and B) are absorbed into circulation and further metabolized to glucuronides [28].

Urolithins A, C, and D have been shown to possess antioxidant activity in a cell-based assay, and urolithin A was found to have significant anti-inflammatory activity in an *in vitro* colon fibroblast model by the inhibition of the activity of eicosanoid generating enzymes and the production of NO [31]. This further suggests that consumption of pomegranate may be of value in inhibiting inflammatory stimuli-induced cartilage breakdown and production of inflammatory mediators in arthritis [31]. Therefore, it is possible that pomegranate polyphenolic compounds may act in multiple ways, with some being absorbed and entering the bloodstream to act directly as antioxidants, and the remainder being digested by the colonic microflora to provide other biologically active substances [28].

Anthocyanins which are also present in abundance in pomegranate, are not bioavailable since these compounds, as well as its metabolites, have not been detected in serum or urine after the consumption of pomegranate juice [26, 71].

Another *in vitro* study [72] examining the outcome of pomegranate juice consumption has shown that pomegranate phenolic compounds are available in increasing amounts during digestion, whereas the anthocyanins are largely metabolized to some no colored forms, oxidized, or degraded into other chemicals [72]. The first report on metabolism and bioavailability of components analyzed punicalagin, an ellagitannin pomegranate juice [72]. In this study, a number of colonic microflora metabolites, ellagic acid derivatives, and ellagitannins were detected in rat urine, plasma, and feces [72].

Notably, the bioavailability and metabolism of pomegranate ellagitannins have been extensively studied in both animals and humans [41]. Interestingly, when pomegranate is consumed, intact punicalagin is not observed in circulation but rather is hydrolyzed to release ellagic acid (which clears from plasma within 4-6 h) and is subsequently converted by gut microflora into further metabolites known as urolithins [41].

The reasons cited for poor bioavailability of punicalagins include the large molecular size which limits its absorption by simple diffusion and extremely low lipid solubility which restricts its permeability through the bilipid layer of the gastrointestinal tract [50]. Moreover, punicalagins are metabolized and made bioavailable but into a relatively poor antioxidant, hydroxy-6H- dibenzopyran-6-one derivatives by the colonic microflora of healthy humans [50].

Although transport and metabolic mechanisms cannot be effectively reproduced, studies can provide a simple predictive instrument to investigate the potential bioavailability of dietary compounds, namely pomegranate polyphenols, by assessing their stability under conditions mimicking the gastrointestinal tract [67].

POMEGRANATE ANTIOXIDANT PROPERTIES

Oxidative stress resulting from an imbalanced defense mechanism of endogenous antioxidants, as it was mentioned above, is an injurious process that can be an important mediator of cell damage and consequently has been reported as being associated with several pathological conditions [2]. The process of ageing has also been referred as being greatly influenced by the damaging actions of free radicals, such as DNA damage, lipid peroxidation and protein oxidation [2]. The DNA damage and lipid peroxidation have been associated with the mechanism of carcinogenesis [2].

Natural antioxidants widely distributed in fruits, vegetables and beverages like tea and red wine have been referred as having interesting health benefits and have received scientific interest in recent years, particularly concerning the action of phenolic compounds [73]. As already described above, pomegranate exhibits phytochemical compounds that are described as possessing antioxidant properties and thus they potentially confer beneficial properties for a wide range of oxidative stress-related diseases [73]. Thus, the antioxidant activity of several extracts from different parts of pomegranates and also from distinct parts of pomegranate tree has been investigated using different experimental approaches and reported in literature [73]. The antioxidant compounds act *in vivo* through distinct mechanisms of action, so no single method of antioxidant activity determination can fully evaluate the antioxidant properties of different natural products [73]. Therefore, results reported in literature, concerning antioxidant capacity of pomegranate extracts use methods that investigate the ability to scavenge free radicals or hydrogen peroxide [73]. Other methods measure the Fe^{2+} chelating activity of extracts, considering that transition metals are involved in oxidative reactions [73]. An important mechanism related with oxidative stress is lipid peroxidation, which can be assessed by the β–carotene bleaching method [73].

The type of solvent used for the sample extractions could influence the results concerning the antioxidant activity as it was pointed out by different authors 25, [74, 75] with methanolic extracts exhibiting the higher antioxidant properties. The obtained extracts had different contents in flavonoid and phenolic compounds and consequently different antioxidant behaviors. Moreover, extracts of different parts of pomegranates exhibited dose dependent antioxidant activities [73, 75, 76].

The extracts of different parts ofpomegranate, peel, seed, leaves and flowers, as well as the fruit juice, have been described as possessing important *in vitro* antioxidant activity depending on the plant part used [11, 25, 32, 73 - 78] and that is cultivar dependent [32, 76, 79]. This activity is probably correlated with the high polyphenol content in general, and with the content in flavonoids in particular [25, 80], as it was also observed for other polyphenol rich fruits [81, 82]. Moreover, the antioxidant activity as well as the polyphenol levels of pomegranate juice are significantly influenced by climatic conditions during fruit ripening and latitude of growing region [32, 79].

An *in vivo* study with young healthy males taking pomegranate juice supplementation during severe physical activity, indicates that the regular intake of this supplementation modulates oxidative stress and thus protects against oxidative damage [11].

EFFECTS OF POMEGRANATE ON OXIDATIVE STRESS-RELATED DISEASES

Inflammatory Processes

Physiological or acute inflammation is a beneficial host response to tissue damage, but when a timely resolution is delayed, it may lead to immune-associated diseases such as rheumatoid arthritis, inflammatory bowel disease, and cancer [22, 83]. Chronic inflammation can lead to cell changes associated with the development of cancer through attraction of soluble pro-inflammatory mediators TNF-α, interleukins (*e.g.* IL-6 and IL-8), transcription activation factors (*e.g.* NF-kB), and bioactive lipids such as eicosanoids (*e.g.* prostaglandin E2 and lipoxygenase derived products) [22, 83]. Hence, anti-inflammatory agents blocking excessive production of these inflammatory mediators could be a major step to reduce inflammation [83]. The study of the anti-inflammatory properties of natural products has gained popularity, aimed at finding new anti-inflammatory agents without the undesirable side effcts and producing synthetic derivatives with enhanced activity has gained increased popularity in the last years [84].

The amount of compelling scientific evidence regarding the therapeutic benefits of pomegranate and its fractions has built a scientific consensus that pomegranate rind methanolic extract has the ability to inhibit inflammation and allergies [85]. The anti-inflammatory components of pomegranate peel, *i.e.*, punicalagin, punicalin, strictinin A and granatin B significantly reduce production of NO and PGE_2 by inhibiting the expression of pro-inflammatory proteins [85]. Nevertheless, inflammatory cells including neutrophils, macrophages and monocytes may inflict damage to nearby tissues, an event thought to be of pathogenic significance in a large number of diseases such as emphysema, acute respiratory distress syndrome, atherosclerosis, reperfusion injury, malignancy and rheumatoid arthritis [85].

Panichayupakaranant *et al.*, 2010 [86] have conducted a study to assess some bioactivities of pomegranate rind extracts [86]. They have concluded that the standardized pomegranate rind extract had marked anti-NO effects, with an IC_{50} value of 10.7 µg/mL [86]. Its potency was the same as that of L-nitroarginine, a NO synthase inhibitor [86]. However, the potency of pomegranate extract was about five fold lower than those of ellagic acid and caffeic acid phenethylester [86].

The anti-inflammatory properties of pomegranate and its metabolites were also evaluated by Larossa *et al.*, 2010 [87]. In this study, it was proven that the effects seen in pomegranate extracts were not solely a result of the accumulation of urolithin A in the colon, when in fact the effect seen in in dextran sodium sulfate (DSS)-urolithin A (DSS is a well-known model of inflammatory bowel disease) was evidently linked to this metabolite [87]. Generally, in terms of the amelioration of inflammation, urolithin A supplementation had more of an effect than pomegranate extract; however, a significant effect was also noted on inflammatory markers, antioxidant status, microbiota and gene expression in the DSS-pomegranate extract group [87]. The effect of this specific outcome could be attributed to either the remaining ellagic acid or the ellagitannin fraction that could have reached the colon together with the small amount of urolithin A produced by the microbiota [87].

The results obtained by Bishayee *et al.*, 2013 [88] clearly demonstrated that the pomegranate bioactive constituents are capable of suppressing DENA-mediated inflammatory cascade by reversing the increased expression of inflammatory mediators, namely, iNOS, 3-NT, HSP70, HSP90, COX-2 and NF-κB, as well as hindering the nuclear translocation of NF-κB during an early stage of experimental hepatic neoplasia [88].

The research of Mo *et al.*, 2013 [89] has for the first time provided firm evidence that topical application of standardized pomegranate rind extract has excellent anti-inflammatory and analgesic properties [89]. Its marker compound, ellagic acid, is the major active constituent responsible for the pharmacological effects, but the pomegranate extract is superior to ellagic acid alone in terms of anti-inflammation and IL-1β modulation [89]. The analgesic activity of pomegranate rind extract is also significant and is better than the steroid anti-inflammatory drug triamcinolone in chronic inflammation even though this effect is milder when compared with diclofenac for acute inflammatory pain [89]. The therapeutic benefits of standardized pomegranate rind extract are probably due to its ability to inhibit the leukocyte infiltration and to block the pro-inflammatory cytokines besides its NO scavenging properties [89]. Therefore, topical standardized pomegranate rind extract formulations are promising phytopharmaceutical candidates and have the potential to be applied in the clinical treatment of inflammatory diseases such as cutaneous inflammation and arthritis as a complementary and alternative therapy [89].

Bekir *et al.*, 2013 [25] have studied for the first time, the 5-lipoxigenase inhibition by pomegranate leaves and have concluded that the greatest anti-inflammatory activity was exerted by the extracts with the most antioxidant activity, which contained the highest amounts of total phenolics [25].

A study by Shukla *et al.*, 2008 [31] provided evidence that bioavailable constituents and/or metabolites of pomegranate extract exert an anti-inflammatory effect by inhibiting the activity of eicosanoid generating enzymes and the production of NO [31]. This further suggests that the consumption of pomegranate may be of value in inhibiting inflammatory stimuli-induced cartilage breakdown and production of inflammatory mediators in arthritis [31].

Since ancient times, pomegranate has been regarded as a "healing food" with numerous beneficial effects against several diseases. The studies carried out today indicate that our ancestors were not entirely wrong and that some of the attributed properties to pomegranate were not unfounded, namely its anti-inflammatory activities [87].

Cancer

Cancer is a difficult disease to manage and treat. For the year 2014, it was estimated that a total of 1.665.540 new cancer cases would be diagnosed in the United States and approximately 585.720 cancer-related deaths would occur [90]. Although these numbers are impressive, it is interesting to note that the cancer-related mortality has declined steadily over last two decades, from 215 deaths per

100.000 inhabitants in 1991 to 172 in 2010 [90]. More than 1.5 million new cases and more than half a million deaths are big numbers that call for sustained efforts in the fight against cancer [90]. One of the biggest challenges in the effective clinical management of human cancers is the absence of validated therapeutic targets, especially when evaluating the activity of natural agents (nutraceuticals) [90]. Several studies have shown that phenolic compounds are the major bioactive phytochemicals with human health benefits [91 - 93]. In fact, many authors have reported a direct relationship between total phenolic content and antioxidant activity in numerous seeds, fruits and vegetables [10, 91].

The unique antioxidant tannins and flavonoids contained in pomegranate extracts have recently drawn the attention of many scientists and the biological activities of pomegranate-derived products, especially their anti-inflammatory and anticancer properties, are under investigation [94]. Recent research has shown that pomegranate extracts selectively inhibit the growth of breast, prostate, colon and lung cancer in cell culture and animal models [94]. Pomegranate juice has shown promising results in a phase II clinical study against prostate cancer [94].

Koyama *et al.*, 2010 [95] have concluded that the results of their study revealed novel interactions between the insulin-like growth factor (IGF) system and pomegranate-induced apoptosis, and suggests that pomegranate products modulate the tumor production and responsiveness to IGFs and the IGF-binding proteins (IGFBP) [95]. As IGFBP-3 is currently being tested in humans as a treatment for prostate cancer and pomegranate supplements are becoming popular as adjuvant nutritional treatments for this disease, they propose that these agents may emerge as useful in the management of prostate cancer [95]. The study performed by Gommersall *et al.*, 2004 [96] demonstrated the significant antitumor activity of pomegranate-derived materials against human prostate cancer [96]. In fact, other researchers have obtained similar results [94, 97, 98]. Malik *et al.*, 2005 [30] in their study, employing human prostate cancer cells, evaluated the antiproliferative and pro-apoptotic properties of pomegranate extract [30]. Pomegranate extract (10–100 µg/mL; 48 h) treatment of highly aggressive human prostate cancer PC3 cells resulted in a dose-dependent inhibition of cell growth/cell viability and induction of apoptosis [30]. They hypothesized the involvement of the cyclin kinase inhibitor-cyclin-cdk network in the antiproliferative effects of pomegranate extract [30]. Oral administration of pomegranate extract (0.1% and 0.2%, wt/vol) to athymic nude mice implanted with androgen-sensitive CWR22Rnul cells resulted in a significant inhibition of tumor growth concomitant with a significant decrease in serum prostate-specific antigen levels [30]. They suggested that pomegranate juice may have cancer-chemopreventive as well as cancer-chemotherapeutic effects against prostate cancer in humans [30]. The work of Hong *et al.*, 2008 [99] showed that pomegranate protects against prostate cancer by down-regulation of genes involved in androgen synthesis. However, the mechanism of the down-regulation is not known [99]. They concluded that pomegranate juice and pomegranate extract as well as their polyphenols showed a capability to arrest proliferation and stimulate apoptosis in human androgen-dependent and androgen-independent prostate cancer cells [99]. Inhibition of gene expression involved in androgen-synthesizing enzymes and the androgen receptor may contribute to the growth-inhibitory effects of pomegranate polyphenols and may provide a molecular target for the inhibition of the emergence of androgen-independent prostate cancer [99]. Wang *et al.*, 2012 [100] have demonstrated that luteolin, ellagic acid, and punicic acid, components of pomegranate juice, can potentially be used as anti-metastatic treatments to deter prostate cancer metastasis [100]. These compounds interfere with multiple biological processes involved in metastasis of cancer cells such as suppression of cell growth, increase in cell adhesion, inhibition of cell migration, and inhibition of chemotaxis toward proteins that are important in prostate cancer metastasis to the bone [100].

Regarding the activity of pomegranate in hapatocellular carcinogenesis, Bishayee *et al.*, 2013 [88] have underlined the importance of simultaneously targeting two interconnected molecular circuits, namely, the Nrf2-mediated redox signaling and NF-κB-regulated inflammatory pathway, to achieve chemoprevention of hapato-cellular carcinogenesis [88]. These interesting attributes together with a safety profile highlight the chemopreventive and therapeutic potential of pomegranate in the management of liver cancer [88].

The potential beneficial effects of pomegranate and its bioavailable metabolites in other types of cancer, namely, colon cancer, melanogenesis/skin cancer, breast cancer and bladder cancer were also previously reported [69, 85, 101].

To summarize, the biological activity of pomegranate-derived products, especially the chemotherapeutic and chemopreventive properties, has been investigated in cell, animal and clinical studies [98]. The pomegranate and its components interfere with multiple biological processes involved in tumor growth, angiogenesis and metastasis of various cancers, and particularly the prostate cancer [98].

Diabetes

Diabetes mellitus is a major endocrine disorder and the global annual cost of treating it and its complications could reach US $ trillion [40]. Improvement of diabetes treatment is a high priority in medical research. Self-management of diabetes is cornerstone to achieving good glycemic control and reducing the risk of developing microvascular (retinopathy, nephropathy, and neuropathy) and macrovascular (cardiovascular and cerebrovascular) complications [40]. Natural extracts have been used to manage of diabetes and its related complications because they are safe and readily available [40]. Many plant species are used in folk medicine for their hypoglycemic properties and therefore potentially used for the treatment of diabetes [40]. Many scientific studies have reported the anti-diabetic activity of pomegranate products [77].

The vast medical potential of pomegranate extracts as well as their active components can potentially provide an effective and safe treatment for type 2 diabetes and its pathological concerns [102]. They affect the type 2 diabetes, mainly by antagonizing the damaging effects of ROS. This can occur directly or indirectly by increasing the activity of specific antioxidant enzymes, such as PON1, SOD, and CAT [102 - 104]. Morever, the pomegranate fractions show properties of metal chelation; including the inhibition or activation of certain transcriptional factors involved in glucose normalization, such as NF-κBand PPAR- γ; as well as the reduction of the resistin formation [102]. The reduction of fasting blood glucose levels were solely decreased by punicic acid, methanolic seed extract, and pomegranate peel extract [102]. Several pomegranate components (punicalagin and ellagic, gallic, oleanolic, ursolic, and uallic acids) were found to have anti-diabetic effects [102]. Additionally, type 2 diabetes variables were potentially impacted by pomegranate juice due to their antioxidant polyphenolic tannins and anthocyanins [102].

Xu *et al.*, 2009 [105] have concluded that pomegranate flower extract ameliorates fatty liver in the rats with type 2 diabetes and obesity [105]. It is possible that the herb-elicited upregulation of hepatic expression of PPAR-α- and SCD-1-regulated genes responsible for fatty acid oxidation results in a decrease in lipid accumulation in the liver [105]. These findings are potentially important for supporting its extension to clinical trials to demonstrate the effectiveness of pomegranate flower extract in the prevention and/or treatment of diabetes and obesity-induced non-alcoholic fatty liver disease, through modulation of abnormal lipid metabolism [105].

Another study [106], dealing with the pomegranate polyphenols in type 2 diabetes, allows to conclude that commercially available pomegranate polyphenols were well tolerated with no adverse effects in healthy volunteers as well as in patients with type 2 diabetes [106]. In addition, pomegranate polyphenols lowered lipid peroxidation only in diabetic patients but had no effects on glucose, lipids, and C-reactive protein in either group of participants [106].

Patel *et al.*, 2014 [107] have demonstrated the antihyperglycemic, anti-hyperlipidemic and antioxidant properties of ethyl acetate fraction of pomegranate leaves in diabetic rats without any hypoglycemic action in normal rats and without evident toxic effects [107]. Therefore the traditional claim was scientifically verified using animal experiments [107]. As can be concluded from the various parameters studied, the possible mechanisms of action by which ethyl acetate fraction of pomegranate leaves has its antihyperglycemic action is by enhancing glycogenesis in liver, increasing glucose utilization by the skeletal muscle, decreasing the absorption of glucose from intestinal lumen, inhibition of α-glucosidase and α-amylase enzymes and regulation of gluconeogenesis [107]. The antihyperglcemic potential of ethyl acetate fraction of pomegranate leaves can be attributed to its flavonoids, tannins and phenolic compounds [107].

The study of Raafat & Samy, 2014 [40] indicated that pomegranate extract as well as its biodegradable polymeric dispersions with casein and chitosan exerted remarkable hypoglycemic activity and improved peripheral nerve function, which might be due to gallic acid, which prevents oxidative stress in diabetic animals beside its insulin-secretagogue action [40]. Therefore, the observed *in vivo* antioxidant potential of the pomegranate extract might possibly be added to the mechanism of action responsible for its antinociceptive effect [40].

There is sufficient scientific evidence to claim that the whole pomegranate, parts of the fruit, and certain extracts have several beneficial health effects against diabetes [26]. The treatment of type 2 diabetes is an interesting and promising property of this fruit achieved through several biological properties, including hypoglycemic and insulin sensitising effects mediated by the inhibition of α-glucosidase, regulation of glucose transporter type 4 function, regulation of adipokine levels, and the improvement of blood lipids (including a reduction of total cholesterol), among others [26].

Cardiovascular Diseases

Cardiovascular disease represents one of the leading causes of death worldwide [108]. Although epidemiological studies have revealed an inverse association between high consumption of fruits and vegetables and a reduced risk of cardiovascular diseases, the role of the compounds providing these protective effects is still under investigation [108]. Edible plants contain numerous bioactive molecules, among which are polyphenols. These molecules are the most abundant dietary antioxidants, and their effects on health have gained huge interest [108]. Epidemiological studies have reported an inverse relationship between flavonoid-rich food consumption and cardiovascular events [108]. This observation is further supported by clinical and preclinical studies of either flavonoid-rich foods or isolated flavonoids demonstrating their beneficial effects on cardiovascular risk factors such as blood pressure, endothelial function, platelet function, and cholesterolemia [108].

Pomegranate is suggested as the "heart-healthy" fruit, and it was indeed shown to attenuate cardiovascular diseases [109]. Measurements of the arterial stiffness of the common carotid arteries in 73 patients with at least one cardiovascular risk factor who consumed pomegranate juice, showed increased arterial elasticity in the juice-treated group *versus* the placebo-treated group (who received beverage of similar caloric content, flavor, and color) [109].

Shema-Didi *et al.*, 2010 [110] in their study about the beneficial effects of pomegranate juice consumption on cardiovascular diseases, have verified that pomegranate juice consumption, but not placebo intake, yielded a significant time response reduction in polymorphonuclear leukocyte priming, protein oxidation and inflammation, all abolished 3 months later by the discontinuation of pomegranate juice intake, validating its beneficial effects [110]. In this study, decrease in blood pressure levels and plasma lipids, concomitant with a significant relative improvement in the atherosclerotic process in the pomegranate juice group, was shown [110]. The incidence rates of first and second infection and cardiovascular events were lower in the pomegranate group *vs.* placebo [110].

Stowe, 2011 [111] have reviewed the effects of pomegranate juice consumption on blood pressure and cardiovascular health and suggested that pomegranate juice has considerable anti-atherosclerotic, anti-hypertensive, antioxidant, and anti-inflammatory effects in human subjects and mouse models [111]. The principal mechanisms of action of pomegranate juice may include decreased systolic blood pressure, thus causing an overall positive effect on the progression of atherosclerosis and the ensuing potential development of coronary heart disease [111]. But with only two clinical trials showing a decrease in blood pressure, it cannot be said with certainty that pomegranate juice should be prescribed for cardiovascular health [111].

The findings from the pilot trial conducted by Asgary *et al.*, 2014 [112] provided evidence for the beneficial impact of pomegranate juice consumption on both systolic and diastolic blood pressures in hypertensive individuals [112]. Besides, consumption of pomegranate juice was found to be associated with significant reductions in serum VCAM-1 which is a biomarker of endothelial function and vascular inflammation [112]. In light of these promising findings, pomegranate juice may be considered as an effective adjunct to the anti-hypertensive medications and also as a constituent of daily regimen for patients who have high risk for hypertension and cardiovascular diseases [112]. Moreover, pomegranate juice with its promising acute hypotensive properties should be considered in the context of both dietary and pharmacological interventions for hypertension [113].

Other authors have reviewed and summarized the potential health benefits of pomegranate consumption regarding the prevention of cardiovascular diseases risk factors, and the main conclusion is that the antioxidant activity of pomegranate polyphenols is the basis for the protective mechanism against cardiovascular disorders [109, 114, 115].

Neurodegenerative Disorders

Aging is the major risk factor for the development of neurodegenerative diseases such as Alzheimer's and Parkinson's diseases, which are the first and the second more prevalent pathologies in elderly people, respectively [2].

A large number of evidence indicates that oxidative stress is involved in the pathophysiology of these diseases [116, 117]. Central nervous system is highly sensitive to oxidative stress and ROS-induced damage on lipids, proteins and DNA, increases with aging [118, 119]. A high level of accumulation of lipid peroxidation or oxidative protein accelerates the progress of aging and neurodegenerative diseases [118, 119]. Polyphenolic antioxidants from fruits and vegetables have been studied for their effectiveness in reducing these harmful effects in several *in vitro* and *in vivo* studies [118, 119]. Increasing number of studies demonstrated the efficacy of these natural antioxidants to prevent or to reduce the progression of neurodegenerative disorders [118, 119].

The high content in polyphenols and the described antioxidant activity of pomegranate extracts and its juice suggest that this species may play an important role in neuroprotection. Kumar *et al.*, 2009 [120] showed that aged and scopolamine treated young mice treated with ethanolic extract of pomegranate seeds exhibited lowered lipid peroxidation level and increased antioxidant glutathione level in brain tissues, which suggest that pomegranate preparations could be used in the treatment of cognitive disorders such as dementia and Alzheimer's disease [120].

A study with pomegranate juice and two pomegranate polyphenols (punicalagin and ellagic acid) performed in transgenic mice model of Alzheimer's disease and in cell cultures showed inhibition of nuclear factor of activated T-cell activity and microglial activation [42]. These results indicate that the intake of pomegranate produces brain anti-inflammatory activities that may attenuate the progression of Alzheimer's disease [42].

Hartman *et al.*, 2009 [58] showed that transgenic mice treated with a long-term administration of pomegranate juice had significantly less (around 50%) accumulation of soluble Aβ42 and amyloid deposition in the hippocampus than the control mice [58]. A subsequent study carried out by other research team [41] indicated that a commercial pomegranate extract altered levels and ratio of the Aβ42 and Aβ40 peptides in Alzheimer's disease mice model by modulation of processing of amyloid-β precursor protein [41].

Concerning Parkinson's disease, they were not found relevant *in vivo* and *in vitro* studies in the literature. However, the results of an epidemiologic study with 805 participants (men and women) performed in order to examine whether higher intakes of different sub-classes of flavonoids, considering five major sources of flavonoid-rich foods (tea, berry fruits, apples, red wine, and orange/orange juice), suggest that the intake of some flavonoids may reduce Parkinson's disease risk, particularly in men, even if a protective effect of other constituents of plant foods cannot be excluded [121]. Moreover, a study with extracts rich in anthocyanins, proanthocyanidins and other polyphenols carried out in primary cell culture model of Parkinson's disease showed that dopaminergic cell death provoked by rotenone was suppressed by those extracts, suggesting that plant extracts rich in anthocyanins and proanthocyanidins may alleviate neurodegeneration in Parkinson's disease *via* enhancement of mitochondrial function [122].

It is noteworthy that pre-administration of standardized pomegranate extract to adult male albino rats, can provide a significant dose-dependent neuroprotective activity against cerebral ischemia/reperfusion-induced brain injury and DNA damage *via* antioxidant, anti-inflammatory, anti-apoptotic and ATP-replenishing activities [123]. These results are promising considering the effectiveness of pomegranate extract in patients predisposed to stroke [123].

CONCLUDING REMARKS

P. granatum extracts and fruit juice have demonstrated promising health effects against oxidative stress and oxidative stress-mediated diseases, such as cancer, diabetes, cardiovascular and neurodegenerative disorders. However, sufficient depth of knowledge on the effectiveness of many of these proposed effects has not yet been attained and more studies *in vitro* and in animal models are needed, as well as human clinical trials to prove evidences of those properties.

CONFLICT OF INTEREST

The author confirms that author has no conflict of interest to declare for this publication.

ACKNOWLEDGEMENTS

Declare none.

REFERENCES

[1]   Li J, O W, Li W, Jiang ZG, Ghanbari HA. Oxidative stress and neurodegenerative disorders. Int J Mol Sci 2013; 14(12): 24438-75.
[http://dx.doi.org/10.3390/ijms141224438] [PMID: 24351827]

[2]   Valko M, Leibfritz D, Moncol J, Cronin MT, Mazur M, Telser J. Free radicals and antioxidants in normal physiological functions and human disease. Int J Biochem Cell Biol 2007; 39(1): 44-84.
[http://dx.doi.org/10.1016/j.biocel.2006.07.001] [PMID: 16978905]

[3]   Parejo I, Viladomat F, Bastida J, *et al.* Investigation of Bolivian plant extracts for their radical scavenging activity and antioxidant activity. Life Sci 2003; 73(13): 1667-81.
[http://dx.doi.org/10.1016/S0024-3205(03)00488-0] [PMID: 12875899]

[4]   Choi DJ, Kim SL, Choi JW, Park YI. Neuroprotective effects of corn silk maysin *via* inhibition of H_2O_2-induced apoptotic cell death in SK-N-MC cells. Life Sci 2014; 109(1): 57-64.
[http://dx.doi.org/10.1016/j.lfs.2014.05.020] [PMID: 24928367]

[5]   Balasubramanian K, Padma PR. Anticancer activity of *Zea mays* leaf extracts on oxidative stress-induced Hep2 cells. J Acupunct Meridian Stud 2013; 6(3): 149-58.
[http://dx.doi.org/10.1016/j.jams.2013.01.015] [PMID: 23787284]

[6]   Aruoma OI. Methodological considerations for characterizing potential antioxidant actions of bioactive components in plant foods. Mutat Res 2003; 523-524: 9-20.
[http://dx.doi.org/10.1016/S0027-5107(02)00317-2] [PMID: 12628499]

[7]   Rosenblat M, Volkova N, Aviram M. Pomegranate phytosterol (β-sitosterol) and polyphenolic antioxidant (punicalagin) addition to statin, significantly protected against macrophage foam cells formation. Atherosclerosis 2013; 226(1): 110-7.
[http://dx.doi.org/10.1016/j.atherosclerosis.2012.10.054] [PMID: 23141585]

[8]   Egea I, Sánchez-Bel P, Romojaro F, Pretel MT. Six edible wild fruits as potential antioxidant additives or nutritional supplements. Plant Foods Hum Nutr 2010; 65(2): 121-9.
[http://dx.doi.org/10.1007/s11130-010-0159-3] [PMID: 20198440]

[9]   Gomes A, Fernandes E, Lima JL, Mira L, Corvo ML. Molecular mechanisms of anti-inflammatory activity mediated by flavonoids. Curr Med Chem 2008; 15(16): 1586-605.
[http://dx.doi.org/10.2174/092986708784911579] [PMID: 18673226]

[10]   Luís A, Domingues F, Duarte AP. Bioactive compounds, RP-HPLC analysis of phenolics, and antioxidant activity of some Portuguese shrub species extracts. Nat Prod Commun 2011; 6(12): 1863-72.
[PMID: 22312726]

[11]   Naghizadeh-Baghi A, Mazani M, Shadman-Fard A, Nemati A. *Punica granatum* juice effects on oxidative stress in severe physical activity. Mater Sociomed 2015; 27(1): 48-51.
[http://dx.doi.org/10.5455/msm.2014.27.48-52] [PMID: 25870532]

[12]   Rodrigues S, Calhelha R, Barreira J, *et al. Crataegus monogyna* buds and fruits phenolic extracts: Growth inhibitory activity on human tumor cell lines and chemical characterization by HPLC–DAD–ESI/MS. Food Res Int 2012; 49(1): 516-23.
[http://dx.doi.org/10.1016/j.foodres.2012.07.046]

[13]   da Silveira CV, Trevisan MT, Rios JB, *et al.* Secondary plant substances in various extracts of the leaves, fruits, stem and bark of *Caraipa densifolia* Mart. Food Chem Toxicol 2010; 48(6): 1597-606.
[http://dx.doi.org/10.1016/j.fct.2010.03.032] [PMID: 20347919]

[14]   Ismail A, Marjan Z, Foong C. Total antioxidant activity and phenolic content in selected vegetables. Food Chem 2004; 87(4): 581-6.
[http://dx.doi.org/10.1016/j.foodchem.2004.01.010]

[15]   Cartea ME, Francisco M, Soengas P, Velasco P. Phenolic compounds in Brassica vegetables. Molecules 2011; 16(1): 251-80.
[http://dx.doi.org/10.3390/molecules16010251] [PMID: 21193847]

[16]   Bohn T. Dietary factors affecting polyphenol bioavailability. Nutr Rev 2014; 72(7): 429-52.
[http://dx.doi.org/10.1111/nure.12114] [PMID: 24828476]

[17]   Crozier A, Del Rio D, Clifford MN. Bioavailability of dietary flavonoids and phenolic compounds. Mol Aspects Med 2010; 31(6): 446-67.
[http://dx.doi.org/10.1016/j.mam.2010.09.007] [PMID: 20854839]

[18]   Patil BS, Jayaprakasha GK, Chidambara Murthy KN, Vikram A. Bioactive compounds: historical perspectives, opportunities, and challenges. J Agric Food Chem 2009; 57(18): 8142-60.
[http://dx.doi.org/10.1021/jf9000132] [PMID: 19719126]

[19]   Dai J, Mumper RJ. Plant phenolics: extraction, analysis and their antioxidant and anticancer properties. Molecules 2010; 15(10): 7313-52.
[http://dx.doi.org/10.3390/molecules15107313] [PMID: 20966876]

[20]   Reddy C, Sreeramulu D, Raghunath M. Antioxidant activity of fresh and dry fruits commonly consumed in India. Food Res Int 2010; 43(1): 285-8.
[http://dx.doi.org/10.1016/j.foodres.2009.10.006]

[21]   Saura-Calixto F, Serrano J, Goñi I. Intake and bioaccessibility of total polyphenols in a whole diet. Food Chem 2007; 101(2): 492-501.
[http://dx.doi.org/10.1016/j.foodchem.2006.02.006]

[22]   Lansky EP, Newman RA. *Punica granatum* (pomegranate) and its potential for prevention and treatment of inflammation and cancer. J Ethnopharmacol 2007; 109(2): 177-206.
[http://dx.doi.org/10.1016/j.jep.2006.09.006] [PMID: 17157465]

[23]   Fernandes L, Pereira J, Lopéz-Cortés I, Salazar D, Ramalhosa E, Casal S. Fatty acid, vitamin E and sterols composition of seed oils from nine different pomegranate (*Punica granatum* L.) cultivars grown in Spain. J Food Compos Anal 2015; 39: 13-22.
[http://dx.doi.org/10.1016/j.jfca.2014.11.006]

[24]   Saad H, Bouhtoury C, Pizzi A, Rode K, Charrier B, Ayed N. Characterization of pomegranate peels tannin extractives. Ind Crops Prod 2012; 40(1): 239-46.
[http://dx.doi.org/10.1016/j.indcrop.2012.02.038]

[25]   Bekir J, Mars M, Souchard JP, Bouajila J. Assessment of antioxidant, anti-inflammatory, anti-cholinesterase and cytotoxic activities of pomegranate (*Punica granatum*) leaves. Food Chem Toxicol 2013; 55: 470-5.
[http://dx.doi.org/10.1016/j.fct.2013.01.036] [PMID: 23380204]

[26]   Medjakovic S, Jungbauer A. Pomegranate: a fruit that ameliorates metabolic syndrome. Food Funct 2013; 4(1): 19-39.
[http://dx.doi.org/10.1039/C2FO30034F] [PMID: 23060097]

[27]   Akhtar S, Ismail T, Fraternale D, Sestili P. Pomegranate peel and peel extracts: chemistry and food features. Food Chem 2015; 174: 417-25.
[http://dx.doi.org/10.1016/j.foodchem.2014.11.035] [PMID: 25529700]

[28]   Johanningsmeier SD, Harris GK. Pomegranate as a functional food and nutraceutical source. Annu Rev Food Sci Technol 2011; 2(1): 181-201.
[http://dx.doi.org/10.1146/annurev-food-030810-153709] [PMID: 22129380]

[29]   Al-Jarallah A, Igdoura F, Zhang Y, *et al.* The effect of pomegranate extract on coronary artery atherosclerosis in SR-BI/APOE double knockout mice. Atherosclerosis 2013; 228(1): 80-9.
[http://dx.doi.org/10.1016/j.atherosclerosis.2013.02.025] [PMID: 23528829]

[30]   Malik A, Afaq F, Sarfaraz S, Adhami V, Syed D, Mukhtar H. Pomegranate fruit juice for chemoprevention and chemotherapy of prostate cancer. Urol Surv 2006; 175: 1171-2.

[31]   Shukla M, Gupta K, Rasheed Z, Khan KA, Haqqi TM. Bioavailable constituents/metabolites of pomegranate (*Punica granatum* L) preferentially inhibit COX2 activity *ex vivo* and IL-1beta-induced PGE2 production in human chondrocytes *in vitro*. J Inflamm (Lond) 2008; 5: 9.
[http://dx.doi.org/10.1186/1476-9255-5-9] [PMID: 18554383]

[32]   Li X, Wasila H, Liu L, *et al.* Physicochemical characteristics, polyphenol compositions and antioxidant potential of pomegranate juices from 10 Chinese cultivars and the environmental factors analysis. Food Chem 2015; 175: 575-84.
[http://dx.doi.org/10.1016/j.foodchem.2014.12.003] [PMID: 25577122]

[33]   Neyrinck AM, Van Hée VF, Bindels LB, De Backer F, Cani PD, Delzenne NM. Polyphenol-rich extract of pomegranate peel alleviates tissue inflammation and hypercholesterolaemia in high-fat diet-induced obese mice: potential implication of the gut microbiota. Br J Nutr 2013; 109(5): 802-9.
[http://dx.doi.org/10.1017/S0007114512002206] [PMID: 22676910]

[34] Viuda-Martos M, Ruiz-Navajas Y, Martin-Sánchez A, *et al.* Chemical, physico-chemical and functional properties of pomegranate (*Punica granatum* L.) bagasses powder co-product. J Food Eng 2012; 110(2): 220-4.
[http://dx.doi.org/10.1016/j.jfoodeng.2011.05.029]

[35] Tehranifar A, Zarei M, Nemati Z, Esfandiyari B, Vazifeshenas M. Investigation of physico-chemical properties and antioxidant activity of twenty Iranian pomegranate (*Punica granatum* L.) cultivars. Sci Hortic (Amsterdam) 2010; 126(2): 180-5.
[http://dx.doi.org/10.1016/j.scienta.2010.07.001]

[36] Li J, He X, Li M, Zhao W, Liu L, Kong X. Chemical fingerprint and quantitative analysis for quality control of polyphenols extracted from pomegranate peel by HPLC. Food Chem 2015; 176: 7-11.
[http://dx.doi.org/10.1016/j.foodchem.2014.12.040] [PMID: 25624199]

[37] Jain V, Viswanatha GL, Manohar D, Shivaprasad HN. Isolation of antidiabetic principle from fruit rinds of *Punica granatum*. Evid Based Complement Alternat Med 2012; 2012(11): 147202.
[http://dx.doi.org/10.1155/2012/147202] [PMID: 22919408]

[38] Seeram NP, Aronson WJ, Zhang Y, *et al.* Pomegranate ellagitannin-derived metabolites inhibit prostate cancer growth and localize to the mouse prostate gland. J Agric Food Chem 2007; 55(19): 7732-7.
[http://dx.doi.org/10.1021/jf071303g] [PMID: 17722872]

[39] Shema-Didi L, Kristal B, Sela S, Geron R, Ore L. Does Pomegranate intake attenuate cardiovascular risk factors in hemodialysis patients? Nutr J 2014; 13: 18.
[http://dx.doi.org/10.1186/1475-2891-13-18] [PMID: 24593225]

[40] Raafat K, Samy W. Amelioration of diabetes and painful diabetic neuropathy by *Punica granatum* L. extract and its spray dried biopolymeric dispersions. Evid Based Complement Alternat Med 2014; 2014(12): 180495.
[http://dx.doi.org/10.155/2014/180495] [PMID: 24982685]

[41] Ahmed AH, Subaiea GM, Eid A, Li L, Seeram NP, Zawia NH. Pomegranate extract modulates processing of amyloid-β precursor protein in an aged Alzheimer's disease animal model. Curr Alzheimer Res 2014; 11(9): 834-43.
[PMID: 25274111]

[42] Rojanathammanee L, Puig KL, Combs CK. Pomegranate polyphenols and extract inhibit nuclear factor of activated T-cell activity and microglial activation *in vitro* and in a transgenic mouse model of Alzheimer disease. J Nutr 2013; 143(5): 597-605.
[http://dx.doi.org/10.3945/jn.112.169516] [PMID: 23468550]

[43] Zarfeshany A, Asgary S, Javanmard SH. Potent health effects of pomegranate. Adv Biomed Res 2014; 3(100): 100.
[PMID: 24800189]

[44] Sengul H, Surek E, Nilufer-Erdil D. Investigating the effects of food matrix and food components on bioaccessibility of pomegranate (*Punica granatum*) phenolics and anthocyanins using an *in vitro* gastrointestinal digestion model. Food Res Int 2014; 62: 1069-79.
[http://dx.doi.org/10.1016/j.foodres.2014.05.055]

[45]   Haidari M, Ali M, Ward Casscells S III, Madjid M. Pomegranate (*Punica granatum*) purified polyphenol extract inhibits influenza virus and has a synergistic effect with oseltamivir. Phytomedicine 2009; 16(12): 1127-36.
[http://dx.doi.org/10.1016/j.phymed.2009.06.002] [PMID: 19586764]

[46]   Modaresi J, Fathi Nasri MH, Rashidi L, Dayani O, Kebreab E. Short communication: effects of supplementation with pomegranate seed pulp on concentrations of conjugated linoleic acid and punicic acid in goat milk. J Dairy Sci 2011; 94(8): 4075-80.
[http://dx.doi.org/10.3168/jds.2010-4069] [PMID: 21787942]

[47]   Minaiyan M, Zolfaghari B, Taheri D, Gomarian M. Preventive Effect of Three Pomegranate (*Punica granatum* L.) Seeds Fractions on Cerulein-Induced Acute Pancreatitis in Mice. Int J Prev Med 2014; 5(4): 394-404.
[PMID: 24829726]

[48]   Vroegrijk IO, van Diepen JA, van den Berg S, *et al.* Pomegranate seed oil, a rich source of punicic acid, prevents diet-induced obesity and insulin resistance in mice. Food Chem Toxicol 2011; 49(6): 1426-30.
[http://dx.doi.org/10.1016/j.fct.2011.03.037] [PMID: 21440024]

[49]   Usta C, Yilmaz B, Tasatargil A, Ozdemir S. Pomegranate seed oil, a rich source of punicic acid, induces endothelium-dependent vasorelaxation in rat thoracic aortic rings. Atherosclerosis 2014; 235(2): e123-4.
[http://dx.doi.org/10.1016/j.atherosclerosis.2014.05.341]

[50]   Vora A, Londhe V, Pandita N. Herbosomes enhance the *in vivo* antioxidant activity and bioavailability of punicalagins from standardized pomegranate extract. J Funct Foods 2015; 12: 540-8.
[http://dx.doi.org/10.1016/j.jff.2014.12.017]

[51]   Yang Y, Xiu J, Zhang L, Qin C, Liu J. Antiviral activity of punicalagin toward human enterovirus 71 *in vitro* and *in vivo*. Phytomedicine 2012; 20(1): 67-70.
[http://dx.doi.org/10.1016/j.phymed.2012.08.012] [PMID: 23146421]

[52]   Yaidikar L, Byna B, Thakur SR. Neuroprotective effect of punicalagin against cerebral ischemia reperfusion-induced oxidative brain injury in rats. J Stroke Cerebrovasc Dis 2014; 23(10): 2869-78.
[http://dx.doi.org/10.1016/j.jstrokecerebrovasdis.2014.07.020] [PMID: 25282190]

[53]   Finegold SM, Summanen PH, Corbett K, Downes J, Henning SM, Li Z. Pomegranate extract exhibits *in vitro* activity against *Clostridium difficile*. Nutrition 2014; 30(10): 1210-2.
[http://dx.doi.org/10.1016/j.nut.2014.02.029] [PMID: 24976424]

[54]   Zhou H, Yuan Q, Lu J. Preparative separation of punicalin from waste water of hydrolysed pomegranate husk by macroporous resin and preparative high-performance liquid chromatography. Food Chem 2011; 126(3): 1361-5.
[http://dx.doi.org/10.1016/j.foodchem.2010.11.100] [PMID: 25214139]

[55]   Wang Y, Zhang H, Liang H, Yuan Q. Purification, antioxidant activity and protein-precipitating capacity of punicalin from pomegranate husk. Food Chem 2013; 138(1): 437-43.
[http://dx.doi.org/10.1016/j.foodchem.2012.10.092] [PMID: 23265509]

[56]   Jaiswal V, DerMarderosian A, Porter J. Anthocyanins and polyphenol oxidase from dried arils of

pomegranate (*Punica granatum* L.). Food Chem 2010; 118(1): 11-6.
[http://dx.doi.org/10.1016/j.foodchem.2009.01.095]

[57] Amyrgialaki E, Makris D, Mauromoustakos A, Kefalas P. Optimisation of the extraction of pomegranate (*Punica granatum*) husk phenolics using water/ethanol solvent systems and response surface methodology. Ind Crops Prod 2014; 59: 216-22.
[http://dx.doi.org/10.1016/j.indcrop.2014.05.011]

[58] Hartman RE, Shah A, Fagan AM, *et al.* Pomegranate juice decreases amyloid load and improves behavior in a mouse model of Alzheimer's disease. Neurobiol Dis 2006; 24(3): 506-15.
[http://dx.doi.org/10.1016/j.nbd.2006.08.006] [PMID: 17010630]

[59] Gawlik-Dziki U, Sugier D, Dziki D, Sugier P. Bioaccessibility *in vitro* of nutraceuticals from bark of selected Salix species. Scientific World Journal 2014; 2014(10): 782763.
[http://dx.doi.org/10.1155/2014/782763] [PMID: 24696660]

[60] Gawlik-Dziki U, Świeca M, Dziki D, *et al.* Anticancer and antioxidant activity of bread enriched with broccoli sprouts. BioMed Res Int 2014; 2014(14): 608053.
[http://dx.doi.org/10.1155/2014/608053] [PMID: 25050366]

[61] Del Rio D, Rodriguez-Mateos A, Spencer JP, Tognolini M, Borges G, Crozier A. Dietary (poly)phenolics in human health: structures, bioavailability, and evidence of protective effects against chronic diseases. Antioxid Redox Signal 2013; 18(14): 1818-92.
[http://dx.doi.org/10.1089/ars.2012.4581] [PMID: 22794138]

[62] Kanellos PT, Kaliora AC, Gioxari A, Christopoulou GO, Kalogeropoulos N, Karathanos VT. Absorption and bioavailability of antioxidant phytochemicals and increase of serum oxidation resistance in healthy subjects following supplementation with raisins. Plant Foods Hum Nutr 2013; 68(4): 411-5.
[http://dx.doi.org/10.1007/s11130-013-0389-2] [PMID: 24114059]

[63] Naselli F, Tesoriere L, Caradonna F, *et al.* Anti-proliferative and pro-apoptotic activity of whole extract and isolated indicaxanthin from Opuntia ficus-indica associated with re-activation of the onco-suppressor p16(INK4a) gene in human colorectal carcinoma (Caco-2) cells. Biochem Biophys Res Commun 2014; 450(1): 652-8.
[http://dx.doi.org/10.1016/j.bbrc.2014.06.029] [PMID: 24937448]

[64] Rubió L, Macià A, Castell-Auví A, *et al.* Effect of the co-occurring olive oil and thyme extracts on the phenolic bioaccessibility and bioavailability assessed by *in vitro* digestion and cell models. Food Chem 2014; 149: 277-84.
[http://dx.doi.org/10.1016/j.foodchem.2013.10.075] [PMID: 24295707]

[65] Thilakarathna SH, Rupasinghe HP. Flavonoid bioavailability and attempts for bioavailability enhancement. Nutrients 2013; 5(9): 3367-87.
[http://dx.doi.org/10.3390/nu5093367] [PMID: 23989753]

[66] Li H, Deng Z, Liu R, Loewen S, Tsao R. Bioaccessibility, *in vitro* antioxidant activities and *in vivo* anti-inflammatory activities of a purple tomato (*Solanum lycopersicum* L.). Food Chem 2014; 159: 353-60.
[http://dx.doi.org/10.1016/j.foodchem.2014.03.023] [PMID: 24767066]

[67] Fazzari M, Fukumoto L, Mazza G, Livrea MA, Tesoriere L, Marco LD. *In vitro* bioavailability of

phenolic compounds from five cultivars of frozen sweet cherries (*Prunus avium* L.). J Agric Food Chem 2008; 56(10): 3561-8.
[http://dx.doi.org/10.1021/jf073506a] [PMID: 18459792]

[68] Costantini S, Rusolo F, De Vito V, *et al.* Potential anti-inflammatory effects of the hydrophilic fraction of pomegranate (*Punica granatum* L.) seed oil on breast cancer cell lines. Molecules 2014; 19(6): 8644-60.
[http://dx.doi.org/10.3390/molecules19068644] [PMID: 24962397]

[69] Jaganathan SK, Vellayappan MV, Narasimhan G, Supriyanto E. Role of pomegranate and citrus fruit juices in colon cancer prevention. World J Gastroenterol 2014; 20(16): 4618-25.
[http://dx.doi.org/10.3748/wjg.v20.i16.4618] [PMID: 24782614]

[70] Syed DN, Afaq F, Mukhtar H. Pomegranate derived products for cancer chemoprevention. Semin Cancer Biol 2007; 17(5): 377-85.
[http://dx.doi.org/10.1016/j.semcancer.2007.05.004] [PMID: 17613245]

[71] Seeram NP, Lee R, Heber D. Bioavailability of ellagic acid in human plasma after consumption of ellagitannins from pomegranate (*Punica granatum* L.) juice. Clin Chim Acta 2004; 348(1-2): 63-8.
[http://dx.doi.org/10.1016/j.cccn.2004.04.029] [PMID: 15369737]

[72] Vini R, Sreeja S.

[73] Lucci P, Pacetti D, Loizzo MR, Frega NG. *Punica granatum* cv. Dente di Cavallo seed ethanolic extract: antioxidant and antiproliferative activities. Food Chem 2015; 167: 475-83.
[http://dx.doi.org/10.1016/j.foodchem.2014.06.123] [PMID: 25149014]

[74] Elfalleh W, Hannachi H, Tlili N, Yhania Y, Nasri N, Ferchichi A. Total phenolic contents and antioxidant activities of pomegranate peel, seed, leaf and flower. J Med Plants Res 2012; 6(32): 4724-30.
[http://dx.doi.org/10.5897/JMPR11.995]

[75] Middha SK, Usha T, Pande V. HPLC evaluation of phenolic profile, nutritive content, and antioxidant capacity of extracts obtained from *Punica granatum* fruit peel. Adv Pharmacol Sci 2013; 2013(6): 296236.
[http://dx.doi.org/10.1155/2013/296236] [PMID: 23983682]

[76] Fawole OA, Makunga NP, Opara UL. Antibacterial, antioxidant and tyrosinase-inhibition activities of pomegranate fruit peel methanolic extract. BMC Complement Altern Med 2012; 12(1): 200.
[http://dx.doi.org/10.1186/1472-6882-12-200] [PMID: 23110485]

[77] Salwe KJ, Sachdev DO, Bahurupi Y, Kumarappan M. Evaluation of antidiabetic, hypolipedimic and antioxidant activity of hydroalcoholic extract of leaves and fruit peel of *Punica granatum* in male Wistar albino rats. J Nat Sci Biol Med 2015; 6(1): 56-62.
[http://dx.doi.org/10.4103/0976-9668.149085] [PMID: 25810635]

[78] Amer OS, Dkhil MA, Hikal WM, Al-Quraishy S. Antioxidant and anti-inflammatory activities of pomegranate (*Punica granatum*) on *Eimeria papillata*-induced infection in mice. BioMed Res Int 2015; 2015(7): 219670.
[http://dx.doi.org/10.1155/2015/219670] [PMID: 25654088]

[79] Borochov-Neori H, Judeinstein S, Tripler E, *et al.* Seasonal and cultivar variations in antioxidant and

sensory quality of pomegranate (*Punica granatum* L.) fruit. J Food Compos Anal 2009; 22(3): 189-95.
[http://dx.doi.org/10.1016/j.jfca.2008.10.011]

[80] Zhang L, Gao Y, Zhang Y, Liu J, Yu J. Changes in bioactive compounds and antioxidant activities in pomegranate leaves. Sci Hortic (Amsterdam) 2010; 123(4): 543-6.
[http://dx.doi.org/10.1016/j.scienta.2009.11.008]

[81] Ballistreri G, Continella A, Gentile A, Amenta M, Fabroni S, Rapisarda P. Fruit quality and bioactive compounds relevant to human health of sweet cherry (*Prunus avium* L.) cultivars grown in Italy. Food Chem 2013; 140(4): 630-8.
[http://dx.doi.org/10.1016/j.foodchem.2012.11.024] [PMID: 23692746]

[82] De Nisco M, Manfra M, Bolognese A, *et al.* Nutraceutical properties and polyphenolic profile of berry skin and wine of *Vitis vinifera* L. (cv. Aglianico). Food Chem 2013; 140(4): 623-9.
[http://dx.doi.org/10.1016/j.foodchem.2012.10.123] [PMID: 23692745]

[83] Liu CL, Cheng L, Ko CH, *et al.* Bioassay-guided isolation of anti-inflammatory components from the root of *Rehmannia glutinosa* and its underlying mechanism *via* inhibition of iNOS pathway. J Ethnopharmacol 2012; 143(3): 867-75.
[http://dx.doi.org/10.1016/j.jep.2012.08.012] [PMID: 23034094]

[84] Susunaga-Notario AdelC, Pérez-Gutiérrez S, Zavala-Sánchez MA, *et al.* Bioassay-guided chemical study of the anti-inflammatory effect of *Senna villosa* (Miller) H.S. Irwin & Barneby (Leguminosae) in TPA-induced ear edema. Molecules 2014; 19(7): 10261-78.
[http://dx.doi.org/10.3390/molecules190710261] [PMID: 25029073]

[85] Ismail T, Sestili P, Akhtar S. Pomegranate peel and fruit extracts: a review of potential anti-inflammatory and anti-infective effects. J Ethnopharmacol 2012; 143(2): 397-405.
[http://dx.doi.org/10.1016/j.jep.2012.07.004] [PMID: 22820239]

[86] Panichayupakaranant P, Tewtrakul S, Yuenyongsawad S. Antibacterial, anti-inflammatory and anti-allergic activities of standardised pomegranate rind extract. Food Chem 2010; 123(2): 400-3.
[http://dx.doi.org/10.1016/j.foodchem.2010.04.054]

[87] Larrosa M, González-Sarrías A, Yáñez-Gascón MJ, *et al.* Anti-inflammatory properties of a pomegranate extract and its metabolite urolithin-A in a colitis rat model and the effect of colon inflammation on phenolic metabolism. J Nutr Biochem 2010; 21(8): 717-25.
[http://dx.doi.org/10.1016/j.jnutbio.2009.04.012] [PMID: 19616930]

[88] Bishayee A, Thoppil RJ, Darvesh AS, Ohanyan V, Meszaros JG, Bhatia D. Pomegranate phytoconstituents blunt the inflammatory cascade in a chemically induced rodent model of hepatocellular carcinogenesis. J Nutr Biochem 2013; 24(1): 178-87.
[http://dx.doi.org/10.1016/j.jnutbio.2012.04.009] [PMID: 22841394]

[89] Mo J, Panichayupakaranant P, Kaewnopparat N, Nitiruangjaras A, Reanmongkol W. Topical anti-inflammatory and analgesic activities of standardized pomegranate rind extract in comparison with its marker compound ellagic acid *in vivo*. J Ethnopharmacol 2013; 148(3): 901-8.
[http://dx.doi.org/10.1016/j.jep.2013.05.040] [PMID: 23743057]

[90] Ahmad A, Ginnebaugh KR, Li Y, Padhye SB, Sarkar FH. Molecular targets of naturopathy in cancer research: bridge to modern medicine. Nutrients 2015; 7(1): 321-34.
[http://dx.doi.org/10.3390/nu7010321] [PMID: 25569626]

[91] Carvalho M, Ferreira PJ, Mendes VS, *et al.* Human cancer cell antiproliferative and antioxidant activities of *Juglans regia* L. Food Chem Toxicol 2010; 48(1): 441-7.
[http://dx.doi.org/10.1016/j.fct.2009.10.043] [PMID: 19883717]

[92] Zhu Y, Soroka DN, Sang S. Synthesis and inhibitory activities against colon cancer cell growth and proteasome of alkylresorcinols. J Agric Food Chem 2012; 60(35): 8624-31.
[http://dx.doi.org/10.1021/jf302872a] [PMID: 22897570]

[93] Barrajón-Catalán E, Fernández-Arroyo S, Saura D, *et al. Cistaceae* aqueous extracts containing ellagitannins show antioxidant and antimicrobial capacity, and cytotoxic activity against human cancer cells. Food Chem Toxicol 2010; 48(8-9): 2273-82.
[http://dx.doi.org/10.1016/j.fct.2010.05.060] [PMID: 20510328]

[94] Ming DS, Pham S, Deb S, *et al.* Pomegranate extracts impact the androgen biosynthesis pathways in prostate cancer models *in vitro* and *in vivo*. J Steroid Biochem Mol Biol 2014; 143: 19-28.
[http://dx.doi.org/10.1016/j.jsbmb.2014.02.006] [PMID: 24565566]

[95] Koyama S, Cobb LJ, Mehta HH, *et al.* Pomegranate extract induces apoptosis in human prostate cancer cells by modulation of the IGF-IGFBP axis. Growth Horm IGF Res 2010; 20(1): 55-62.
[http://dx.doi.org/10.1016/j.ghir.2009.09.003] [PMID: 19853487]

[96] Albrecht M, Jiang W, Kumi-Diaka J, *et al.* Pomegranate extracts potently suppress proliferation, xenograft growth, and invasion of human prostate cancer cells. J Med Food 2004; 7(3): 274-83.
[http://dx.doi.org/10.1089/jmf.2004.7.274] [PMID: 15383219]

[97] Chrubasik-Hausmann S, Vlachojannis C, Zimmermann B. Pomegranate juice and prostate cancer: importance of the characterisation of the active principle. Phytother Res 2014; 28(11): 1676-8.
[http://dx.doi.org/10.1002/ptr.5181] [PMID: 24895232]

[98] Wang L, Martins-Green M. Pomegranate and its components as alternative treatment for prostate cancer. Int J Mol Sci 2014; 15(9): 14949-66.
[http://dx.doi.org/10.3390/ijms150914949] [PMID: 25158234]

[99] Hong MY, Seeram NP, Heber D. Pomegranate polyphenols down-regulate expression of androgen-synthesizing genes in human prostate cancer cells overexpressing the androgen receptor. J Nutr Biochem 2008; 19(12): 848-55.
[http://dx.doi.org/10.1016/j.jnutbio.2007.11.006] [PMID: 18479901]

[100] Wang L, Ho J, Glackin C, Martins-Green M. Specific pomegranate juice components as potential inhibitors of prostate cancer metastasis. Transl Oncol 2012; 5(5): 344-55.
[http://dx.doi.org/10.1593/tlo.12190] [PMID: 23066443]

[101] Zhou B, Yi H, Tan J, Wu Y, Liu G, Qiu Z. Anti-proliferative effects of polyphenols from pomegranate rind (*Punica granatum* L.) on EJ bladder cancer cells *via* regulation of p53/miR-34a axis. Phytother Res 2015; 29(3): 415-22.
[http://dx.doi.org/10.1002/ptr.5267] [PMID: 25572695]

[102] Banihani S, Swedan S, Alguraan Z. Pomegranate and type 2 diabetes. Nutr Res 2013; 33(5): 341-8.
[http://dx.doi.org/10.1016/j.nutres.2013.03.003] [PMID: 23684435]

[103] Khateeb J, Gantman A, Kreitenberg AJ, Aviram M, Fuhrman B. Paraoxonase 1 (PON1) expression in hepatocytes is upregulated by pomegranate polyphenols: a role for PPAR-gamma pathway.

Atherosclerosis 2010; 208(1): 119-25.
[http://dx.doi.org/10.1016/j.atherosclerosis.2009.08.051] [PMID: 19783251]

[104] Fuhrman B, Volkova N, Aviram M. Pomegranate juice polyphenols increase recombinant paraoxonase-1 binding to high-density lipoprotein: studies *in vitro* and in diabetic patients. Nutrition 2010; 26(4): 359-66.
[http://dx.doi.org/10.1016/j.nut.2009.05.003] [PMID: 19762215]

[105] Xu KZ, Zhu C, Kim MS, Yamahara J, Li Y. Pomegranate flower ameliorates fatty liver in an animal model of type 2 diabetes and obesity. J Ethnopharmacol 2009; 123(2): 280-7.
[http://dx.doi.org/10.1016/j.jep.2009.03.035] [PMID: 19429373]

[106] Basu A, Newman ED, Bryant AL, Lyons TJ, Betts NM. Pomegranate polyphenols lower lipid peroxidation in adults with type 2 diabetes but have no effects in healthy volunteers: a pilot study. J Nutr Metab 2013; 2013(7): 708381.
[http://dx.doi.org/10.1155/2013/708381] [PMID: 23936637]

[107] Patel A, Bandawane D, Mhetre N. Pomegranate (*Punica granatum* Linn.) leaves attenuate disturbed glucose homeostasis and hyperglycemia mediated hyperlipidemia and oxidative stress in streptozotocin induced diabetic rats. Eur J Integr Med 2014; 6(3): 307-21.
[http://dx.doi.org/10.1016/j.eujim.2014.03.009]

[108] Chanet A, Milenkovic D, Manach C, Mazur A, Morand C. Citrus flavanones: what is their role in cardiovascular protection? J Agric Food Chem 2012; 60(36): 8809-22.
[http://dx.doi.org/10.1021/jf300669s] [PMID: 22574825]

[109] Aviram M, Rosenblat M. Pomegranate for your cardiovascular health. Rambam Maimonides Med J 2013; 4(2): e0013.
[http://dx.doi.org/10.5041/RMMJ.10113] [PMID: 23908863]

[110] Shema-Didi L, Sela S, Geron R, Sapiro G, Ore L, Kristal B. The Beneficial Effects of One Year Pomegranate Juice Consumption on Traditional and Nontraditional Risk Factors for Cardiovascular Diseases. Free Radic Biol Med 2010; 49
[http://dx.doi.org/10.1016/j.freeradbiomed.2010.10.572]

[111] Stowe CB. The effects of pomegranate juice consumption on blood pressure and cardiovascular health. Complement Ther Clin Pract 2011; 17(2): 113-5.
[http://dx.doi.org/10.1016/j.ctcp.2010.09.004] [PMID: 21457902]

[112] Asgary S, Sahebkar A, Afshani MR, Keshvari M, Haghjooyjavanmard S, Rafieian-Kopaei M. Clinical evaluation of blood pressure lowering, endothelial function improving, hypolipidemic and anti-inflammatory effects of pomegranate juice in hypertensive subjects. Phytother Res 2014; 28(2): 193-9.
[http://dx.doi.org/10.1002/ptr.4977] [PMID: 23519910]

[113] Asgary S, Keshvari M, Sahebkar A, Hashemi M, Rafieian-Kopaei M. Clinical investigation of the acute effects of pomegranate juice on blood pressure and endothelial function in hypertensive individuals. ARYA Atheroscler 2013; 9(6): 326-31.
[PMID: 24575134]

[114] Lynn A, Hamadeh H, Leung WC, Russell JM, Barker ME. Effects of pomegranate juice supplementation on pulse wave velocity and blood pressure in healthy young and middle-aged men and women. Plant Foods Hum Nutr 2012; 67(3): 309-14.

[http://dx.doi.org/10.1007/s11130-012-0295-z] [PMID: 22648092]

[115] Aviram M, Rosenblat M. Pomegranate Protection against Cardiovascular Diseases. Evid Based Complement Alternat Med 2012; 2012(10): 382763.
[http://dx.doi.org/10.1155/2012/382763] [PMID: 23243442]

[116] Ramassamy C. Emerging role of polyphenolic compounds in the treatment of neurodegenerative diseases: a review of their intracellular targets. Eur J Pharmacol 2006; 545(1): 51-64.
[http://dx.doi.org/10.1016/j.ejphar.2006.06.025] [PMID: 16904103]

[117] Hybertson BM, Gao B, Bose SK, McCord JM. Oxidative stress in health and disease: the therapeutic potential of Nrf2 activation. Mol Aspects Med 2011; 32(4-6): 234-46.
[http://dx.doi.org/10.1016/j.mam.2011.10.006] [PMID: 22020111]

[118] Radak Z, Zhao Z, Goto S, Koltai E. Age-associated neurodegeneration and oxidative damage to lipids, proteins and DNA. Mol Aspects Med 2011; 32(4-6): 305-15.
[http://dx.doi.org/10.1016/j.mam.2011.10.010] [PMID: 22020115]

[119] Reed TT. Lipid peroxidation and neurodegenerative disease. Free Radic Biol Med 2011; 51(7): 1302-19.
[http://dx.doi.org/10.1016/j.freeradbiomed.2011.06.027] [PMID: 21782935]

[120] Kumar S, Maheshwari KK, Singh V. Protective effects of *Punica granatum* seeds extract against aging and scopolamine induced cognitive impairments in mice. Afr J Tradit Complement Altern Medicines 2008; 6(1): 49-56.
[PMID: 20162041]

[121] Gao X, Cassidy A, Schwarzschild MA, Rimm EB, Ascherio A. Habitual intake of dietary flavonoids and risk of Parkinson disease. Neurology 2012; 78(15): 1138-45.
[http://dx.doi.org/10.1212/WNL.0b013e31824f7fc4] [PMID: 22491871]

[122] Strathearn KE, Yousef GG, Grace MH, *et al.* Neuroprotective effects of anthocyanin- and proanthocyanidin-rich extracts in cellular models of Parkinson's disease. Brain Res ;2014 :1555 60-77.
[http://dx.doi.org/10.1016/j.brainres.2014.01.047] [PMID: 24502982]

[123] Ahmed MA, El Morsy EM, Ahmed AA. Pomegranate extract protects against cerebral ischemia/reperfusion injury and preserves brain DNA integrity in rats. Life Sci 2014; 110(2): 61-9.
[http://dx.doi.org/10.1016/j.lfs.2014.06.023] [PMID: 25010842]

CHAPTER 6

Nutritional and Functional Properties of Edible Berries: Implications For Health Claims

Amadeo Gironés-Vilaplana[1], Cristina García-Viguera[1,*], Diego A. Moreno[1], Ral Domínguez-Perles[2]

[1] *Research Group on Quality, Safety and Bioactivity of Plant Foods, Department of Food Science and Technology, CEBAS (CSIC), P.O. Box 164, 30100 Campus University Espinardo, Murcia, Spain*

[2] *Centre for the Research and Technology for Agro-Environment and Biological Sciences, Universidade de Trás-os-Montes e Alto Douro (CITAB-UTAD), Quinta de Prados, 5001-801 Vila Real, Portugal*

Abstract: The data reported over the last decades on the relevance of fruits and vegetables for human health promotion has increased their use as nutraceutical ingredients. In this connection, berries display an interesting content in a wide variety of nutrients that contribute to a balanced diet such as sugars, essential oils, carotenoids, vitamins, and minerals as well as bioactive non-nutrients namely flavonoids, phenolic acids, stilbenes, and tannins. Bioactive compounds from berries have potent biological activities namely antioxidant, anticancer, antimutagenic, antimicrobial, anti-inflammatory, and antineurodegenerative, which are supported by *in vitro* and, to a lesser extent, by *in vivo* models. This is a comprehensive review on nutritional and non-nutritional compounds present in berries, intended to provide rational information on the health benefits of integrating berries in balanced diets

Keywords: Absorption, Berries, Bioavailaility, Biological activity, Flavonoids, Health, Intestinal microbiota, Metabolism, Nutrients, Phenolic acids, (Poly)phenols.

---

* **Corresponding author Cristina Garca Viguera:** Research Group on Quality, Safety and Bioactivity of Plant Foods, Department of Food Science and Technology, CEBAS (CSIC); Espinardo, Murcia (Spain). Email: cgviguera@cebas.csic.es.

Luís Rodrigues da Silva and Branca Silva (Eds.)

INTRODUCTION

Berry fruits, small fruits or berries are generally referred to any small fruit that lacks seeds and can be eaten whole (Fig. **1**), which has been promoted for dietary consumption with resort to their content in bioactive nutrients and non-nutrients. This composition turns berries into very special and healthy foods.

Dietary Guidelines of international experts recommend to increase the fruit intake as part of a healthy diet. Thus, although this dietary habit can be achieved by a large group of forms of fruit, including fresh, frozen, and canned, as well as fruit juices and dried fruit updated and accurate evaluations of the nutritional and phytochemical composition of processing foods is required for rational advice. In this sense, nowadays many of the berry species and subspecies present in the market are understudied, whilst the current trend on health promotion through balanced diets requires further investigations to identify new challenges for dietary interventions towards improved human health [1].

Fig. (1). Common berry fruits.

Berries are widely distributed and include blackberry (*Rubus* spp.), black raspberry (*Rubus occidentalis*), blueberry (*Vaccinium corymbosum*), cranberry (*Vaccinium oxycoccus*), red raspberry (*Rubus idaeus*), chokeberry (*Aronia* spp.), and strawberry (*Fragaria* spp.), among others fruits.

Over the last year these kind of fruits have been related with an interesting content in a wide variety of nutrients such as sugars, essential oils, carotenoids, vitamins, and minerals as well as healthy non-nutrients such as flavonoids, phenolic acids, stilbenes, and tannins. An overwhelming body of research has now steadily established that their dietary intake exerts a positive and depth impact on health and human performance [2]. Thus, bioactive compounds from berries have potent biological activities namely antioxidant, antimicrobial, anti-inflammatory, anticancer, antimutagenic, and neuroprotector, which have been reported both *in vitro* and *in vivo*. Therefore, an overview about their beneficial effects on health related to their nutritional and bioactive composition is needed.

Berry fruits are usually consumed not only in fresh and frozen shapes but also as derived and processed products including canned fruits, beverages, jams, yogurts, and jellies. Besides to these traditional uses, over the last years there is a growing trend regarding the application of berry parts as ingredients in the development of functional foods and dietary supplements, which can or cannot be combined with other colorful vegetable extracts [3]. Hence, in addition to their consumption in fresh form, berries have been used in the production of several manufactured commodities including juices, marmalades, spirits, or infusions, which consumption has also been promoted based on their preventive properties against degenerative processes [4]. However, in comparison with fresh fruits significant losses of bioactive (poly)phenols have been identified in manufactured products namely blueberry juices [5, 6], raspberry purée and jam [7], strawberry jams [8], canned fruit, and nectar [9]. Hence, Hager *et al.*, reported that processing of canned berries in water, purée or syrup decreased the anthocyanin content up to 51% from the fresh berry, on average [10].

CHEMICAL COMPOSITION

The information available on the properties of berries as a source of nutrients and phytochemical compounds has allowed to promote their consumption, which has grown considerably during the last years. Indeed, berries have reported to contain over 190 compounds, with concentrations varying significantly between distinct species and subspecies, according to the origin, climate, and time of harvesting [11]. The composition of berries includes carbohydrates, proteins and amino acids, lipids and fatty acids, organic acids, vitamins (A, B1, B2, B9, C, K, and E) and minerals, phenolic acids, flavanoids, carotenoids, and terpenes [12, 13].

Basic Chemical Composition

Moisture, Dietary Fibre, and Total Soluble Solids (TSS)

The moisture content of berries ranges from 82.0 to 91.0%, whilst the concentration of dietary fibre content of berries varies from 2.0 to 6.5 g/100 g, respectively (USDA data base, May 2015). Dietary fibre has demonstrated to possess many positive health effects namely contributing to appetite regulation and weight loss and reducing the risk of develop some diseases such as heart disease and type 2 diabetes [14]. Total soluble solids of berries have been described for Chinese and Indian cultivars of buckthorn ranging from 9.30 to 22.74 [13, 15, 16].

Carbohydrates

The content of carbohydrates in raw blueberries, blackberries, blackcurrant, cranberries, strawberries, and raspberries, calculated by difference is 14.49, 9.61, 15.38, 12.20, 7.68, and 11.94 g/100g, respectively (USDA data base, May 2015). The carbohydrates profile is composed by monosaccharides, oligosaccharides, and polysaccharides being glucose, fructose, and sucrose the main oligosaccharides and cellulose and starches the major polysaccharides [17], although the specific composition is strongly dependent on the subspecies considered. Thus, in strawberry fructose has been described as the most abundant sugar, whilst glucose predominates in blackthorn. This composition allows to promote berries as appropriate fruits for satisfying hunger. Actually, they have been highlighted within local daily diets, specifically during famine periods [17].

Using EtOH to precipitate neutral sugars has allowed to describe 14.4% galacturonic acid, 5.0% arabinose, 1.6% rhamanose, and 3.0% galactose, suggesting that homogalacturonan is not the major polysaccharide in the ethanolic fraction. The achievement of the content of arabinose and galactose in the EtOH precipitate suggests the presence of arabinans and/or arabinogalactans [18].

Table 1. Nutritional composition of berries

Nutrient (Unit)	Blueberry (*Vaccinium myrtillus*)	Blackberry (*Rubus villosus*)	Blackcurrant (*Ribes nigrum*)	Cranberry (*Vaccinium macrocarpon*)	Strawberry (*Fragaria virginiana*)	Raspberry (*Rubus idaeus*)
Basic chemical composition						
Moisture (g/100g)	84.210	88.150	81.960	87.130	90.950	85.750
Dietary fibre (g/100g)	2.400	5.300	---	4.600	2.000	6.540
Carbohydrates (g/100g) (by difference)	14.490	9.610	15.380	12.200	7.680	11.940
Total sugars (g/100g)	9.960	4.880	---	4.040	4.890	4.420
Total Protein (g/100g)	0.740	1.390	1.400	0.390	0.670	1.200
Total lipids (g/100g)	0.330	0.490	0.410	0.130	0.300	0.650
Fatty acids, total saturated	0.028	0.014	0.034	0.011	0.015	0.019
Fatty acids, total monounsaturated	0.047	0.047	0.058	0.018	0.043	0.064
Fatty acids, total polyunsaturated	0.146	0.280	0.179	0.055	0.155	0.375
Micronutrients						
Vitamins						
Vitamin C (mg/100g)	9.700	21.000	181.000	13.300	58.800	26.200
Thiamine (Vitamin B1) (mg/100g)	0.037	0.020	0.050	0.012	0.024	0.032
Riboflavin (Vitamin B2) (mg/100g)	0.041	0.026	0.050	0.020	0.022	0.038
Niacin (Vitamin B3) (mg/100g)	0.418	0.646	0.300	0.101	0.386	0.598
Pyridoxine (Vitamin B6) (mg/100g)	0.052	0.030	0.066	0.057	0.047	0.055
Folate (Vitamin B9) (μg/100g)	6.000	25.000	---	1.000	24.000	21.000

(Table 1) contd.....

Nutrient (Unit)	Blueberry (*Vaccinium myrtillus*)	Blackberry (*Rubus villosus*)	Blackcurrant (*Ribes nigrum*)	Cranberry (*Vaccinium macrocarpon*)	Strawberry (*Fragaria virginiana*)	Raspberry (*Rubus idaeus*)
Vitamin A (IU)	54.000	214.000	230.000	60.000	12.000	33.000
α-tocopherol (Vitamin E) (mg/100g)	0.570	1.170	1.000	1.200	0.290	0.870
Phylloquinone (Vitamin K) (µg/100g)	19.300	19.800	---	5.100	2.200	7.800
Minerals						
Calcium, Ca (mg/100g)	6.000	29.000	55.000	8.000	16.000	25.000
Iron, Fe (mg/100g)	0.280	0.620	1.540	0.250	0.410	0.690
Magnesium, Mg (mg/100g)	6.000	20.000	24.000	6.000	13.000	22.000
Phosphorus, P (mg/100g)	12.000	22.000	59.000	13.000	24.000	29.000
Potassium, K (mg/100g)	77.000	162.000	322.000	85.000	153.000	151.000
Sodium, Na (mg/100g)	1.000	1.000	2.000	2.000	1.000	1.000
Zinc, Zn (mg/100g)	0.160	0.530	0.270	0.100	0.140	0.420
United States Department of Agriculture (USDA, http://fnic.nal.usda.gov/food-composition, May 2015)						

During fruit ripening, the insoluble wall-bound protopectin, of high molecular weight, is converted to water-soluble pectin and, through this process, becomes partially demethoxylated and depolymerized. In addition, the presence of glucose and xylose in the EtOH precipitate indicates the probable existence of xyloglucan [19]. Since glucuronic acid is found in several types of hemicellulose, its presence in the EtOH precipitate may respond to its breakdown [20]. Thus, the sugar composition of this fraction evidences that it contains a mixture of pectic and hemicellulosic polymers. In this connection, the presence of different polymers suggests great potential to be used as a source of valuable polysaccharides non-food industries, besides their role in diets [18].

Organic Acids

Berries contain organic acids, mostly represented by quinic and malic acids, which together constitute about 90% of all the acids from the fruit. Large variations in the concentration of these acids have been reported in berries, which range between 2.1 and 9.1 g/100 g depending on the species, subspecies, origins, and climatic conditions, specially concerning temperature [13, 21]. Besides these factors, the total amount of organic acids decreases during ripening [22].

Protein Content and Amino Acids

Protein content in berries has been reported to be in low levels, ranging from 0.39 to 1.40 g/100 g in Blueberry, blackberry, black currant, cranberry, strawberry, and raspberry (Table **1**). Thus, according to the USDA database the highest concentration corresponds to blackberry and black currant (1.390 and 1.400 g/100 g, respectively).

With respect to the nutritional value of berries as sources of amino acids, the scarce number of reports available describing the amino acid content in these plant foods have pointed out the presence of 18 essential and nonessential amino acids in buckthorn fruit [13, 23]. This is of special relevance because of their crucial role in several biological processes including energy production, building cells, fat loss, and neurological functions.

Oils and Fatty Acids

The concentration of total fats as well as the content in total saturated fatty acids, monounsaturated fatty acids, and polyunsaturated fatty acids of berries are shown in Table **1**. So far, twenty-four fatty acids were identified and quantified, being stressed the concentration of polyunsaturated fatty acids in raspberries. The major fatty acids found in these fruits are α-linolenic acid (C18:3) and linoleic acid (C18:2). These are essential fatty acids as they cannot be synthesised by humans due to the lack of desaturase enzymes and must be obtained by the diet. After consumption these fatty acids originate the omega-3 and omega-6, respectively [24], which constitute biosynthetic precursors of eicosanoids, and thus, make part of diverse metabolic functions. Furthermore, they can also decrease the total amount of fat in blood (cholesterol), reducing the risk of cardiovascular diseases [24]. As an exception, in blackthorn fruits monounsaturated predominated over polyunsaturated fatty acids due to the abundance of oleic acid (C18:1).

Micronutrients

Plant foods in general and especially regarding berries, have been promoted as valuable sources of vitamins and minerals. Thus, vitamin C, thiamine, riboflavin, niacin, pyridoxine, folate, vitamin A, α-tocopherol, and phylloquinone as well as the minerals calcium, iron, magnesium, phosphorus, potassium, sodium, and zinc are present in berries in high concentrations, although considerable differences were recorded between species and subspecies (Table **1**). From the information available on the content in micronutrients it can be stressed the valuable contribution of these fruits to achieve the recommended dairy allowance (RDA) that guarantee health status in humans. The differential content in vitamins observed in the separate berries allows to emphasize the most relevant species in order to achieve the RDA of these nutrients.

Vitamins

Berries are particularly valuable sources of vitamin C (9.7-181.0 mg/100 g fw), being stressed its concentration in blackcurrant and strawberry (Table **1**). Thus, the regular consumption of these berries allows to reach its RDA in health adults (45 mg/day in healthy adults) and also under especial physiologic situations (pregnancy and lactation, 55 and 70 mg/day, respectively) (www.fao.org, May 2015).

Regarding the B-complex vitamins, niacin (vitamin B3) is at the highest concentration in blackberry and raspberry (0.646 and 0.598 mg/100g, respectively), whilst additional berries remain in similar lower concentration (0.368 mg/100 g fw), contributing to fulfil the diary allowance pointed out by the scientific community to guarantee health status (USDA data base, Table **1**)). From this vitamin group, can be also stressed the concentration of folate (Vitamin B9), which is at the highest level in blackberry, strawberry, and raspberry (23.333 µg/100 g fw), contributing interestingly to achieve the RDA (400 µg/day).

Vitamin A (retinol) is an essential nutrient needed in small amounts by humans (RDA 285.000 µg/day) for the correct functioning of the visual system; growth and development; and upkeep of epithelial cellular wholeness, reproduction, and immune function. The dietary needs for vitamin A are normally supplied as retinol (mainly as retinol ester) and pro-vitamin A carotenoids. Berries provide diverse concentrations of retinol, which range from 12 to 214 International Units per 100 g fw.

As a relevant part of the antioxidant machinery of plant foods, vitamin E, which is represented by diverse tocopherols, exerts a much powerful effect to supress reactive oxygen species. Thus, α-tocopherol is considered as a reference regarding the content in vitamin E of plant foods due to its high biological activity. The data on the content in α-tocopherol provided by the USDA data base allow to remark blackberry and cranberry as the species with the highest amounts (1185.0 mg/100 g fw, on average), although all berry fruits contribute interestingly to the RDA for this vitamin (10.0 mg/day), by providing from 3 to 12% of this amount by a service. This concentration is strongly dependent on the cultivar, year, agro-climatic conditions, and ripening stage [14].

Concerning vitamin K, this is present in low concentration in berries, ranging from 2.2 to 19.9 µg/100 g fw, being highlighted their concentration in blueberries and blackberry.

Minerals

The information available on the mineral composition of berries allows to point out their interest as valuable sources of potassium, phosphorous, magnesium, and calcium, although several differences between them are noticed as a consequence of the diverse agro-climatic conditions. Thus, potassium, which constitutes the major mineral nutrient described in these fruits, is a vital part of the ionic balance and maintains the tissue excitability *in vivo* [25]. Thus, data available on the differential mineral composition of berries pointed out blackcurrant as the best one for providing dietary potassium (162.000 mg/100 g fw) followed by blackberry, strawberry, and raspberry (155.500 mg/100 g fw, on average) (Table 1). Besides potassium, other abundant minerals present in these fruits are calcium, phosphorus, and magnesium, which again appear at the highest concentration in blackcurrant (59.000, 24.000, and 25.000 mg/100 g fw, respectively). Thus, whilst calcium is related with the maintenance of the skeletal integrity, phosphorus and magnesium are related with several metabolic functions namely protein synthesis, energy metabolism, RNA and DNA synthesis, and the maintenance of the electrical potential of cell membranes and nervous tissues. Thus, the confirmation of the capacity of berries to achieve up to 10% of the RDA has prompted to promote these fruits as a valuable part of balanced diets to achieve the recommended amounts of micro nutrients.

Given the nutritional composition of berries and the strong dependence on the climatic conditions of wild berries, their growth under greenhouse environment conditions could result in a constant source of nutrients and phytochemicals fitting adequately the market and consumers requirements [21].

Phytochemical Compounds of Berries

Phytochemicals found in berries include (poly)phenols, with high proportions of flavonoids, highlighting anthocyanins as the most characteristic compounds. Other phenolic compounds found are flavonols, procyanidins, ellagitannins, and phenolic acids, among others. Additionally, other non-phenolic bioactive compounds have been identified previously, such as vitamin C and E [26], as it was explained before.

Phenolic compounds

Plant foods, including berries, are one of the most important sources of dietary phenolic compounds. These bioactive phytochemicals are secondary metabolites in plants, commonly found in fruits and vegetables, responsible for beneficial pharmacological properties [27]. They are essential to plant's physiology, playing a key role in diverse and crucial biological functions such as structure maintenance, pigmentation, pathogen and predator resistance, and plants' performance [28].

Phenolic compounds possess at least one aromatic ring attaching one or more hydroxyl groups that have diverse molecular complexity, from simple, with low molecular weight, and single aromatic-ring compounds to the large and complex tannins and polyphenols [29]. They are usually found conjugated to sugars and are classified into different groups: flavonoids, phenolic acids, stilbenes, tannins, and lignans, all of them with demonstrated health benefits (Fig. **2**).

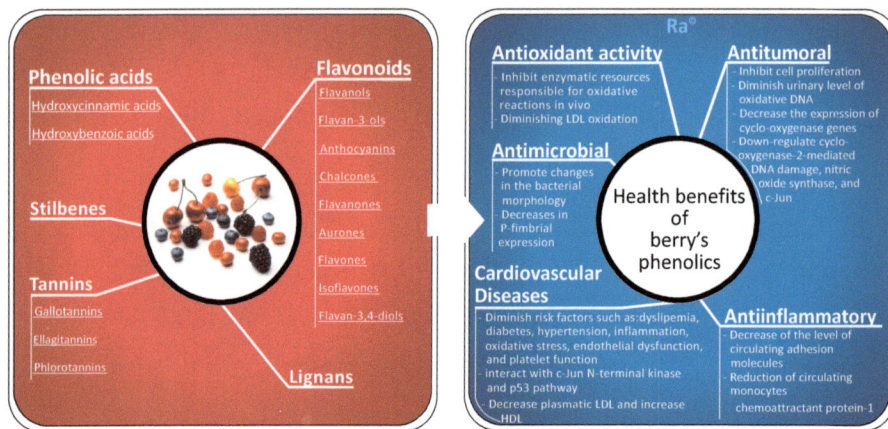

Fig. (2). Common berry´s phytochemicals and their health benefits.

Polyphenols

Flavonoids

Flavonoids are one of the most extensive groups of secondary plant metabolites present in edible vegetables and fruits [30]. Flavonoid skeleton consists of two aromatic rings (A and B), connected through a pyrone or hydropyrone ring (C) [31]. The main subclasses of these C6–C3–C6 compounds are flavones, flavonols, flavan-3-ols, isoflavones, flavanones, and anthocyanidins. Other flavonoids are the chalcones, dihydrochalcones, dihydroflavonols, flavan-3,4-diols, coumarins, and aurones (Fig. **3**).

The basic structure of flavonoids may have diverse substituents. The mostly of flavonoids occurs naturally as glycosides rather than aglycones [32]. Thus, this section deals with the major flavonoids in berries.

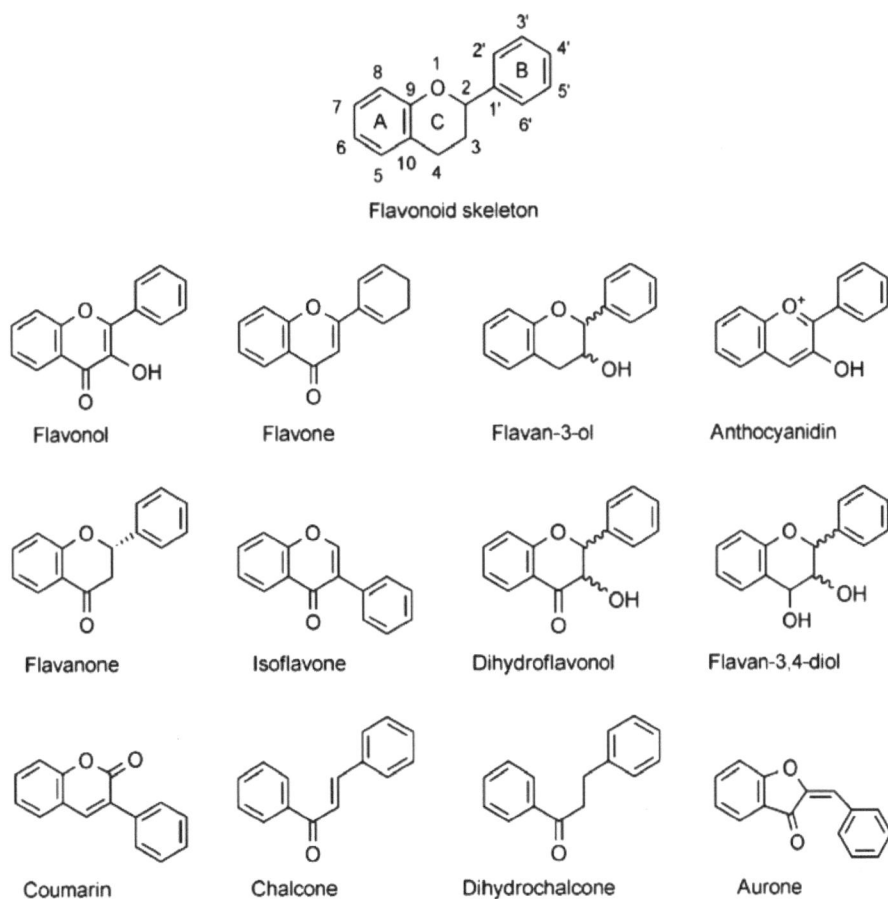

Fig. (3). Flavonoids. Skeleton and types. Extracted from Crozier *et al.*, 2009.

The main flavonoids present in berries are the anthocyanins, which comprises the biggest group of water soluble pigments in the plant kingdom and are characteristic of the flowering plants, which themselves provide our major source of food crops. In food vegetables, anthocyanins have been identified in at least 27 families, 73 genera, and a large number of species [33]. These phenolic compounds appear in plants as glycosides in which the anthocyanidin (pigment free of sugar) molecule is linked with a sugar. Anthocyanins can be break down into different sub-types, such as different substituent groups on the B ring, number and type of conjugated sugars, and the presence or not of acyl group. There are six main anthocyanins: pelargonidins, cyanidins, delphinidins, peonidins, petunidins, and malvidins (Fig. **4**).

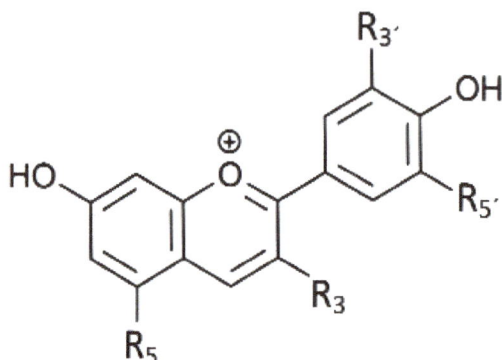

Antocyanin	Abbreviature	$R_{3'}$	$R_{5'}$
Pelargonidin	Pg	H	H
Cyanidin	Cy	OH	H
Delphinidin	Dp	OH	OH
Peonidin	Pn	OCH_3	H
Petunidin	Pt	OCH_3	OH
Malvidin	Mv	OCH_3	OCH_3

Fig. (4). General anthocyanins structure.

Anthocyanins are of growing interest, not only because of technological reasons and organoleptic properties but also due to their health-promoting potential, as suggested by available experimental and epidemiological evidences [34 - 36]. Nevertheless, anthocyanins are rather unstable compounds, being influenced by pH, oxygen, light, temperature, concentration, and enzymes, as well as due to the presence of ascorbic acid, sugars, metal ions, sulphur compounds, and copigments [34, 37 - 39].

Other phenolic compounds quite prevalent in berries are flavonols, which constitute the most widespread flavonoids, being found throughout the plant kingdom except of algae species [29]. They have substitutions in A and/or B ring with hydroxylation in the 5 and 7 or 3′ and 4′ positions, respectively [40] (Fig. **5**).

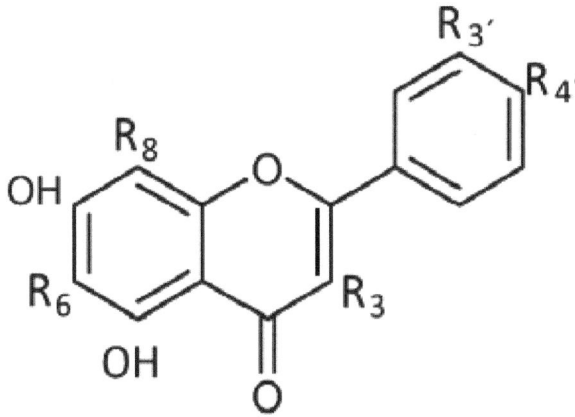

Fig. (5). General flavonols structure.

The flavonols predominant in berries are quercetin, kaempferol, myricetin, and isorhamnetin, and appear most commonly as *O*-glycosides derivatives or, in a lesser extent, as *C*-glycosides, in which one or more of the hydroxyl groups are bound to a suggar by an acid-labile hemiacetal bond.

More flavonoid subclasses have been found in berries. Among these, it is important to note the flavanols or flavan-3-ols, which constitutes the most complex sub-group of flavonoids. The molecular structure of this flavonoid group is divided into the simple monomers (such as catechin and its isomer epicatechin) to oligomers (from dimers to decamers), polymers (>10mers), and other derived compounds (*e.g.* theaflavins and thearubigins). The oligomers and polymers of flavanols are also named as proanthocyanidins or condensed tannins [41]. Specifically regarding berries, they contain the flavanol monomers (+)-catechin and (-)-epicatechin as well as dimers, trimmers, and polymeric proanthocyanidins.

Other flavonoid subclasses identified in berries in lesser amounts are chalcones, flavanones, aurones, flavones, isoflavones, and flavan-3,4-diols [42].

Non-Flavonoids

Ellagitannins are also a class of tannins and thereby share common properties with proanthocyanidins, differing only in their chemical structures [43]. They likely derive from a common gallotannin biosynthetic precursor, penta-*O*-galloyl-β-D-glucose, by the oxidative formation of one or several biphenyl bonds between two or more galloyl residues [43]. Ellagitannins can be found in many kinds of berries, as strawberries, raspberries, blueberries or blackberries [26].

Simple Phenols

Phenolic acids

Within the non-flavonoid phenolic compounds, highlight the phenolic acids that may be classified in two wide groups: benzoic acid derivatives and cinnamic acid derivatives, being this last group the most common in nature [44] (Fig. **6**).

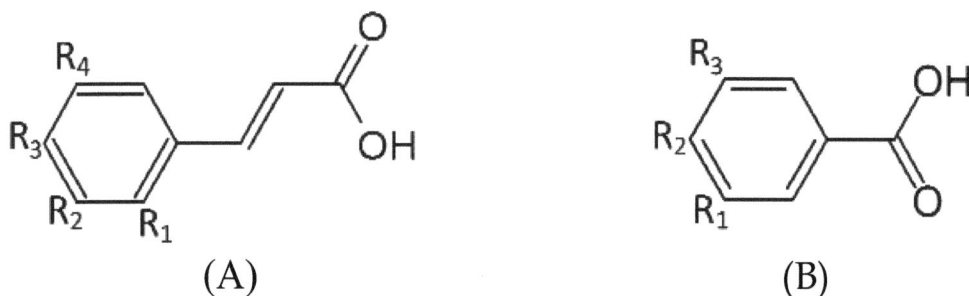

Fig. **(6)**. General structure of cinnamic (A) and benzoic (B) acids.

Multitude of phenolic compounds have been detected in berries, divided into cinnamic acids (mainly represented by *p*-coumaric, caffeic, and ferulic acids) and benzoic acids (mainly represented by *p*-hydroxybenzoic, gallic and ellagic acids) [45].

Coumarins

Coumarins (2H-1-benzopyran-2-one) consist of a widespread sub-group of phenolic compounds identified in plants and are made of fused benzene and α-pyrone rings [46]. Over 1300 different coumarins have been found as secondary metabolites from plants, bacteria, and fungi [47]. The prototypical structure compound is known as 1,2-benzopyrone or as o-hydroxycinnamic acid and lactone. These phenolic compounds were initially identified in tonka bean (Dipteryx odorata Wild) and, currently, have been described in about 150 different species, including berries, distributed over nearly 30 different families, of which a few relevant ones are Rutaceae, Umbelliferae, Clusiaceae, Guttiferae, Caprifoliaceae, Oleaceae, Nyctaginaceae, and Apiaceae.

Other non-flavonoids phytochemicals found in berries are stilbenes and lignans, although in a lesser degree [42].

ABSORPTION AND BIOAVAILABILITY OF HEALTHY NUTRIENTS AND NONNUTRIENTS

The main phytochemicals found in berries have demonstrated to exert beneficial health effects *in vitro* and *in vivo*. Therefore, it is important to examine the gastrointestinal absorption and systemic bioavailability of these compounds, responsible for this beneficial activity concerning health promotion in humans.

Bioavailability of Flavonoids and Phenolic Compounds

After the ingestion of dietary flavonoids, absorption of most of compounds but not all into the circulatory system happens in the small intestine [48]. Normally, this is associated with hydrolysis reactions, releasing the aglycone, by the action of lactase phloridizin hydrolase in the brush border of the epithelial cells in the small intestine. Before to passage into the blood stream the aglycones are subjected to metabolism, forming glucuronide, sulfate and/or methylated metabolites through the respective action of sulfotransferases (SULT), uridine-50-diphosphate glucuronosyltransferases (UGTs), and catechol-*O*-methyltransferases (COMT) [49]. Once in the bloodstream, metabolites undergo to phase II metabolism with more conversions that occur in the liver, where enterohepatic transport in the bile results in some recycling back to the small intestine. Flavonoids and their metabolites not absorbed in the small intestine reach colon and undergo the metabolic action of the local microflora, being absorbed in the large intestine as metabolic derivatives, responsible for different biological actions *in vivo* [1].

Anthocyanins

Evidences on the beneficial activity of anthocyanins have been reported over the last years that has allowed to recognize the therapeutic protective and therapeutic power of these phenolics. With respect to their bioavailability, the key difference compared to the other flavonoid glycosides is the re-arrangements suffered by anthocyanins in response to pH and temperature. The constraints regarding stability of anthocyanins contribute significantly to their low bioavailability, which does not justify all the biological activities previously associated with the huge consumption of these compounds.

Following ingestion, anthocyanin glycosides may be quickly absorbed in stomach by a process that can involve bilitranslocase, entering the systemic circulation after pass through the liver. Anthocyanin glycosides that are not absorbed in the stomach move to the small intestine where, because of the higher pH, are converted into different structures as chalcone, hemiketal, and quinonoidal forms [50]. Further absorption appears to take place in the jejunum. The transport mechanism has not been found, but if similar to flavonols, can involve hydrolysis of the glycosides by various hydrolases and absorption of the corresponding aglycone [51].

Anthocyanins that reach the colon can be degraded to sugar and other simple phenolic components by the microbial population, with the phenolic components later degraded by disruption of the C-ring to yield phenolic acids and aldehydes. These metabolites, derived from the ingested anthocyanins, contribute to the positive health effects of anthocyanins either directly in the gastrointestinal tract or systemically after absorption [51].

Flavonols

The main studies on flavonols bioavailability are focused on flavonols glucosides and diglucosides, mainly in the form of quercetin-4'-*O*-glucoside, quercetin-3-*O*-rutinoside and quercetin-3,4'-*O*-diglucoside, also present in berries, in onion, and tomato [52, 53]. These phenolics derive in quercetin-3'-*O*-sulfate, quercetin-3-*O*-glucuronide, isorhamnetin-3-*O*-glucuronide, and two partially identified metabolites, a quercetin-*O*-diglucuronide and a quercetin-*O*-glucuronide-*O*-

sulfate, in the circulatory system around 30 minutes after ingestion [52, 53].

It is important to emphasize that quercetin liberated in the small intestine by split of quercetin glucosides is converted to glucuronide, methylated, and sulfated metabolites [52], while quercetin produced in the colon from quercetin-3-*O*-rutinoside is derived to methyl and glucuronide metabolites, but not sulfate metabolites [53]. This suggests that sulfation of quercetin is a characteristic of sulfotransferases in the wall of the small intestine rather than the colon or the liver.

Flavanols

Bioavailability of flavanols is in great measure influenced by their degree of polymerization; being monomers quickly absorbed in the small intestine. On the other hand, oligomers and polymers need to be subjected by the action of colonic microbiota before absorption [54]. Glucuronide, sulfate, and methyl conjugate derivatives of (epi)catechin and (epi)gallocatechin reach peak of nanomolar plasma concentrations range 1.6 to 2.3 hours after intake, suggestion that their absorption occurs in the small intestine. These quantities decrease until trace amounts 8 h after ingestion. After green tea consumption, urinary excretion of metabolites over a 24-h period reports to almost 28% of the ingested (epi)catechin and 11% of (epi)gallocatechin, indicating higher absorption in comparison with other flavonoids [55]. Phase II metabolites synthesized as a consequence of the small intestine and hepatic metabolism appear as conjugated derivatives (glucuronic acid or sulfate esters, methyl ether, or their combined forms) of monomeric flavanols and are usually excreted in the bile [54]. Otherwise, about two-thirds of the ingested flavanols pass from the small to the large intestine where the bioaction of the microbiota becomes to C-6-C-5 phenylvalerolactones and phenylvaleric acids, which are subjected to side-chain shortening to produce C-6-C-1 phenolic and aromatic acids that are absorbed into the bloodstream and are excreted in urine in amounts equivalent to around 36% of flavanol intake [55].

Phenolic Acids

With respect to phenolic acids, over 30% of the compounds ingested are absorbed in the small intestine. Thus, around the 70% of the ingested phenolic acids pass from the small to the large intestine, where they undergo to the metabolism of the colonic microbiota [56], demonstrating the importance of the biotransformation taking place in the colon with respect to the bioavailability of phenolic acids.

BENEFITS FOR HEALTH

Although bioactivity of polyphenols of dietary berries has been evaluated by animal models and in human subjects, it is important to remark that dosages assayed in experimental animals are frequently higher than in humans and, in addition, the metabolism of phenolic compounds may differ in both models. Thus, the assessment of the actual biological activity of berries *in vivo* requires of complementary determinations using diverse models [1].

With respect to antitumoral activity of berries, anthocyanins extracts of bilberry have proved their capacity to reduce the number of colonic aberrant crypt foci, colonic cell proliferation, urinary level of oxidative DNA, and expression of cyclo-oxygenase genes as biomarkers of colon cancer in rats, supporting the protective role of berry extracts against this type of cancer [57]. In the same way, raspberries, blackberries, and strawberries have been demonstrated to reduce the initiation and progression of oesophageal cancer throughout the influence of their phenolic compounds on the mechanism of N-nitrosomethylbenzylamine, causing a down-regulation of cyclo-oxygenase-2-mediated DNA damage [58]. In this issue, a similar study allowed to point out the ability of berry's phenolics to reduce the levels of cyclo-oxygenase-2, nitric oxide synthase, and c-Jun, which are related with cancer progression as well as oesophagus preneoplasic lesions [59]. However, berries also have shown protector activity against non-digestive cancers. Thus, trials focused on the effect of berries supplemented diets in animal models have allowed to evidence their capacity to decrease, in a dose-dependent form, the size of haemangioendothelioma, urinary bladder, and breast tumours [60]. Although the accumulation of *in vitro* results suggests the protective role of dietary berries, to date there exist scarce evidences on the capacity of dietary berries to reduce the progression of cancer in humans, which are mainly restricted to the association of anthocyanins with the reduction of the risk of non-Hodgkin lymphoma [61].

The anti-inflammatory power of berry's phenolics has been also demonstrated with resort to diverse *in vivo* studies. De Furia, *et al.* demonstrated that the supplementation of rat diets with lyophilized blueberry might protect against adipose tissue-related inflammation [62]. Human trials have allowed to describe the decrease of the level of circulating adhesion molecules in sedentary humans and in patients affected by hypercholesterolemia as well as the reduction of circulating monocytes chemoattractant protein 1 (an inflammatory biomarker related with risk of cardiovascular disease) as a consequence of diets supplemented with berries [63].

The increase of berries intake also displays preventive activity against cardiovascular diseases based on their capacity to diminish risk factors such as hypertension, inflammation, dyslipemia, oxidative stress, diabetes, endothelial dysfunction, and platelet function, which is closely related with cardiovascular diseases progression [64]. This protection has been further supported by diverse animal models. Thus, blueberry enriched diets protected rats against ischemic damage and diminished post-myocardial infarction chronic heart failure [65] due to the potential of berries to increase myocardial gluthatione levels and, thus, modulate antioxidant defences [65]. Also in rats, dietary berries decreased brain infarct after focal cerebral ischaemic injury by interacting with c-Jun N-terminal kinase and p53 pathway [66]. Diets supplemented with berries (raspberry, strawberry, and bilberry) also demonstrated to be efficient inhibiting aortic lipid deposition by almost 90% in rats and triggering activity of hepatic antioxidant enzymes [67]. Regarding the application of berries in animal models, has been verified a decrease of plasmatic low density lipoproteins and an increase of high density lipoprotein [68]. This effect has been proved also in patients affected by metabolic diseases although it is required to take into account the effect of baseline markers and background medications that could condition the capacity of berries intake on cardiovascular prevention, providing confusing results [69].

Diabetes has been proposed as a clinical entity, which incidence and severity could be affected by berry's phenolics. However, although this effect has been supported by diverse animal assays, focused on the evaluation of the capacity of pelargonidin to normalize glycemia, improve serum levels of insulin, superoxide disumatse, and malonaldehyde and fructosamine [70, 71], to date the obtained results have not been reproduced in humans.

Oxidative stress constitutes an additional risk factor associated to diverse degenerative diseases that can be prevented by including berries in diets. Thus, bioactive compounds from these fruits are able to inhibit oxidative reactions *in vivo* and diminishing of low density lipoproteins oxidation in healthy and sedentary men [72, 73]. In contrast to results supporting the antioxidant activity of berry extracts, some works failed to provide evidences regarding an actual decrease of oxidative stress monitored by determining eicosanoids levels, which suggests that further research is required to rational comprehension of the pathways involved in the antioxidant attributions of berries *in vivo* [64].

Several clinical trials have allowed to establish the reduction of the prevalence of urinary tract infections due to the ingestion of berries and berry-based products [64]. Although some controversial data have been published on the efficiency of these dietary interventions. The diverse mechanisms proposed to explain the preventive activity related to urinary tract infections enclosed to berries intake are mainly related to the interference of bioactive compounds in berries with bacterial adhesion to urinary tract receptors by promoting changes in the bacterial morphology or genetically based decreases in P-fimbrial expression in bacteria [74].

Evidences suggesting that berries intake may decrease the prevalence of metabolic and cardiovascular diseases, and contribute to the microbial homeostasis in the gastrointestinal tract encourage the daily consumption of these fruits to achieve a healthy dietary pattern, implement recommendations for micronutrients intake, and promote the ingestion of a great diversity of bioactive phytochemicals.

CONFLICT OF INTEREST

The author confirms that author has no conflict of interest to declare for this publication.

ACKNOWLEDGEMENTS

The authors express their gratitude to the Spanish Ministry of Economy and Competitiveness for the CYTED Program (Ref. 112RT0460) CORNUCOPIA

Thematic Network (URL: redcornucopia.org).

REFERENCES

[1]   Del Rio D, Borges G, Crozier A. Berry flavonoids and phenolics: bioavailability and evidence of protective effects. Br J Nutr 2010; 104(3) (Suppl. 3): S67-90.
[http://dx.doi.org/10.1017/S0007114510003958] [PMID: 20955651]

[2]   Seeram NP. Berry fruits: compositional elements, biochemical activities, and the impact of their intake on human health, performance, and disease. J Agric Food Chem 2008; 56(3): 627-9.
[http://dx.doi.org/10.1021/jf071988k] [PMID: 18211023]

[3]   Seeram NP. Berry fruits for cancer prevention: current status and future prospects. J Agric Food Chem 2008; 56(3): 630-5.
[http://dx.doi.org/10.1021/jf072504n] [PMID: 18211019]

[4]   González-Tejero MR, Casares-Porcel M, Sánchez-Rojas CP, *et al.* Medicinal plants in the Mediterranean area: synthesis of the results of the project Rubia. J Ethnopharmacol 2008; 116(2): 341-57.
[http://dx.doi.org/10.1016/j.jep.2007.11.045] [PMID: 18242025]

[5]   Lee J, Durst RW, Wrolstad RE. Impact of juice processing on blueberry anthocyanins and polyphenolics: Comparison of two pretreatments. J Food Sci 2002; 67: 1660-7.
[http://dx.doi.org/10.1111/j.1365-2621.2002.tb08701.x]

[6]   Srivastava A, Akoh CC, Yi W, Fischer J, Krewer G. Effect of storage conditions on the biological activity of phenolic compounds of blueberry extract packed in glass bottles. J Agric Food Chem 2007; 55(7): 2705-13.
[http://dx.doi.org/10.1021/jf062914w] [PMID: 17348670]

[7]   Ochoa MR, Kesseler AG, Vullioud MB, Lozano JE. Physical and chemical characteristics of raspberry pulp: Storage effect on composition and color. LWT -. Food Sci Technol (Campinas) 1999; 32: 149-53.

[8]   García-Viguera C, Zafrilla P, Romero F, Abellán P, Artés F, Tomás-Barberán FA. Color stability of strawberry jam as affected by cultivar and storage temperature. J Food Sci 1999; 64: 243-7.
[http://dx.doi.org/10.1111/j.1365-2621.1999.tb15874.x]

[9]   Klopotek Y, Otto K, Böhm V. Processing strawberries to different products alters contents of vitamin C, total phenolics, total anthocyanins, and antioxidant capacity. J Agric Food Chem 2005; 53(14): 5640-6.
[http://dx.doi.org/10.1021/jf047947v] [PMID: 15998127]

[10]   Hager A, Howard LR, Prior RL, Brownmiller C. Processing and storage effects on monomeric anthocyanins, percent polymeric color, and antioxidant capacity of processed black raspberry products. J Food Sci 2008; 73(6): H134-40.
[http://dx.doi.org/10.1111/j.1750-3841.2008.00855.x] [PMID: 19241590]

[11]   Yang B, Karlsson RM, Oksman PH, Kallio HP. Phytosterols in sea buckthorn (Hippophaë rhamnoides L.) berries: identification and effects of different origins and harvesting times. J Agric Food Chem 2001; 49(11): 5620-9.

[http://dx.doi.org/10.1021/jf010813m] [PMID: 11714369]

[12]   Mironov VA. Chemical composition of Hippophaė rhamnoides of different populations of the USSR.

[13]   Zhang W, Yan J, Duo J, Ren B, Guo J. Preliminary study of biochemical constitutions of berry of sea buckthorn growing in Shanxi province and their changing trend.

[14]   Anderson JW, Baird P, Davis RH Jr, *et al.* Health benefits of dietary fiber. Nutr Rev 2009; 67(4): 188-205.
        [http://dx.doi.org/10.1111/j.1753-4887.2009.00189.x] [PMID: 19335713]

[15]   Arimboor R, Venugopalan VV, Sarinkumar K, Arumughan C, Sawhney RC. Integrated processing of fresh Indian sea buckthorn (Hippophae rhamnoides) berries and chemical evaluation of products. J Sci Food Agric 2006; 86: 2345-53.
        [http://dx.doi.org/10.1002/jsfa.2620]

[16]   Dhyani D, Maikhuri RK, Rao KS, Kumar L, Purohit VK, Sundriyal M, *et al.* Basic nutritional attributes of Hippophae rhamnoides (Seabuckthorn) populations from Uttarakhand Himalaya, India. Curr Sci 2007; 92: 1148-52.

[17]   Barros L, Carvalho AM, Morais JS, Ferreira IC. Strawberry-tree, blackthorn and rose fruits: Detailed characterisation in nutrients and phytochemicals with antioxidant properties. Food Chem 2010; 120: 247-54.
        [http://dx.doi.org/10.1016/j.foodchem.2009.10.016]

[18]   Chidouh A, Aouadi S, Heyraud A. Extraction, fractionation and characterization of water-soluble polysaccharide fractions from myrtle *(Myrtus communis L.)* fruit. Food Hydrocoll 2014; 35: 733-9.
        [http://dx.doi.org/10.1016/j.foodhyd.2013.08.001]

[19]   Hayashi T. Xyloglucans in the primary cell wall. Annu Rev Plant Physiol 1989; 40: 139-68.
        [http://dx.doi.org/10.1146/annurev.pp.40.060189.001035]

[20]   Selvendran RR, ONeil MA. Isolation and analysis of cell walls from plant material. Methods of Biochemical Analysis. 2006; 32: pp. 25-153.

[21]   Jaakkola M, Korpelainen V, Hoppula K, Virtanen V. Chemical composition of ripe fruits of Rubus chamaemorus L. grown in different habitats. J Sci Food Agric 2012; 92(6): 1324-30.
        [http://dx.doi.org/10.1002/jsfa.4705] [PMID: 22083544]

[22]   Haila K, Kumpulainen J, Häkkinen U, Tahvonen R. Sugars and organic acids in berries and fruits consumed in Finland during 1987-1989. J Food Compos Anal 1992; 5: 108-11.
        [http://dx.doi.org/10.1016/0889-1575(92)90025-F]

[23]   Bal LM, Meda V, Naik SN, Satya S. Sea buckthorn berries: A potential source of valuable nutrients for nutraceuticals and cosmoceuticals. Food Res Int 2011; 44: 1718-27.
        [http://dx.doi.org/10.1016/j.foodres.2011.03.002]

[24]   Voet D, Voet JG. 2004.

[25]   Indrayan AK, Sharma S, Durgapal D, Kumar N, Kumar M. Determination of nutritive value and analysis of mineral elements for some medicinally valued plants from Uttaranchal. Curr Sci 2005; 89: 1252-5.

[26]   Pimpão RC, Dew T, Oliveira PB, Williamson G, Ferreira RB, Santos CN. Analysis of phenolic

compounds in Portuguese wild and commercial berries after multienzyme hydrolysis. J Agric Food Chem 2013; 61(17): 4053-62.
[http://dx.doi.org/10.1021/jf305498j] [PMID: 23530973]

[27]   Parr AJ, Bolwell GP. Phenols in the plant and in man. The potential for possible nutritional enhancement of the diet by modifying the phenols content or profile. J Sci Food Agric 2000; 80: 985-1012.
[http://dx.doi.org/10.1002/(SICI)1097-0010(20000515)80:7<985::AID-JSFA572>3.0.CO;2-7]

[28]   Croteau R, Kutchan TM, Lewis NG. Natural products (secondary metabolites). Biochem Mol Biol Plant 2000; pp. 1250-318.

[29]   Crozier A, Jaganath IB, Clifford MN. Dietary phenolics: chemistry, bioavailability and effects on health. Nat Prod Rep 2009; 26(8): 1001-43.
[http://dx.doi.org/10.1039/b802662a] [PMID: 19636448]

[30]   Robards K, Prenzler PD, Tucker G, Swatsitang P, Glover W. Phenolic compounds and their role in oxidative processes in fruits. Food Chem 1999; 66: 401-36.
[http://dx.doi.org/10.1016/S0308-8146(99)00093-X]

[31]   Gattuso G, Barreca D, Gargiulli C, Leuzzi U, Caristi C. Flavonoid composition of Citrus juices. Molecules 2007; 12(8): 1641-73.
[http://dx.doi.org/10.3390/12081641] [PMID: 17960080]

[32]   Del Rio D, Rodriguez-Mateos A, Spencer JP, Tognolini M, Borges G, Crozier A. Dietary (poly)phenolics in human health: structures, bioavailability, and evidence of protective effects against chronic diseases. Antioxid Redox Signal 2013; 18(14): 1818-92.
[http://dx.doi.org/10.1089/ars.2012.4581] [PMID: 22794138]

[33]   Bridle P, Timberlake CF. Anthocyanins as natural food colours - Selected aspects. Food Chem 1997; 58: 103-9.
[http://dx.doi.org/10.1016/S0308-8146(96)00222-1]

[34]   Castañeda-Ovando A, Pacheco-Hernández ML, Páez-Hernández ME, Rodríguez JA, Galán-Vidal CA. Chemical studies of anthocyanins: A review. Food Chem 2009; 113: 859-71.
[http://dx.doi.org/10.1016/j.foodchem.2008.09.001]

[35]   De Pascual-Teresa S, Sanchez-Ballesta MT. Anthocyanins: From plant to health. Phytochem Rev 2008; 7: 281-99.
[http://dx.doi.org/10.1007/s11101-007-9074-0]

[36]   Tsuda T. Dietary anthocyanin-rich plants: biochemical basis and recent progress in health benefits studies. Mol Nutr Food Res 2012; 56(1): 159-70.
[http://dx.doi.org/10.1002/mnfr.201100526] [PMID: 22102523]

[37]   García-Viguera C, Bridle P. Influence of structure on colour stability of anthocyanins and flavylium salts with ascorbic acid. Food Chem 1999; 64: 21-6.
[http://dx.doi.org/10.1016/S0308-8146(98)00107-1]

[38]   Jurd L. Anthocyanidins and related compounds-XI. Catechin-flavylium salt condensation reactions. Tetrahedron 1967; 23: 1057-64.
[http://dx.doi.org/10.1016/0040-4020(67)85056-7]

[39]   Parisa S, Reza H, Elham G, Rashid J. Effect of heating, UV irradiation and pH on stability of the anthocyanin copigment complex. Pak J Biol Sci 2007; 10(2): 267-72.
[http://dx.doi.org/10.3923/pjbs.2007.267.272] [PMID: 19070027]

[40]   Herrmann K. Flavonols and flavones in food plants: a review. J Food Technol 1976; 11: 443-8.
[http://dx.doi.org/10.1111/j.1365-2621.1976.tb00743.x]

[41]   Knaze V, Zamora-Ros R, Luján-Barroso L, *et al.* Intake estimation of total and individual flavan-3-ols, proanthocyanidins and theaflavins, their food sources and determinants in the European Prospective Investigation into Cancer and Nutrition (EPIC) study. Br J Nutr 2012; 108(6): 1095-108.
[http://dx.doi.org/10.1017/S0007114511006386] [PMID: 22186699]

[42]   Nile SH, Park SW. Edible berries: bioactive components and their effect on human health. Nutrition 2014; 30(2): 134-44.
[http://dx.doi.org/10.1016/j.nut.2013.04.007] [PMID: 24012283]

[43]   Clifford MN, Scalbert A. Ellagitannins - Nature, occurrence and dietary burden. J Sci Food Agric 2000; 80: 1118-25.
[http://dx.doi.org/10.1002/(SICI)1097-0010(20000515)80:7<1118::AID-JSFA570>3.0.CO;2-9]

[44]   Manach C, Scalbert A, Morand C, Rémésy C, Jiménez L. Polyphenols: food sources and bioavailability. Am J Clin Nutr 2004; 79(5): 727-47.
[PMID: 15113710]

[45]   Häkkinen S, Heinonen M, Kärenlampi S, Mykkänen H, Ruuskanen J, Törrönen R. Screening of selected flavonoids and phenolic acids in 19 berries. Food Res Int 1999; 32: 345-53.
[http://dx.doi.org/10.1016/S0963-9969(99)00095-2]

[46]   Aoyama Y, Katayama T, Yamamoto M, Tanaka H, Kon K. A new antitumor antibiotic product, demethylchartreusin. Isolation and biological activities. J Antibiot 1992; 45(6): 875-8.
[http://dx.doi.org/10.7164/antibiotics.45.875] [PMID: 1500353]

[47]   Iranshahi M, Askari M, Sahebkar A, Hadjipavlou-Litina D. Evaluation of antioxidant, anti-inflammatory and lipoxygenase inhibitory activities of the prenylated coumarin umbelliprenin. Daru 2009; 17: 99-103.

[48]   Donovan JL, Manach C, Faulks RM, Kroon PA. Absorption and Metabolism of Dietary Plant Secondary Metabolites. Plant Secondary Metabolites: Occurrence. Structure and Role in the Human Diet 2007; pp. 303-51.

[49]   Crozier A, Del Rio D, Clifford MN. Bioavailability of dietary flavonoids and phenolic compounds. Mol Aspects Med 2010; 31(6): 446-67.
[http://dx.doi.org/10.1016/j.mam.2010.09.007] [PMID: 20854839]

[50]   Fernandes I, Faria A, Calhau C, de Freitas V, Mateus N. Bioavailability of anthocyanins and derivatives. J Funct Foods 2013.

[51]   McGhie TK, Walton MC. The bioavailability and absorption of anthocyanins: towards a better understanding. Mol Nutr Food Res 2007; 51(6): 702-13.
[http://dx.doi.org/10.1002/mnfr.200700092] [PMID: 17533653]

[52]   Mullen W, Edwards CA, Crozier A. Absorption, excretion and metabolite profiling of methyl-, glucuronyl-, glucosyl- and sulpho-conjugates of quercetin in human plasma and urine after ingestion

of onions. Br J Nutr 2006; 96(1): 107-16.
[http://dx.doi.org/10.1079/BJN20061809] [PMID: 16869998]

[53]   Jaganath IB, Mullen W, Edwards CA, Crozier A. The relative contribution of the small and large intestine to the absorption and metabolism of rutin in man. Free Radic Res 2006; 40(10): 1035-46.
[http://dx.doi.org/10.1080/10715760600771400] [PMID: 17015248]

[54]   Monagas M, Urpi-Sarda M, Sánchez-Patán F, *et al.* Insights into the metabolism and microbial biotransformation of dietary flavan-3-ols and the bioactivity of their metabolites. Food Funct 2010; 1(3): 233-53.
[http://dx.doi.org/10.1039/c0fo00132e] [PMID: 21776473]

[55]   Clifford MN, van der Hooft JJ, Crozier A. Human studies on the absorption, distribution, metabolism, and excretion of tea polyphenols. Am J Clin Nutr 2013; 98(6) (Suppl.): 1619S-30S.
[http://dx.doi.org/10.3945/ajcn.113.058958] [PMID: 24172307]

[56]   Stalmach A, Edwards CA, Wightman JD, Crozier A. Identification of (poly)phenolic compounds in concord grape juice and their metabolites in human plasma and urine after juice consumption. J Agric Food Chem 2011; 59(17): 9512-22.
[http://dx.doi.org/10.1021/jf2015039] [PMID: 21812481]

[57]   Lala G, Malik M, Zhao C, *et al.* Anthocyanin-rich extracts inhibit multiple biomarkers of colon cancer in rats. Nutr Cancer 2006; 54(1): 84-93.
[http://dx.doi.org/10.1207/s15327914nc5401_10] [PMID: 16800776]

[58]   Wang LS, Hecht SS, Carmella SG, *et al.* Anthocyanins in black raspberries prevent esophageal tumors in rats. Cancer Prev Res (Phila) 2009; 2(1): 84-93.
[http://dx.doi.org/10.1158/1940-6207.CAPR-08-0155] [PMID: 19139022]

[59]   Chen T, Hwang H, Rose ME, Nines RG, Stoner GD. Chemopreventive properties of black raspberries in N-nitrosomethylbenzylamine-induced rat esophageal tumorigenesis: down-regulation of cyclooxygenase-2, inducible nitric oxide synthase, and c-Jun. Cancer Res 2006; 66(5): 2853-9.
[http://dx.doi.org/10.1158/0008-5472.CAN-05-3279] [PMID: 16510608]

[60]   Gordillo G, Fang H, Khanna S, Harper J, Phillips G, Sen CK. Oral administration of blueberry inhibits angiogenic tumor growth and enhances survival of mice with endothelial cell neoplasm. Antioxid Redox Signal 2009; 11(1): 47-58.
[http://dx.doi.org/10.1089/ars.2008.2150] [PMID: 18817478]

[61]   Frankenfeld CL, Cerhan JR, Cozen W, *et al.* Dietary flavonoid intake and non-Hodgkin lymphoma risk. Am J Clin Nutr 2008; 87(5): 1439-45.
[PMID: 18469269]

[62]   DeFuria J, Bennett G, Strissel KJ, *et al.* Dietary blueberry attenuates whole-body insulin resistance in high fat-fed mice by reducing adipocyte death and its inflammatory sequelae. J Nutr 2009; 139(8): 1510-6.
[http://dx.doi.org/10.3945/jn.109.105155] [PMID: 19515743]

[63]   Youdim KA, McDonald J, Kalt W, Joseph JA. Potential role of dietary flavonoids in reducing microvascular endothelium vulnerability to oxidative and inflammatory insults (small star, filled). J Nutr Biochem 2002; 13(5): 282-8.
[http://dx.doi.org/10.1016/S0955-2863(01)00221-2] [PMID: 12015158]

[64]   Blumberg JB, Camesano TA, Cassidy A, *et al.* Cranberries and their bioactive constituents in human health. Adv Nutr 2013; 4(6): 618-32.
[http://dx.doi.org/10.3945/an.113.004473] [PMID: 24228191]

[65]   Toufektsian MC, de Lorgeril M, Nagy N, *et al.* Chronic dietary intake of plant-derived anthocyanins protects the rat heart against ischemia-reperfusion injury. J Nutr 2008; 138(4): 747-52.
[PMID: 18356330]

[66]   Shin WH, Park SJ, Kim EJ. Protective effect of anthocyanins in middle cerebral artery occlusion and reperfusion model of cerebral ischemia in rats. Life Sci 2006; 79(2): 130-7.
[http://dx.doi.org/10.1016/j.lfs.2005.12.033] [PMID: 16442129]

[67]   Yamanouchi J, Nishida E, Itagaki S, Kawamura S, Doi K, Yoshikawa Y. Aortic atheromatous lesions developed in APA hamsters with streptozotocin induced diabetes: a new animal model for diabetic atherosclerosis. 1. Histopathological studies. Exp Anim 2000; 49(4): 259-66.
[http://dx.doi.org/10.1538/expanim.49.259] [PMID: 11109551]

[68]   Yung LM, Tian XY, Wong WT, *et al.* Chronic cranberry juice consumption restores cholesterol profiles and improves endothelial function in ovariectomized rats. Eur J Nutr 2013; 52(3): 1145-55.
[http://dx.doi.org/10.1007/s00394-012-0425-2] [PMID: 22836513]

[69]   Chu YF, Liu RH. Cranberries inhibit LDL oxidation and induce LDL receptor expression in hepatocytes. Life Sci 2005; 77(15): 1892-901.
[http://dx.doi.org/10.1016/j.lfs.2005.04.002] [PMID: 15982671]

[70]   Sasaki R, Nishimura N, Hoshino H, *et al.* Cyanidin 3-glucoside ameliorates hyperglycemia and insulin sensitivity due to downregulation of retinol binding protein 4 expression in diabetic mice. Biochem Pharmacol 2007; 74(11): 1619-27.
[http://dx.doi.org/10.1016/j.bcp.2007.08.008] [PMID: 17869225]

[71]   Shabrova EV, Tarnopolsky O, Singh AP, Plutzky J, Vorsa N, Quadro L. Insights into the molecular mechanisms of the anti-atherogenic actions of flavonoids in normal and obese mice. PLoS One 2011; 6(10): e24634.
[http://dx.doi.org/10.1371/journal.pone.0024634] [PMID: 22016761]

[72]   Basu A, Betts NM, Ortiz J, Simmons B, Wu M, Lyons TJ. Low-energy cranberry juice decreases lipid oxidation and increases plasma antioxidant capacity in women with metabolic syndrome. Nutr Res 2011; 31(3): 190-6.
[http://dx.doi.org/10.1016/j.nutres.2011.02.003] [PMID: 21481712]

[73]   Hollman PC, Cassidy A, Comte B, *et al.* The biological relevance of direct antioxidant effects of polyphenols for cardiovascular health in humans is not established. J Nutr 2011; 141(5): 989S-1009S.
[http://dx.doi.org/10.3945/jn.110.131490] [PMID: 21451125]

[74]   Liu Y, Black MA, Caron L, Camesano TA. Role of cranberry juice on molecular-scale surface characteristics and adhesion behavior of Escherichia coli. Biotechnol Bioeng 2006; 93(2): 297-305.
[http://dx.doi.org/10.1002/bit.20675] [PMID: 16142789]

CHAPTER 7

Bioactive Compounds of Tropical Fruits as Health Promoters

Iris Feria Romero[1], Christian Guerra-Araiza[2], Hermelinda Salgado Ceballos[2], Juan M. Gallardo[3], Julia J. Segura-Uribe[1], Sandra Orozco-Suárez[1,*]

[1] *Unidad de Investigación Médica en Enfermedades Neurológicas. Centro Médico Nacional "Siglo XXI". Instituto Mexicano del Seguro Social. México, D.F., México*

[2] *Unidad de Investigación Médica en Farmacología. Centro Médico Nacional "Siglo XXI". Instituto Mexicano del Seguro Social. México, D.F., México*

[3] *Unidad de Investigación Médica en Enfermedades Nefrológicas. Hospital de Especialidades. Centro Médico Nacional "Siglo XXI". Instituto Mexicano del Seguro Social. México, D.F., México*

Abstract: Currently, it is known that ingestion of fruits and vegetables decreases the incidence of several degenerative and aging-related disorders. Plants contain mostly antioxidant compounds that preserve the balance between oxidants and antioxidants in the body. Excess of reactive oxygen species [ROS] can be produced after the so-called "oxidative stress", a process that can damage and even kill the cells. Although antioxidant substances represent one of the most important mechanisms of defense against free radicals, the endogenous antioxidant molecules alone are not effective enough to counteract the injuries caused by ROS, particularly in current times when lifestyles based on smoking, consumption of drugs and/or alcohol, unbalanced diet, pollution, and exposure to solar radiation, among others, can facilitate free radical formation. For this reason, increasing the intake of dietary antioxidants is of great importance for a good health, as it is evidenced by studies on food with high antioxidant contents as well as with anti-inflammatory properties. Among these foods are tropical fruits which include a large number of plants that grow in tropical and subtropical climates above 4°C.

* **Corresponding author Sandra Orozco Suárez:** Unidad de Investigación Médica en Enfermedades Neurológicas, Centro Médico Nacional,"Siglo XXI", Instituto Mexicano del Seguro Social, México, D.F., México; Email: sorozco5@hotmail.com

As Latin America and Asia are the main regions that produce this kind of fruits for human consumption, we described some of the phytochemical properties, biological activities and bioactive compounds of several tropical fruits from the Latin American region. The aim of this review was to present them as potential health promoters and their use for prevention and treatment of several neurological diseases.

Keywords: Antioxidants, Health promoters, Phytochemicals, Tropical fruits.

INTRODUCTION

The influence of diet in the prevention and treatment of diseases is becoming stronger to date. This is attributed to many substances present in food that act synergistically on intermediary and xenobiotic metabolism. Among these substances are those with antioxidant capacity which eliminate or inactivate free radicals, preventing the development of oxidative stress related diseases as diabetes mellitus II, cardiovascular disease (CVD), neurodegenerative diseases and cancer. The antioxidant benefit of fruits and vegetables has been supported by many analytical (bioactive), epidemiological (protector) and interventional (dose-response) studies, where the consumption of fruits, particularly tropical fruits and berries increased these antioxidant effects. The beneficial effects of plants have been associated with the presence of essential and non-essential bioactive compounds, which are found in nature and are part of the food chain, and also have effects on human health [1].

Tropical fruits are known as fruits native to the tropical or subtropical climate. They have in common not withstanding the cold or temperature falls below 4°C. Tropical fruits are not characterized by the geographical area where they grow but rather the surrounding climate. Many tropical fruits are grown in areas that are not classified as tropical or subtropical, but enjoy a warm, constant temperature at an average of 27°C [2]. In regard to tropical fruits and their products attention is growing mainly because their appearance and interesting taste to consumers worldwide (Fig. **1**). The tropical fruit exporting countries are in Asia, Middle East, Latin America, the Caribbean and, to a lesser extent, in Africa. The five fruits with the major export volume are banana, mango, pineapple, papaya and avocado. Major tropical fruit production was reported by the Food and Agriculture Organization (FAO) to be over 184 million tons in 2013 [3]. Bananas are extensively available in all markets, with a production of 93 million tons in 2008 and 100 million tons in 2013. Major banana producers are the countries of Central America and the Philippines. On the other hand, mango production rose by 25% since 2007, when it was less than a third of that of bananas [3, 4].

Fig. (1). As global consumers demand and are willing to pay for new appealing and exotic foods, tropical fruits are being intensively investigated for their several properties and bioactive compounds, such as carotenoids that produce their attractive colors.

Many of the tropical fruits remain unknown to consumers due to a lack of communication, poor storage and bad transportation that causes a short useful life.

However, because of the demand for new food products in all economies tropical fruits could be a source of bioactive compounds to improve health as several of them have been used in traditional medicine and ethnopharmacology [5, 6], especially for diseases that occur in tropical and subtropical regions [7].

Bioactive Fruit Compounds

Bioactive food components are obtained from fruits and are known as phytochemicals. Most of these phytochemicals have antioxidant properties, which can remove and counteract the formation of reactive oxygen species (ROS). It is known that ROS may contribute to degenerative diseases, so it has been postulated that providing different plant antioxidants can contribute to the observed protective effects of a plant-based diet. In contrast to the hypothesis that high doses of supplementary antioxidants may be beneficial, it has been suggested that low concentrations of a mix of different phytochemicals with redox activity may have effects [8].

Table 1. Main compounds in tropical fruits as health promoters.

Phenolic class	Chemical structure	Main compounds	Tropical fruits
Flavonoids			
Flavonols		quercetin kaempferol myricetin isorhamnetin	*Theobroma cacao* (cocoa) *Psidium guajava* (guava) *Mangifera indica* (mango)
Flavones		apigenin luteneolin 6-hydroxiflavone	*Mangifera indica* (mango)
Flavans		cathechin epichathechin	*Averrhoa carambola L.* (carambola)

(Table 1) contd.....

Phenolic class	Chemical structure	Main compounds	Tropical fruits
Anthocyanidins		cyanidyn malvidin	Berries *Musa paradisiaca* (banana) *Punicum granatum* (grenade)
Isoflavones		daizein ginestein	Peas
Lignans		pinoresinol podophyllotoxin steganacin	*Coffea arabica* (coffee) Kiwi
Diterpenes		cafestol kahweal	*Coffea arabica* (coffee)
Carotenoids		lutein lycopene β carotene	*Mangifera indica* (mango) *Carica papaya* (papaya)
Vitamins		vit A vit C vit D vit E	*Persea americana (avocado)* *Averrhoa carambola L.* (carambola) *Ananas comosus* (pineapple) *Psidium guajava* (guava)

There are several hundred antioxidants, some of them from the carotenoid group and a very large class of compounds called phenols or polyphenols. Their main structural feature is one or more hydroxyl groups (-OH) attached to one or more aromatic rings (phenolic group). Between polyphenols it is possible to distinguish two subtypes of compounds: flavonoids, which structure (propane diphenyl, C6-C3-C6) comprises two aromatic rings linked together by a heterocycle formed by three carbon atoms and oxygen. There have been described over 5000 compounds. At the same time flavonoids are subdivided into six groups of compounds: anthocyanidins, flavonols, flavanones, flavanols, flavones and isoflavones. The other subtype compounds are called non-flavonoids. Phenolic compounds comprise mostly mono-alcohols (*e.g.* hydroxytyrosol), simple phenolic acids such as benzoic acid derivatives and stilbene (*e.g.* rasveratrol) (Table **1**) [9, 10].

Flavonols are the most widely represented flavonoids in food. Fruits often contain 5-10 different flavonol glycosides (glycosylated form of flavonols). Flavonols accumulate in the skin of the fruit and leaves because their biosynthesis is stimulated by light. Thus the flavonol concentration of fruits and vegetables depends on the exposure to light. Flavonoids are commonly found in nature as conjugated compounds, namely attached to different sugars (as glucose, fructose) or in the form of free compounds (aglycon). The proportions of free and conjugated flavonoids depend on the type of food they are located in. In turn, the gastrointestinal tract is exposed to a different ratio of free/conjugated flavonoids, depending on how cooked is the food, and the action taken along the digestion process. Several enzymes are capable of hydrolyzing sugars, however, the absorption is not a fundamental process when the action of these compounds is exerted directly in the lumen of the gastrointestinal tract, for example, modulating the activity of any digestive enzyme or acting as an antioxidant directly on ROS present in the lumen.

The ability of polyphenols to act as antioxidants, both flavonoids as those that are not, depends primarily on the presence of -OH groups in their structure. When attached to a benzene ring, polyphenol hydroxyl groups function either as hydrogen atom donors (HAT) or as electron donor (SET) to a free radical (or to other reactive species). Some flavonoids in particular may also act as antioxidants through a mechanism that involves their ability to react with (chelate) certain transition metals such as copper and iron [9, 10].

Another class of bioactive compounds are carotenoids. Carotenoids or tetra-terpenoids are terpenoid pigments with 40 carbon atoms derived biosynthetically from two geranylgeranyl units. Most are soluble in nonpolar solvents and colors range from yellow (*e.g.* beta-carotene) to red (*e.g.* lycopene). They are widely distributed in the plant kingdom, in bacteria, and very few have been reported in animals. In humans, β-carotene is an essential dietary requirement because it is a precursor of vitamin A [10, 11]. The most often found carotene is β-carotene, and usually constitutes 25-30% of total carotenoids in plants. Lutein is the most abundant xanthophyll (40-45%), but in a smaller proportion than the β-carotene. Carotenoids, chlorophylls and flavonoids are the most distributed plant pigments. In particular, carotenoids give yellow, orange, red and violet colorations to plant tissues and animal organs [10, 11].

Tropical Fruits in Health

At the present time these bioactive or chemopreventive compounds are the spotlight of pharmaceutical research laboratories and food industry. In the scientific literature this research field is known as functional foods. While no one can consider them as essential substances required for our metabolism, they are essential for long term health since they are involved and exert protective effects on the cardiovascular system reducing blood pressure, and act as blood glucose and cholesterol regulators, reducing the risk of cancer and improving the immune defense response of the body. It is worth noting that some tropical fruits with these characteristics in Latin America could be useful in central nervous system (CNS) neuroprotection, and against cancer and vascular diseases.

Cancer and Bioactive Compounds

The time between the development of a cancerous tumor and its clinical manifestation is considered a slow evolution process because it can take decades, from the cancer cell developing a niche until the tumor diagnosis which can be benign or malignant. This provides enough time for the application of different preventive treatments. Nevertheless, cancer remains a leading cause of death worldwide, which also can be interpolated to different geographical regions and countries. According to GLOBOCAN, about 14.07 million people are diagnosed with cancer each year, causing about 8.2 million deaths [12].

The National Cancer Institute of the United States defines cancer as collection of related diseases. However, categorization of cancer is very complicated. First, it can be present in virtually any tissue which is considered tissue of origin or primary tumor. This is important because the cellular structure of the tumor may be similar in different tissues sharing the same cell lines (*e.g.* epithelial cells), which will influence in making the best decision for the course of the treatment. Another aspect to consider is the invasive behavior of the tumor. If tumor infiltration is observed in healthy tissue, it is termed malignant but if the tumor is defined, it is considered a benign tumor.

Epidemiological and experimental studies have provided additional insights into cancer development. These studies have found that lifestyle is one of the most important aspects. Due to studies in individuals with chronic diseases (*e.g.* obese and diabetic patients) it has been observed that these individuals have an increased susceptibility to develop cancer compared to those watching their weight and maintaining their biochemical levels within the normal range. Thus, a healthy lifestyle that includes exercise, proper nutrition and the use of natural or synthetic agents delays or prevents the carcinogenic process. These are considered preventive or chemopreventive actions [13].

Conversely, a diet rich in red meat and animal fat as well as low consumption of fresh fruits, vegetables and whole grains has been associated with an increased risk of developing some types of cancer. Interestingly, in the current nutrimental research potentially chemopreventive dietary components have been identified to use as functional foods or nutraceuticals [14, 15].

Persea americana, which common names are avocado and alligator pear, is originated in Puebla, Mexico. This fruit has been grown in different climates (tropical and Mediterranean) around the world, and its consumption has been known since 10,000 BC. The avocado is used in dip "guacamole", in sushi and smoothies. Among its main components are monounsaturated fats. It is also rich in vitamins B3, C and E, and persin and persenone A, that are considered its bioactive compounds. The main biological activity experimentally demonstrated was anticancer activity against breast, oral and prostate cancers. Among several studies with avocado fruit there has been reported that acetone extract (100 µg/mL) significantly inhibited (60.8%) the proliferation of a human prostate cancer cell (LNCaP). Chloroform-soluble extract isolated from the mesocarp selectively inhibited the growth (IC_{50} value of 18.8 µg/mL) of human oral cancer cell line (83-01-82CA) by targeting the EGFR/RAS/RAF/MEK/ERK1/2 pathway; and significant decrease of mean cell viability using 20 µg/mL of ethanol, ethyl acetate, petroleum and chloroform extracts on esophageal cancer cells (42.0 ± 3.3, 32.7 ± 3.3, 31.7 ± 3.2 and 33.0 ± 3.3, respectively) and colon cancer cells (70.3 ± 3.3, 56.0 ± 3.3, 56.0 ± 3.3 and 52.0 ± 3.3, respectively), compared with mononuclear blood cells (96.7 ± 3.3, 96.8 ± 3.2, 96.7 ± 3.3 and 96.8 ± 3.2, respectively) [16 - 18].

Annona muricata, (soursop) also known as graviola or *guanabana*, is native to Mexico, Colombia, Cuba, Peru, Brazil, Venezuela and Ecuador, as well as Southeast Asia, Africa and the Pacific. Until now there are not enough studies that relate the soursop's pulp with the inhibition of cancer development. Recently, Sun *et al.* found three acetogenins as the main bioactive compounds with inhibitory activities against a human prostate cancer cell line (PC-3) [19]. On the other hand, more than 40 acetogenins are present in the stems, leaves, and seeds of the soursop showing anti-proliferative properties.

Theobroma cacao seeds are known because once dried and fermented are used to make beverages and the famous chocolate. Ancient Mesoamerican civilizations like Aztec and Mayan consumed cacao since 1,100 B.C. [20]. But it was not until the 16th century that Hernan Cortes carried some seeds to Europe for consumption, which was initially banned [21]. The antioxidant, detoxifying, anti-inflammatory and protective effects on cell cultures are attributed to polyphenolic compounds present in cacao [22]. Proanthocyanidins found in cocoa liquor (CLPR) shown to have inhibitory effect on mutagenicity, which is caused by different heterocyclic amines. In an experimental work where a CLPR diet was administered during the onset of mutagenesis it was able to inhibit the initiation stage of the pancreas but not mamary carcinogenesis [23]. When CLPR was integrated into the basal diet —one week after carcinogen treatment in a lung and thyroid carcinogenesis model induced by different mutagenic compounds—, adenocarcinomas in the lung showed a significant reduction of development and a tendency for decrease in thyroid lesions [24]. Similarly, a significant reduction in the incidence of prostate lesions was observed in rats when orally administered 24 mg/kg of a rich polyphenolic cocoa compounds extract before the initiating and promoting phases of prostate cancer. Cocoa diet (16% Natural Forastero cocoa powder) completely inhibited liver damage induced by a mutagenic (N-nitorosodiethylamine) in both pre-and post-treated Sprague-Dawley, as enzymes of phase I and II rats [25]. In a model of genetically induced multiple intestinal neoplasia in C57BL/6J-Min/+ mice it was shown that a diet with (+) - catechin from cocoa using concentrations of 0.1 and 1% reduced the formation of intestinal tumors by 75 and 71%, respectively [26]. When a diet rich in cocoa (12%) was administered two weeks before an azoxymethane-induced colon cancer model and throughout the experimental time of 8 weeks, it was possible to prevent the development of early stage of this colon cancer [27]. In humans, Kuna people has supported the chemopreventive role of cocoa in cancer development. In this population cocoa is the main drink, which favors consumption over 900 mg/day of flavanols. A survey on mortality conducted between 2000 and 2004 revealed that the rates of diabetes mellitus, cardiovascular diseases and cancer were lower among the residents of the island of Kuna compared with the people of Panama [28]. Supporting the anticancer effect of cocoa, in Spain a case-control study reported that consumption of cocoa flavonoids decreased the incidence of gastric cancer [29]. Also, the Iowa Women's Study in postmenopausal women recognized that the catechin consumption from cacao decreased the incidence of rectal cancer [30]. Conversely, studies with humans who could not prove cocoa consumption have reported a beneficial effect in reducing the rate of occurrence of cancer. Also, adverse effects about cocoa consumption on the incidence of different cancers have been reported [31 - 34].

Coffea arabica, which common name is coffee, is consumed worldwide as a beverage, with an annual consumption of 1.5-8 kg in industrialized countries [35]. Among the bioactive compounds in coffee with antitumor properties is caffeine that is the main alkaloid; the diterpenes kahweal and cafestol have been attributed anticancer properties because they induce detoxification of carcinogens by induction of phase II enzymes, inhibiting the phase I enzymes that activate carcinogens and stimulate antioxidant defense mechanisms; also polyphenols are considered anticarcinogenic, which provide color and coffee flavor [36 - 39]. Finally, caffeic acid besides inhibiting DNA methylation in human cancer cell cultures also inhibited the inflammatory response, stress, cell cycle regulation and apoptosis, pathways involved in tumorigenic processes [36]. In North America,

Europe and Asia-Pacific epidemiological studies related to coffee consumption have tried to associate it or reject it as a risk factor in the development of different types of cancer. Among the most outstanding results, people regardless of sex, who consume 1-6 cups of coffee a day may reduce bladder, breast, oral and pharyngeal, colorectal, endometrial, esophageal, hepatocellular, pancreatic and prostate cancers and leukemia. However, no apparent association was observed with stomach, lung, non-melanoma, ovarian and kidney cancers [41].

Averrhoa carambola L. The star fruit (Oxalidaceae) also known as carambola is rich in antioxidants (vitamin C, catechines and epicatechines), which act against ROS [42]. The ripe star fruit is used to treat headache, vomiting, coughing, hangovers, eczemas, and has digestive and biliousness properties. Regarding the study of its anticancer properties, an ethanol extract of star fruit (25 mg/kg of administrated orally for 5 days) was tested in a liver cancer induced by the hepatocarcinogen diethylnitrosamine (DENA) in a mouse model. There was a significant reduction in the LPO levels elicited by carambola and an increase in GSH, SOD, and catalase protein levels. The authors suggested a protective role of the extract in the structural integrity of hepatocyte cell membranes and a preventive role in the progression of DENA-induced hepatic carcinogenesis [43].

Cocos nucifera (coconut) is a member of the family Arecaceae and is part of the daily diet of many people. It is different from any other fruit because, when immature, it contains a large quantity of "water" and may be harvested for drinking. In this stage, there are known as tender-nuts or jelly-nuts. When mature, it still contains some water and can be used as seednuts or processed to give oil from the kernel. *Cn*-AMP2 is an anionic host defense peptide (AHDP) found in green coconut water [44]. This peptide inactivates human glioma cell lines 1321N1 and U87MG. It was found that increasing concentrations up to 2.0 mM of the *Cn*-AMP2 peptide led to the progressive decrease in the viability of cell lines (approximately 70%). In contrast, this peptide at the same concentration range achieved only reductions in the viability of SVGp12 cell line (non-cancerous glial cells) at approximately 25% [45].

Psidium guajava (guava). Although the guava is native from America, it has been cultivated in the different tropics and subtropics of the world in Europe, Africa,

South Asia, and Oceania. Among its bioactive compounds oleanolic acid, ursolic acid, glucuronic acid and arjunolic acid, morin-3-O-a-L-lyxopyranoside and morin-3-O-a-L-arabinopyranoside; pentane-2-thiol; and flavonoids guaijavarin and quercetin have been found [46, 47]. Anti-proliferative and apoptotic activity was observed upon administration of a concentration of 1.5 mg/mL of acetone fruit extract for 24 and 48 h in NB4 promyelocitic leukaemia cells [48].

Litchi chinensis (lychee). The lychee is native to Southern China, Taiwan, Bangladesh, and Southeast Asia, but it is cultivated in semitropical areas. Chemical analysis on mature and pre-ripe lychee pericarp extracts, also known as LFPs, identified flavonoids and phenolic compounds with important antioxidant activity when exposed to fat oxidation [49]. Fresh pericarp extract contains polymeric proanthocyanidins (condensed tannins), epicatechin and procyanidin A2 [50]. Regarding to its biological activity, a concentration of 320 g/mL reached maximum cell growth inhibition (> 80%) in human breast cancer cells (MCF-7 and MDA-MB-231) and an inhibition rate growth of 40.7% in invasive ductal breast carcinoma [51].

Mangifera indica (mango). Mango is originated to South Asia although it is considered the national fruit of India and commonly cultivated in different subtropical and tropical regions. The fruit is used to prepare different beverages, meals and desserts. It contains natural dyes such as carotenes (α- and β-carotene) and xanthophylls (lutein), antioxidant compounds such as flavonoids (quercetin, kaempferol and catechins) as well as caffeic and gallic acids; furthermore, branches, leaves and green skin of the fruit contains a rich resin, mangiferina, mangiferol and mangiferic acid that can cause contact dermatitis or blisters in some people [52, 53]. The chemical analysis performed on extracts from the pulp of different varieties of mango such as Kent, Ataulfo, Tommy Atkins, and Haden from Mexico, and Francis from Haiti showed that the Ataulfo variety contains the highest amount of total phenols (56.7 ± 0.3 mg GAE/100 g). Values were normalized to gallic acid equivalents (GAE) [54]. Mango exhibited antineoplastic transformation activity in approximately 50% relative to the positive controls using 0.01% of whole mango juice, whereas the addition of 0.1% whole mango juice reduced the number of foci in approximately 70%. Interestingly, the ability of the cell mango nectar inhibition was observed in the late stage of carcinogenesis using an *in vitro* model, possibly in the promotion of carcinogenesis stage. This evidence suggests that compounds on the fruit do not only have antioxidant activity but also may have anticancer functions [55]. The lupeol on mango pulp extract (0.18 mg/100 g of mango pulp) possess protective effects against testosterone induced alterations in mouse prostate (1.3-1.4 fold compared to testosterone treated group). In an *in vitro* experimental model with LNCaP cells it exhibited antiproliferative effect by ROS mediated apoptosis in a dose-dependent (10–125 μM) as well as in a time-dependent (24, 48, 72 h) manner [56]. About mango varieties, Ataulfo and Haden cultivars had antiproliferative effects in Molt-4 leukemia, A-549 lung, MDA-MB-231 breast, LnCap prostate, and SW-480 colon cancer cells, where the SW-480 colon cancer cell line seemed to be the most susceptible to the treatment of 42 mg GAE/L [54]. Besides, in recent studies of essential oils β-pinene (40.7%) and terpinolene (28.3%) from the pulp of *M. indica* var. Rosa tumor cell growth inhibition was identified in different cancer cell lines [57].

Opuntia ficus-indica. Cactus pear is a popular fruit largely consumed in the North America, Central America and Mediterranean basin. Betalains are water-soluble nitrogenous vacuolar pigments present in flowers and fruits of cactus pear [58]. The antioxidant properties of betalains from cactus pear have been demonstrated [59]. The main dietary betacyanin found in cactus pear is betanin. This compound has been shown to inhibit the growth of stomach, breast, CNS, colon and lung cancer cells, and induce apoptosis in K562, a human myeloid leukemia precedence cell line [60 - 62].

Bertholletia excelsa (Brazil nut). The Brazil nut is native to the Guianas, Venezuela, Brazil, Eastern Colombia, Peru and Bolivia. Nutritionally, Brazil nuts are rich in thiamin, vitamin E, magnesium, phosphorus, manganese, zinc and it is considered among the main sources of selenium for human consumption because 28g or about six nuts provide 774% of daily value [63]. The anti-tumoral properties have been associated with selenium content, which restores the selenoproteins in selenium-depleted animals. Laboratory assays observed a lower tumor development (approximately 50%) in a mammary tumor rat model supplemented with a Brazil nut diet (12.5–18.7%) [64].

Carica papaya (papaya). Papaya is a tropical fruit native to America, but has been cultivated in Africa and Asia. The lycopene, a carotenoid of papaya, is endowed with powerful anticancer properties and is now potentially considered to be of importance for cancer prevention. Experimental studies of antiproliferative activity on liver cancer cell lines reported cell death in the liver cancer line Hep G2 using pure lycopene and papaya juice. The half maximal inhibitory concentration (IC_{50}) of pure lycopene and papaya juice were 22.8 µg/mL and 20 mg/mL, respectively [65]. The antioxidants from papaya flesh extracts such as β-carotene, polyphenols, and flavonoids contributed in the inhibition of breast cancer cell line MCF-7 proliferation. In the same cell model, the extract at concentrations of 2 and 4% produced 30 and 53% inhibition of cell proliferation, respectively [66]. Another study demonstrated that higher sources of polyphenols and flavonoids obtained from an ethanolic extract of papaya fruit administered to an *in vitro* model (MCF-7) showed protective effects against proliferation induced by nitric oxide [67].

Ananas comosus (pineapple). The pineapple is a tropical fruit indigenous to South America. In traditional medicine it is known to enhance digestion and it has been used to treat inflammatory disorders. Pineapple contains the bromelain enzyme obtained from the stem (80%) and fruit (10%). The anti-cancer activity is attributed by observations on traditional medicine in Southeast Asia and data obtained in animal and cell models [68]. However, apparently the inhibitory effect of bromelain in the growth of cancer cell lines and modification of their genetic profile has been only observed in gastric carcinoma cells but not in glioma cells [69, 70].

Anacardium occidentale (cashew tree). The cashew tree is native to the Amazon rainforest of Brazil, and was unveiled by Portuguese explorers. Currently it is grown and marketed by Brazil, Vietnam, India and several African countries. The cashew nut and its apple contain anacardic acids and cardols with antioxidant activity that is associated with a lowered incidence of some forms of cancer and CVD. However, this nut exhibited moderate cytotoxicity in a murine model of B16-F10 melanoma cells [71].

Effects of Fruit Consumption in Neurodegenerative Diseases

Degenerative diseases are initially characterized by causing progressive neurological deterioration accompanied by a decrease of personal functionality and independence. They are classified according to their predominant clinical characteristics in different groups: dementia syndromes (Alzheimer's disease); disorders of movement and posture (Parkinson's disease); progressive ataxia (olivopontocerebellar atrophy); weakness and muscle atrophy (amyotrophic lateral sclerosis, ALS and sclerosis). The etiology of neurodegenerative diseases is still unknown. In Parkinson's disease clinical manifestations are due to a loss of nigrostriatal dopaminergic neurons. It is believed to be multifactorial, linked to aging, genetic and environmental factors [72]. Alzheimer's dementia (AD), Parkinson's disease (PD) and stroke (CVA) are the most frequent causes of physical and mental disability in elderly people.

These diseases cause great personal and family impact, as they have consequences in the workplace and in social relationships. The progression of clinical symptoms leads the patient to incapacity and a different role in the family. Therefore, we must continue the research on the pathophysiology of neurodegenerative diseases and, based on these findings, develop and introduce new therapies to help prevent these diseases by introducing dietary bioactive compounds that provide neuroprotection.

Using different therapeutic strategies or natural neuroprotection produces relative preservation of structure and/or function of neurons according to the pathology or insult. It constitutes a strategy aimed at preventing the ultimate result of different process of neuronal damage [72].

Even though it is not clear whether a diet rich in vegetables reduces the risk of neurodegenerative diseases such as AD and PD in humans, recent studies in animal models that resemble part of the pathophysiology of these diseases suggest that bioactive compounds present in some fruits can be neuroprotective [73, 74]. On the other hand, in a prospective study that lasted an average of 6.3 years with 1,836 elderly Japanese Americans it was found that the regular consumption of fruit and vegetable juices was associated with a lower risk of developing these diseases [75].

Dementia and Alzheimer's Disease

Among the most abundant antioxidants in the diet are ascorbic acid, vitamin E, carotenoids, and polyphenols. Polyphenols are abundant micronutrients in our diet and exhibit antioxidant properties. These compounds are biosynthesized by plants (fruit, leaves, stems, roots, seeds or other parts) and account for most of the antioxidant activity exhibited by fruits, vegetables and certain herbal teas and natural drinks usually consumed by the population [76, 77]. It has been reported that intake of flavonoids in the diet reduces cognitive impairment and risk of dementia [77, 78].

According to the latest numbers from the US Food and Drug Administration (FDA), the Western countries consume 300 mg of caffeine a day. The main sources of the compound are coffee, tea, chocolate. The major active components of coffee are similar molecules belonging to the family of methylxanthines, which main mechanism of action is blocking adenosine. Adenosine and adenosine triphosphate is obtained after hydrolyzing its three phosphates. Adenosine is distributed throughout our body. It is considered as a hormone outside the nervous system and in the central nervous system, as a neurotransmitter that helps to reduce neural activity. The effect of methylxanthines would block this soothing effect and exert a stimulatory/excitatory effect as a non-selective A1 and A2A adenosine receptor antagonist. It acts as a brain stimulant most likely by increasing cerebral energy metabolism, cortical activity, and extracellular levels of acetylcholine, thus promoting increased alertness and arousal. Roughly, this would be the explanation of why coffee helps to fight sleep [80]. Additionally it is known that caffeine is a calcium channel inhibitor and phosphodiesterases, but its main effect is not considered because of these inhibitions. Caffeine is able to cross the blood-brain barrier (BBB), so its "brain" levels will be significant in normal people who consume coffee. The effect of caffeine on cognitive ability is also known: is moderate under normal conditions, but it is significant when the capacities are decreased due to illness, and neurodegenerative disease [81 - 83]. Long-term caffeine or A1 receptor antagonist administration (up to 2 weeks) to normal rodents has not been shown to provide cognitive benefits [84, 85], although no truly long-term caffeine administration studies have been done in rodents that might relate to habitual caffeine use in humans.

Epidemiological studies showed that AD patients consumed less caffeine during the 20 years preceding diagnosis of AD compared with individuals of the same age without AD, indicating that caffeine intake was associated with a significantly lower risk developing AD [86].

An additional study in APPsw transgenic mice showed that long-term caffeine administration protected against certain cognitive impairment, while also reduced brain βA levels. That study demonstrated that the effects of caffeine do not involve brain adenosine receptor, but rather a caffeine-induced suppression of βA production due to reduced expression of both Presenilin 1 (PS1) and β-secretase (BACE) [87]. All these data indicate that moderate amounts of caffeine consumed daily have a neuroprotective effect for neurodegenerative diseases such as AD.

While all polyphenols exhibit antioxidant properties, it has been established that some of these compounds also exhibit, among others, anti-inflammatory, anti-aggregating (platelets), anti-bacterial, estrogenic activity and modulating the activity of numerous enzymes, including of certain digestive enzymes. Also the capacity to protect neurons from death induced by several injuries and stimulate memory, learning and cognitive functions. These effects are related to two common processes. First, as detailed earlier, flavonoids interact with signaling pathways involving kinases that lead to the inhibition of neuronal apoptosis induced by neurotoxins. Also, increasing evidence has noted that the consumption of foods rich in polyphenols, such as cocoa (in the form of black/dark chocolate), green tea (drinks/juices containing it) or red wine (moderate) bring effects that would be potentially favorable for conservation and/or normalization of physiological parameters relevant or indicative of mental health. It has been suggested that cocoa flavanols might be beneficial in conditions with reduced cerebral blood flow, involving vascular dementia, AD and stroke [88 - 90].

Flavonoids found in high concentrations in cocoa, and therefore in chocolate, are called flavanols. Cocoa flavanols are presented in two structural forms, as single entities or monomers, or oligomeric structures (polymers). Among the most important flavanols monomers found in cocoa and their products are the (−)-epicatechin and (+)-catechin, and among the polymeric products, the procyanidins have been shown in different experimental models the ability to inhibit LDL oxidation and to increase the production of NO by stimulating the activity of the endothelial nitric oxide synthase in endothelial tissue, which effect results in a vasodilator and vascular relaxation and improves cerebrovascular perfusion [90, 91]. Furthermore the consumption of antioxidants from other plant sources appears to have no association with the risk to avoid the development of AD [92]. However, several studies have reported a decreased cerebral blood flow in patients with dementia [92]. Cerebrovascular injury is also known to lead to mild cognitive impairment and subsequently to vascular dementia (VD). It is therefore conceivable that beneficial properties of cocoa flavanols may slow down the transition from mild cognitive impairment to AD [93, 94]. Commenges *et al.* [95] showed that intake of antioxidant flavonoids is inversely related to the risk of dementia. However, in this study the flavonoids came mainly from fruits, vegetables, wine and tea instead of cocoa.

Cocos nucifera **L.** is a tree grown to provide a large number of products, although it is mainly grown for its nutritional values and regional consumption. Coconut oil, derived from coconut fruit has been recognized to contain high levels of saturated fats. Unlike most other dietary fats high in long chain fatty acids, coconut oil comprises medium chain fatty acids (MCFA). MCFA are unique in that they are easily absorbed and metabolized by the liver and can be converted into ketones. Ketones are an important source of alternative energy in the brain, and can be beneficial for people with cognitive impairment, such as AD. Coconut is classified as a highly nutritious "functional food". It is rich in dietary fiber, vitamins and minerals; however, in particular, there is increasing evidence to support the concept that coconut may be beneficial in the treatment of obesity, dyslipidemia, high LDL, insulin resistance and hypertension; these are risk factors for cardiovascular disease and type 2 diabetes, and for AD. Moreover, phenolic compounds and hormones (cytokinins) found in coconut may aid in preventing the aggregation of beta-amyloid (βA) peptide, *i.e.* to potentially inhibit a key step in the pathogenesis of AD [96].

Mangifera indica or mango, a plant from the family of Anacardiaceae, has been used in traditional folklore for a long term. The medicinal properties of mango appear to be varied depending on the parts of mango tree and its cultivar. According to the traditional folklore, ripe mango fruit is believed to be invigorating and freshening. Its juice is reputed for restorative tonic and antiheat stroke [97]. In addition, mango is also regarded as a valuable dietary source of many phytochemical compounds which provide health benefit for the nervous system [98]. Recently, ethanolic extract of mango fruit is reported to improve age-related memory deficit and memory deficit induced by scopolamine [99]. Based on what is known about the cholinergic dysfunction and oxidative stress in memory impairment and the protective effect of the ethanol extract of mango fruit, the effect of ethanolic extract of *Mangifera indica* (var. Nam Dok Mai) has shown to improve memory deficit, cholinergic dysfunction, and oxidative stress in an animal model of mild cognitive impairment (MCI). A recent study showed that different doses of the extract improved memory with decreased MDA level and increased SOD and GSH-Px activities of enzymes. It also preserved the density of cholinergic neurons in CA1 and CA3 of the hippocampus in rats treated with the extract at doses of 50 and 200 mg/kg body weight. Based on these results it is suggested that *M. indica* is a neuroprotective cholinergic agent and reduced oxidative stress which in turn produces improved memory [100].

Parkinson's Disease

Parkinson's disease (PD) is the second most common neurodegenerative disease affecting approximately 1% of the adult population. It is considered to be a disease of the motor system and its diagnosis is based on the presence of a set of motor signs (rigidity, bradykinesia, rest tremor) that are consequences of the death of dopaminergic neurons in the substantia nigra pars compacta, although it has been observed that non-dopaminergic degeneration also occurs in other areas of the brain that appear to be responsible for alterations in olfactory, emotional and memory functions that precede the classical motor symptoms in PD. Epidemiologic and preclinical data suggest that caffeine may confer neuroprotection against the degeneration of dopaminergic neurons and influence the onset and progression of PD. The data suggest that caffeine can improve motor symptoms in PD and antagonists of A2A adenosine receptors as istradefylline reduces dyskinesia associated with standard therapies of dopamine replacement. Finally, recent experimental results have indicated caffeine as a potential aid in the management of non-motor symptoms of PD, which do not improve the current potential with dopaminergic drugs [101].

The mechanisms by which caffeine produces motor activation mainly depend on its ability to release the braking presynaptic and postsynaptic that adenosine imposed on striatal dopaminergic neurotransmission by acting on different heteromeric adenosine receptors. However, dopamine depletion reduces the effect on motor activity of caffeine in PD. On the other hand, several prospective epidemiologic studies have shown that the incidence of PD declines with increasing intake of caffeine [102]. Caffeine and other A2A receptor antagonists are neuroprotective against dopaminergic neuronal loss in the MPTP mouse model of PD [103] and can attenuate βA neurotoxicity *in vitro* [104].

In PD, the abnormal action of adenosine neuromodulator fails to suppress unwanted motor activity in the basal ganglia *via* striatopallidal neurons, and has been associated with impaired motor function [105]. Adenosine antagonists, including caffeine and chocolate, have been considered good candidates to improve parkinsonian motor dysfunction. Parkinson's patients report an increase in chocolate consumption, independent of concomitant depressive symptoms [106]. However, in a crossover double-blind controlled trial in 26 patients with moderate PD —assessed by Unified Parkinson's Disease Rating Scale motor score— [107] it was observed that a single dose of black chocolate failed to improve motor function over white flavanol-free chocolate. The poor outcome might have been due to the dose of chocolate, its flavanol content, the duration of the treatment, since it has been shown that chronic exposure to caffeine differently modified motor effects dependent of the blockade of A1 receptors. These studies provide strong evidence that caffeine may represent a promising therapeutic tool in PD, thus being the first compound to restore both motor and non-motor early symptoms of PD together with its neuroprotective potential.

Stroke

Brain stroke is the major cause of death worldwide. Each year more people die from stroke than from any other cause. It often results in irreversible brain damage and subsequent loss of neuronal function. It is known that neuronal loss is generated by an ischemic process, in wich neural cells and microglia are activated and release a number of cytotoxic ROS or products including H_2O_2, NO and the synthesis of interleukins (IL-1, IL-6), tumor necrosis factor alpha (TNF-α) and interferon gamma (ITF-γ) [108].

Some studies have shown that flavonoid intake reduces neuronal loss and improves function after stroke. A meta-analysis with a sample of 114,009 participants reported a 29% reduction in the risk of stroke in high consumers of chocolate compared with low consumers [109]. In another study, an inverse association between chocolate consumption and stroke was observed and was even stronger than that for myocardial infarction [110]. A recent human study examined the relationship between consumption of total antioxidants (including fruits, vegetables, tea, coffee and chocolate) and risk of stroke in women of the Swedish Mammography Cohort. This study included 31,035 of women with no history of cardiovascular disease (CVD) and 5,680 women with a history of CVD. The authors reported that the total antioxidant capacity diet was inversely associated with stroke in women-free CVD (risk reduction 17%) and hemorrhagic stroke in women with a history of cardiovascular disease (risk reduction of 45%) [111]. Furthermore, mice pretreated orally with 5, 15 or 30 mg/kg of epicatechin (from cacao) 90 min before occlusion of the middle cerebral artery (MCAO) had less volume lesion and improved neurological scores compared to the control group. Mice were post-treated with 30 mg/kg of epicatechin 3.5 h after MCAO also had infarct volumes significantly smaller and better scores in neurological tests.

Other Degenerative Diseases: Diabetes, Dyslipidemia and Other Complications

Diabetes mellitus (DM) can be classified as type 1 (DM1), type 2 (DM2), gestational (DG) and other specific types of diabetes mellitus (DM3). Diabetes mellitus type 2 (DM2) is the most common form of diabetes, including more than 90% of all diabetics. DM2 is frequently associated with reduced sensitivity to insulin in target tissues that is identified as insulin resistance [112].

DM2 is one of the major health problems worldwide and its consequences are ranked as one of the most common causes of morbidity and mortality. The International Diabetes Federation estimates that 366 million people around the world have DM. It is expected that by 2030 these figures will increase to 552 million, equivalent to approximately 14 million new cases each year. DM2 is a complex disease where a global metabolic disorder that includes carbohydrates, lipids and proteins coexist. It is a multisystemic organic syndrome with different geno- and phenotypic characteristics with a genetic predisposition and defects in secretion and insulin action. The main feature of the DM is hyperglycemia, which results from defects in insulin secretion, cellular resistance to the action thereof or both, generated by an inflammatory alteration of the β-cells of the pancreas or resistance to insulin action in different tissues [113]. Microvascular (retinopathy and nephropathy) and macrovascular (atherosclerotic) disorders are the leading causes of morbidity and mortality in diabetic patients. Therefore, emphasis on diabetes care and management is on optimal blood glucose control to avert these adverse outcomes.

Recent clinical studies have shown that diabetes is associated with a state of chronic low-grade inflammation [114, 115]. Inflammatory cytokines and fatty acids produce insulin resistance. Activated intracellular signals in response to inflammation inhibit signaling pathways of insulin by binding to its active receptor in the cascade "downstream" to MAP kinases; these same kinases can be activated independently of insulin by external agents, such as microbial products, or internal agents, such as cytokines. When activated by pathways independent of insulin, these protein kinases generate the same resistance [116 - 120].

Dietary interventions to reduce complications of the DM2 have focused on diets rich in fruits and vegetables that provide natural sources of bioactive compounds and have shown beneficial effects on body weight and blood glucose control [121, 122]. On the other hand, a series of phyto and natural products with antiglucemic traditional medicine or insulin-sensitizing properties have been reported, and some of them have been used in the clinic [114, 122, 123]. Medicinal plants are important alternatives to conventional medicine for developing countries, especially in poor communities living in rural areas and lack access to health services [124, 125]. A large number of native medicinal plants of the tropics have traditionally high value for indigenous communities not only for their healing properties, but also because of other uses [126, 127].

Antidiabetic and Hypocholesterolemic Activities

It has been shown that regular consumption of avocado can help to maintain normal total cholesterol [128]. Clinical trials have consistently showed avocado effects on blood lipids in a variety of diets in healthy individuals with hypercholesterolemia and DM2 subjects [129 - 133]. In subjects with hypercholesterolemia, who had an enriched avocado diet enhance blood lipid profile by reducing LDL-cholesterol and triglycerides and increasing HDL cholesterol compared to high-carbohydrate diets or diets without avocado. In patients with DM2 diabetes, beneficial effects on blood lipids were observed with a diet rich in monounsaturated fats derived from avocado compared to a low-fat diet rich in complex carbohydrates [130].

Annona muricata, *A. glabra* and *A. cherimola* have been reported as hypoglycemic species [134]. *A. macroprophyllata* would be an interesting species to study since it may have the same metabolic pathways that synthesize the compounds as the other members of the genus and thus may show some therapeutic properties.

The **Opuntia** genus is also used by the Pima Indians for the treatment of diabetes and hyperlipidemia. Previous results suggest that daily consumption of the fruit, known in Mexico as *tuna* (prickly pear), would be beneficial to cardiovascular health through a reduction in total cholesterol and LDL-C [135], while the intake of the cactus pad could potentially improve blood glucose levels [136 - 139].

Mangifera indica L. contains a variety of bioactive compounds that include carotenoids, tocopherols, ascorbic acid, dietary fiber, and the phenolic compounds mangiferin, gallic acid, and quercetin [140]. The results of a recent study show that treatment with the lyophilized pulp has beneficial effects on body composition and improved glucose and lipid profile in mice fed with a high-fat diet [141]. In human studies, the consumption of mango pulp, compared with other fruits, produced a better response of postprandial glucose and insulin in people with DM2 [142, 143]. Besides pulp, other parts of the plant also have hypoglycemic properties. For example, the extract of the stem bark and foliage with a high concentration of mangiferin reduced effectively blood glucose in diabetic rats and streptozotocin-induced hyperglycemia induced by glucose in rodent models [144, 145].

Effects on Cardiovascular System and Blood

Diabetic complications, including damage to large and small blood vessels, can lead to coronary heart disease, stroke and hypertension, the latter being a well-established risk factor for stroke that contributes to worsening renal disease (WRD). Therefore, reduction of blood pressure (BP) is an effective way to prevent or slow the progression of WRD. It is therefore conventional renoprotection — which is accomplished through the reduction in BP with antihypertensive regimens and diet changes — important [146]. Antihypertensive therapy in diabetes not only improves the quality of life, but also reduces renal complications [147].

CONCLUDING REMARKS

The opening of the world market to new tastes and textures of tropical fruits and vegetables makes the characterization of these "new fruits" necessary in terms of their phytochemical constitution, nutritional values, levels of bioactivity and food security based mainly in the traditional uses of each plant. Phytochemical composition analysis shows that tropical fruits are similar in its constitution as are temperate fruits. A great value of their antioxidant composition has been identified. If most of the bioactive compounds present in fruits this identified, are could help to develop a new category defined by the content of nutrients.

Phytochemical studies usually report high activity of the extracts or fractions of fruits and vegetables as total extracts instead of pure compounds, suggesting a complementary and synergy of different activities. In several studies, tropical fruits are promoted for their health benefits, especially for their antioxidant properties in cardiovascular protection and diseases derived from it, nervous system diseases, and mainly their anti-cancer properties, which put fruits with a high added value. Clinical studies should be conducted to support experimental work on chronic diseases.

CONFLICT OF INTEREST

The authors confirm that they have no conflict of interest to declare for this publication.

ACKNOWLEDGEMENTS

We would like to thank the CYTED programme (Action 112RT0460. "CORNOCUPIA" thematic network) for the support to the accomplishment of this work.

REFERENCES

[1]   Kalra EK. Nutraceutical--definition and introduction. AAPS PharmSci 2003; 5(3): E25.
        [http://dx.doi.org/10.1208/ps050325] [PMID: 14621960]

[2]   de Souza Sant'Ana A. Special issue on exotic fruits. Food Res Int 2011; 44: 1657.
        [http://dx.doi.org/10.1016/j.foodres.2011.06.017]

[3]   FAOSTAT. Food and Agriculture Organization of the United Nations. Statistics Division. Available from: http://faostat.fao.org/ 2015 May 25;

[4]   Pierson JT, Dietzgen RG, Shaw PN, Roberts-Thomson SJ, Monteith GR, Gidley MJ. Major Australian tropical fruits biodiversity: bioactive compounds and their bioactivities. Mol Nutr Food Res 2012; 56(3): 357-87.
        [http://dx.doi.org/10.1002/mnfr.201100441] [PMID: 22147637]

[5]   Scartezzini P, Speroni E. Review on some plants of Indian traditional medicine with antioxidant activity. J Ethnopharmacol 2000; 71(1-2): 23-43.
        [http://dx.doi.org/10.1016/S0378-8741(00)00213-0] [PMID: 10904144]

[6]   Rahman S, Hasnat A, Hasan C, Rashid M, Ilias M. Pharmacological evaluation of Bangladeshi medicinal plants—a review. Pharm Biol 2001; 39: 1-6.
        [http://dx.doi.org/10.1076/phbi.39.1.1.5939]

[7]   Valadeau C, Pabon A, Deharo E, *et al.* Medicinal plants from the Yanesha (Peru): evaluation of the leishmanicidal and antimalarial activity of selected extracts. J Ethnopharmacol 2009; 123(3): 413-22.
        [http://dx.doi.org/10.1016/j.jep.2009.03.041] [PMID: 19514108]

[8]   Blomhoff R. Dietary antioxidants and cardiovascular disease. Curr Opin Lipidol 2005; 16(1): 47-54.
        [http://dx.doi.org/10.1097/00041433-200502000-00009] [PMID: 15650563]

[9]   Manach C, Williamson G, Morand C, Scalbert A, Rémésy C. Bioavailability and bioefficacy of polyphenols in humans. I. Review of 97 bioavailability studies. Am J Clin Nutr 2005; 81(1) (Suppl.): 230S-42S.
        [PMID: 15640486]

[10]   Williamson G, Manach C. Bioavailability and bioefficacy of polyphenols in humans. II. Review of 93 intervention studies. Am J Clin Nutr 2005; 81(1) (Suppl.): 243S-55S.

[PMID: 15640487]

[11]   Lindsay DG, Astley SB. European research on the functional effects of dietary antioxidants - EUROFEDA. Mol Aspects Med 2002; 23(1-3): 1-38.
[http://dx.doi.org/10.1016/S0098-2997(02)00005-5] [PMID: 12079769]

[12]   World Health Organization. International Agency for Research on Cancer. GLOBOCAN 2012: Estimated Cancer Incidence, Mortality and Prevalence Worldwide in 2012 Available from: http://globocan.iarc.fr/Default.aspx 2012 May 21;

[13]   Kelloff GJ, Lippman SM, Dannenberg AJ, *et al.* Progress in chemoprevention drug development: the promise of molecular biomarkers for prevention of intraepithelial neoplasia and cancer--a plan to move forward. Clin Cancer Res 2006; 12(12): 3661-97.
[http://dx.doi.org/10.1158/1078-0432.CCR-06-1104] [PMID: 16778094]

[14]   Ramos S. Cancer chemoprevention and chemotherapy: dietary polyphenols and signalling pathways. Mol Nutr Food Res 2008; 52(5): 507-26.
[http://dx.doi.org/10.1002/mnfr.200700326] [PMID: 18435439]

[15]   Surh YJ. Cancer chemoprevention with dietary phytochemicals. Nat Rev Cancer 2003; 3(10): 768-80.
[http://dx.doi.org/10.1038/nrc1189] [PMID: 14570043]

[16]   Lu QY, Arteaga JR, Zhang Q, Huerta S, Go VL, Heber D. Inhibition of prostate cancer cell growth by an avocado extract: role of lipid-soluble bioactive substances. J Nutr Biochem 2005; 16(1): 23-30.
[http://dx.doi.org/10.1016/j.jnutbio.2004.08.003] [PMID: 15629237]

[17]   D'Ambrosio SM, Han C, Pan L, Kinghorn AD, Ding H. Aliphatic acetogenin constituents of avocado fruits inhibit human oral cancer cell proliferation by targeting the EGFR/RAS/RAF/MEK/ERK1/2 pathway. Biochem Biophys Res Commun 2011; 409(3): 465-9.
[http://dx.doi.org/10.1016/j.bbrc.2011.05.027] [PMID: 21596018]

[18]   Vahedi Larijani L, Ghasemi M, Abedian Kenari S, Naghshvar F. Evaluating the effect of four extracts of avocado fruit on esophageal squamous carcinoma and colon adenocarcinoma cell lines in comparison with peripheral blood mononuclear cells. Acta Med Iran 2014; 52(3): 201-5.
[PMID: 24901722]

[19]   Sun S, Liu J, Kadouh H, Sun X, Zhou K. Three new anti-proliferative Annonaceous acetogenins with mono-tetrahydrofuran ring from graviola fruit (*Annona muricata*). Bioorg Med Chem Lett 2014; 24(12): 2773-6.
[http://dx.doi.org/10.1016/j.bmcl.2014.03.099] [PMID: 24780120]

[20]   Hurst WJ, Tarka SM Jr, Powis TG, Valdez F Jr, Hester TR. Cacao usage by the earliest Maya civilization. Nature 2002; 418(6895): 289-90.
[http://dx.doi.org/10.1038/418289a] [PMID: 12124611]

[21]   Rössner S. Chocolate--divine food, fattening junk or nutritious supplementation? Eur J Clin Nutr 1997; 51(6): 341-5.
[http://dx.doi.org/10.1038/sj.ejcn.1600409] [PMID: 9192189]

[22]   Lee KW, Kundu JK, Kim SO, Chun KS, Lee HJ, Surh YJ. Cocoa polyphenols inhibit phorbol ester-induced superoxide anion formation in cultured HL-60 cells and expression of cyclooxygenase-2 and activation of NF-kappaB and MAPKs in mouse skin *in vivo*. J Nutr 2006; 136(5): 1150-5.

[PMID: 16614396]

[23]   Natsume M, Osakabe N, Yamagishi M, *et al.* Analyses of polyphenols in cacao liquor, cocoa, and chocolate by normal-phase and reversed-phase HPLC. Biosci Biotechnol Biochem 2000; 64(12): 2581-7.
[http://dx.doi.org/10.1271/bbb.64.2581] [PMID: 11210120]

[24]   Yamagishi M, Natsume M, Osakabe N, *et al.* Chemoprevention of lung carcinogenesis by cacao liquor proanthocyanidins in a male rat multi-organ carcinogenesis model. Cancer Lett 2003; 191(1): 49-57.
[http://dx.doi.org/10.1016/S0304-3835(02)00629-8] [PMID: 12609709]

[25]   Granado-Serrano AB, Martín MA, Bravo L, Goya L, Ramos S. A diet rich in cocoa attenuates N-nitrosodiethylamine-induced liver injury in rats. Food Chem Toxicol 2009; 47(10): 2499-506.
[http://dx.doi.org/10.1016/j.fct.2009.07.007] [PMID: 19602430]

[26]   Weyant MJ, Carothers AM, Dannenberg AJ, Bertagnolli MM. (+)-Catechin inhibits intestinal tumor formation and suppresses focal adhesion kinase activation in the min/+ mouse. Cancer Res 2001; 61(1): 118-25.
[PMID: 11196148]

[27]   Rodríguez-Ramiro I, Ramos S, López-Oliva E, *et al.* Cocoa-rich diet prevents azoxymethane-induced colonic preneoplastic lesions in rats by restraining oxidative stress and cell proliferation and inducing apoptosis. Mol Nutr Food Res 2011; 55(12): 1895-9.
[http://dx.doi.org/10.1002/mnfr.201100363] [PMID: 21953728]

[28]   Bayard V, Chamorro F, Motta J, Hollenberg NK. Does flavanol intake influence mortality from nitric oxide-dependent processes? Ischemic heart disease, stroke, diabetes mellitus, and cancer in Panama. Int J Med Sci 2007; 4(1): 53-8.
[http://dx.doi.org/10.7150/ijms.4.53] [PMID: 17299579]

[29]   Garcia-Closas R, Gonzalez CA, Agudo A, Riboli E. Intake of specific carotenoids and flavonoids and the risk of gastric cancer in Spain. Cancer Causes Control 1999; 10(1): 71-5.
[http://dx.doi.org/10.1023/A:1008867108960] [PMID: 10334645]

[30]   Arts IC, Jacobs DR Jr, Gross M, Harnack LJ, Folsom AR. Dietary catechins and cancer incidence among postmenopausal women: the Iowa Women's Health Study (United States). Cancer Causes Control 2002; 13(4): 373-82.
[http://dx.doi.org/10.1023/A:1015290131096] [PMID: 12074507]

[31]   McKelvey W, Greenland S, Sandler RS. A second look at the relation between colorectal adenomas and consumption of foods containing partially hydrogenated oils. Epidemiology 2000; 11(4): 469-73.
[http://dx.doi.org/10.1097/00001648-200007000-00018] [PMID: 10874557]

[32]   Peterson J, Lagiou P, Samoli E, *et al.* Flavonoid intake and breast cancer risk: a case--control study in Greece. Br J Cancer 2003; 89(7): 1255-9.
[http://dx.doi.org/10.1038/sj.bjc.6601271] [PMID: 14520456]

[33]   Boutron-Ruault MC, Senesse P, Faivre J, Chatelain N, Belghiti C, Méance S. Foods as risk factors for colorectal cancer: a case-control study in Burgundy (France). Eur J Cancer Prev 1999; 8(3): 229-35.
[http://dx.doi.org/10.1097/00008469-199906000-00011] [PMID: 10443952]

[34]   Giannandrea F. Correlation analysis of cocoa consumption data with worldwide incidence rates of

testicular cancer and hypospadias. Int J Environ Res Public Health 2009; 6(2): 568-78.
[http://dx.doi.org/10.3390/ijerph6020578] [PMID: 19440400]

[35]   United Nations Conference on Trade and Development (UNCTAD). Beverages. Coffee. Market. Coffee production over the period 1852-2011 in millions of bags. 2015 May 21. Available from: http://www.unctad.info/en/Infocomm/Beverages/Coffee-French-version-only/Market/ 2015.

[36]   Rosendahl AH, Perks CM, Zeng L, *et al.* Caffeine and caffeic acid inhibit growth and modify estrogen receptor and insulin-like growth factor I receptor levels in human breast cancer. Clin Cancer Res 2015; 21(8): 1877-87.
[http://dx.doi.org/10.1158/1078-0432.CCR-14-1748] [PMID: 25691730]

[37]   Huber WW, Scharf G, Nagel G, Prustomersky S, Schulte-Hermann R, Kaina B. Coffee and its chemopreventive components Kahweol and Cafestol increase the activity of O6-methylguanine-DNA methyltransferase in rat liver--comparison with phase II xenobiotic metabolism. Mutat Res 2003; 522(1-2): 57-68.
[http://dx.doi.org/10.1016/S0027-5107(02)00264-6] [PMID: 12517412]

[38]   Cavin C, Marin-Kuan M, Langouët S, *et al.* Induction of Nrf2-mediated cellular defenses and alteration of phase I activities as mechanisms of chemoprotective effects of coffee in the liver. Food Chem Toxicol 2008; 46(4): 1239-48.
[http://dx.doi.org/10.1016/j.fct.2007.09.099] [PMID: 17976884]

[39]   Oleaga C, Ciudad CJ, Noé V, Izquierdo-Pulido M. Coffee polyphenols change the expression of STAT5B and ATF-2 modifying cyclin D1 levels in cancer cells. Oxid Med Cell Longev 2012; 2012: 390385.
[http://dx.doi.org/10.1155/2012/390385] [PMID: 22919439]

[40]   Feng R, Lu Y, Bowman LL, Qian Y, Castranova V, Ding M. Inhibition of activator protein-1, NF-kappaB, and MAPKs and induction of phase 2 detoxifying enzyme activity by chlorogenic acid. J Biol Chem 2005; 280(30): 27888-95.
[http://dx.doi.org/10.1074/jbc.M503347200] [PMID: 15944151]

[41]   Yu X, Bao Z, Zou J, Dong J. Coffee consumption and risk of cancers: a meta-analysis of cohort studies. BMC Cancer 2011; 11: 96.
[http://dx.doi.org/10.1186/1471-2407-11-96] [PMID: 21406107]

[42]   Shui G, Leong LP. Analysis of polyphenolic antioxidants in star fruit using liquid chromatography and mass spectrometry. J Chromatogr A 2004; 1022(1-2): 67-75.
[http://dx.doi.org/10.1016/j.chroma.2003.09.055] [PMID: 14753772]

[43]   Singh R, Sharma J, Goyal PK. Prophylactic role of *Averrhoa carambola* (star fruit) extract against chemically induced hepatocellular carcinoma in Swiss albino mice. Adv Pharmacol S 2014 2014. 158936

[44]   Mandal SM, Dey S, Mandal M, Sarkar S, Maria-Neto S, Franco OL. Identification and structural insights of three novel antimicrobial peptides isolated from green coconut water. Peptides 2009; 30(4): 633-7.
[http://dx.doi.org/10.1016/j.peptides.2008.12.001] [PMID: 19111587]

[45]   Prabhu S, Dennison SR, Mura M, Lea RW, Snape TJ, Harris F. *Cn*-AMP2 from green coconut water is an anionic anticancer peptide. J Pept Sci 2014; 20(12): 909-15.

[http://dx.doi.org/10.1002/psc.2684] [PMID: 25234689]

[46]   Paniandy JC, Chane MJ, Pieribattesti JC. Chemical composition of the essential oil and headspace solid-phase microextraction of the guava fruit (Psidium guajava L.). J Essent Oil Res 2000; 12: 153-8.
        [http://dx.doi.org/10.1080/10412905.2000.9699486]

[47]   Hwang JS, Yen YP, Chang MC, Liu CY. Extraction and identification of volatile components of guava fruits and their attraction to Oriental fruit fly, *Bactrocera dorsalis* (Hendel). Plant Protection Bull (Taipei) 2002; 44: 279-302.

[48]   Bontempo P, Doto A, Miceli M, *et al. Psidium guajava* L. anti-neoplastic effects: induction of apoptosis and cell differentiation. Cell Prolif 2012; 45(1): 22-31.
        [http://dx.doi.org/10.1111/j.1365-2184.2011.00797.x] [PMID: 22172154]

[49]   Sarni-Manchado P, Le Roux E, Le Guernevé C, Lozano Y, Cheynier V. Phenolic composition of litchi fruit pericarp. J Agric Food Chem 2000; 48(12): 5995-6002.
        [http://dx.doi.org/10.1021/jf000815r] [PMID: 11312772]

[50]   Zheng G, Yi Z, Zhang J, Zhong D. Studies on the antioxidative effect of extract from mature and premature litchi pericarp. Nat Prod Res Dev 2003; 15: 341-4.

[51]   Wang X, Yuan S, Wang J, *et al.* Anticancer activity of litchi fruit pericarp extract against human breast cancer *in vitro* and *in vivo.* Toxicol Appl Pharmacol 2006; 215(2): 168-78.
        [http://dx.doi.org/10.1016/j.taap.2006.02.004] [PMID: 16563451]

[52]   Schieber A, Ullrich W, Carle R. Characterization of polyphenols in mango puree concentrate by HPLC with diode array and mass spectrometric detection. Innov Food Sci Emerg Technol 2000; 1: 161-6.
        [http://dx.doi.org/10.1016/S1466-8564(00)00015-1]

[53]   Schieber A, Berardini N, Carle R. Identification of flavonol and xanthone glycosides from mango (*Mangifera indica* L. Cv. "Tommy Atkins") peels by high-performance liquid chromatography-electrospray ionization mass spectrometry. J Agric Food Chem 2003; 51(17): 5006-11.
        [http://dx.doi.org/10.1021/jf030218f] [PMID: 12903961]

[54]   Noratto GD, Bertoldi MC, Krenek K, Talcott ST, Stringheta PC, Mertens-Talcott SU. Anticarcinogenic effects of polyphenolics from mango (*Mangifera indica*) varieties. J Agric Food Chem 2010; 58(7): 4104-12.
        [http://dx.doi.org/10.1021/jf903161g] [PMID: 20205391]

[55]   Percival SS, Talcott ST, Chin ST, Mallak AC, Lounds-Singleton A, Pettit-Moore J. Neoplastic transformation of BALB/3T3 cells and cell cycle of HL-60 cells are inhibited by mango (*Mangifera indica* L.) juice and mango juice extracts. J Nutr 2006; 136(5): 1300-4.
        [PMID: 16614420]

[56]   Prasad S, Kalra N, Shukla Y. Induction of apoptosis by lupeol and mango extract in mouse prostate and LNCaP cells. Nutr Cancer 2008; 60(1): 120-30.
        [http://dx.doi.org/10.1080/01635580701613772] [PMID: 18444143]

[57]   Ramos EH, Moraes MM, Nerys LL, *et al.* Chemical composition, leishmanicidal and cytotoxic activities of the essential oils from *Mangifera indica* L. var. Rosa and Espada Biomed Res Int 2014; 2014: 734946.
        [http://dx.doi.org/10.1155/2014/734946] [PMID: 25136617]

[58]   Stintzing FC, Schieber A, Carle R. Phytochemical and nutritional significance of cactus pear. Eur Food Res Technol 2001; 212: 396-407.
[http://dx.doi.org/10.1007/s002170000219]

[59]   Gentile C, Tesoriere L, Allegra M, Livrea MA, D'Alessio P. Antioxidant betalains from cactus pear (*Opuntia ficus-indica*) inhibit endothelial ICAM-1 expression. Ann N Y Acad Sci 2004; 1028: 481-6.
[http://dx.doi.org/10.1196/annals.1322.057] [PMID: 15650274]

[60]   Reddy MK, Alexander-Lindo RL, Nair MG. Relative inhibition of lipid peroxidation, cyclooxygenase enzymes, and human tumor cell proliferation by natural food colors. J Agric Food Chem 2005; 53(23): 9268-73.
[http://dx.doi.org/10.1021/jf051399j] [PMID: 16277432]

[61]   Sreekanth D, Arunasree MK, Roy KR, Chandramohan Reddy T, Reddy GV, Reddanna P. Betanin a betacyanin pigment purified from fruits of *Opuntia ficus-indica* induces apoptosis in human chronic myeloid leukemia Cell line-K562. Phytomedicine 2007; 14(11): 739-46.
[http://dx.doi.org/10.1016/j.phymed.2007.03.017] [PMID: 17482444]

[62]   Keller J, Camaré C, Bernis C, *et al.* Antiatherogenic and antitumoral properties of *Opuntia* cladodes: inhibition of low density lipoprotein oxidation by vascular cells, and protection against the cytotoxicity of lipid oxidation product 4-hydroxynonenal in a colorectal cancer cellular model. J Physiol Biochem 2015; 71(3): 577-87.
[http://dx.doi.org/10.1007/s13105-015-0408-x] [PMID: 25840808]

[63]   Moodley R, Kindness A, Jonnalagadda SB. Elemental composition and chemical characteristics of five edible nuts (almond, Brazil, pecan, macadamia and walnut) consumed in Southern Africa. J Environ Sci Health B 2007; 42(5): 585-91.
[http://dx.doi.org/10.1080/03601230701391591] [PMID: 17562467]

[64]   Ip C, Lisk DJ. Bioactivity of selenium from Brazil nut for cancer prevention and selenoenzyme maintenance. Nutr Cancer 1994; 21(3): 203-12.
[http://dx.doi.org/10.1080/01635589409514319] [PMID: 8072875]

[65]   Rahmat A, Rosli R, Zain WN, Endrini S, Sani HA. Antiproliferative activity of pure lycopene compared to both extracted lycopene and juices from watermelon (*Citrullus vulgaris*) and papaya (*Carica papaya*) on human breast and liver cancer cell lines. J Med Sci 2002; 2: 55-8.
[http://dx.doi.org/10.3923/jms.2002.55.58]

[66]   García-Solís P, Yahia EM, Morales-Tlalpan V, Díaz-Muñoz M. Screening of antiproliferative effect of aqueous extracts of plant foods consumed in México on the breast cancer cell line MCF-7. Int J Food Sci Nutr 2009; 60 (Suppl. 6): 32-46.
[http://dx.doi.org/10.1080/09637480802312922] [PMID: 19468947]

[67]   Jayakumar R, Kanthimathi MS. Inhibitory effects of fruit extracts on nitric oxide-induced proliferation in MCF-7 cells. Food Chem 2011; 126: 956-60.
[http://dx.doi.org/10.1016/j.foodchem.2010.11.093]

[68]   Chobotova K, Vernallis AB, Majid FA. Bromelain's activity and potential as an anti-cancer agent: Current evidence and perspectives. Cancer Lett 2010; 290(2): 148-56.
[http://dx.doi.org/10.1016/j.canlet.2009.08.001] [PMID: 19700238]

[69]   Taussig SJ, Szekerczes J, Batkin S. Inhibition of tumour growth *in vitro* by bromelain, an extract of the pineapple plant (*Ananas comosus*). Planta Med 1985; 51(6): 538-9.
[http://dx.doi.org/10.1055/s-2007-969596] [PMID: 17345291]

[70]   Tysnes BB, Maurer HR, Porwol T, Probst B, Bjerkvig R, Hoover F. Bromelain reversibly inhibits invasive properties of glioma cells. Neoplasia 2001; 3(6): 469-79.
[http://dx.doi.org/10.1038/sj.neo.7900196] [PMID: 11774029]

[71]   Kubo I, Nitoda T, Tocoli FE, Green IR. Multifunctional cytotoxic agents from *Anacardium occidentale*. Phytother Res 2011; 25(1): 38-45.
[http://dx.doi.org/10.1002/ptr.3109] [PMID: 20623613]

[72]   Hernandez R. Demography of ageing. Bold 1992; 2(4): 8-12.
[PMID: 12288978]

[73]   Joseph JA, Denisova NA, Arendash G, *et al.* Blueberry supplementation enhances signaling and prevents behavioral deficits in an Alzheimer disease model. Nutr Neurosci 2003; 6(3): 153-62.
[http://dx.doi.org/10.1080/1028415031000111282] [PMID: 12793519]

[74]   Suganuma H, Hirano T, Arimoto Y, Inakuma T. Effect of tomato intake on striatal monoamine level in a mouse model of experimental Parkinson's disease. J Nutr Sci Vitaminol (Tokyo) 2002; 48(3): 251-4.
[http://dx.doi.org/10.3177/jnsv.48.251] [PMID: 12350086]

[75]   Dai Q, Borenstein AR, Wu Y, Jackson JC, Larson EB. Fruit and vegetable juices and Alzheimer's disease: the Kame Project. Am J Med 2006; 119(9): 751-9.
[http://dx.doi.org/10.1016/j.amjmed.2006.03.045] [PMID: 16945610]

[76]   Manach C, Scalbert A, Morand C, Rémésy C, Jiménez L. Polyphenols: food sources and bioavailability. Am J Clin Nutr 2004; 79(5): 727-47.
[PMID: 15113710]

[77]   Ramassamy C. Emerging role of polyphenolic compounds in the treatment of neurodegenerative diseases: a review of their intracellular targets. Eur J Pharmacol 2006; 545(1): 51-64.
[http://dx.doi.org/10.1016/j.ejphar.2006.06.025] [PMID: 16904103]

[78]   Letenneur L, Proust-Lima C, Le Gouge A, Dartigues JF, Barberger-Gateau P. Flavonoid intake and cognitive decline over a 10-year period. Am J Epidemiol 2007; 165(12): 1364-71.
[http://dx.doi.org/10.1093/aje/kwm036] [PMID: 17369607]

[79]   de Mendonça A, Sebastião AM, Ribeiro JA. Adenosine: does it have a neuroprotective role after all? Brain Res Brain Res Rev 2000; 33(2-3): 258-74.
[http://dx.doi.org/10.1016/S0165-0173(00)00033-3] [PMID: 11011069]

[80]   Fredholm BB, Bättig K, Holmén J, Nehlig A, Zvartau EE. Actions of caffeine in the brain with special reference to factors that contribute to its widespread use. Pharmacol Rev 1999; 51(1): 83-133.
[PMID: 10049999]

[81]   Jarvis MJ. Does caffeine intake enhance absolute levels of cognitive performance? Psychopharmacology (Berl) 1993; 110(1-2): 45-52.
[http://dx.doi.org/10.1007/BF02246949] [PMID: 7870897]

[82]   Hameleers PA, Van Boxtel MP, Hogervorst E, *et al.* Habitual caffeine consumption and its relation to memory, attention, planning capacity and psychomotor performance across multiple age groups. Hum

Psychopharmacol 2000; 15(8): 573-81.
[http://dx.doi.org/10.1002/hup.218] [PMID: 12404609]

[83] van Boxtel MP, Schmitt JA, Bosma H, Jolles J. The effects of habitual caffeine use on cognitive change: a longitudinal perspective. Pharmacol Biochem Behav 2003; 75(4): 921-7.
[http://dx.doi.org/10.1016/S0091-3057(03)00171-0] [PMID: 12957237]

[84] Von Lubitz DK, Paul IA, Bartus RT, Jacobson KA. Effects of chronic administration of adenosine A1 receptor agonist and antagonist on spatial learning and memory. Eur J Pharmacol 1993; 249(3): 271-80.
[http://dx.doi.org/10.1016/0014-2999(93)90522-J] [PMID: 8287914]

[85] Molinengo L, Scordo I, Pastorello B. Action of caffeine, L-PIA and their combination on memory retention in the rat. Life Sci 1994; 54(17): 1247-50.
[http://dx.doi.org/10.1016/0024-3205(94)00851-5] [PMID: 8164506]

[86] Maia L, de Mendonça A. Does caffeine intake protect from Alzheimer's disease? Eur J Neurol 2002; 9(4): 377-82.
[http://dx.doi.org/10.1046/j.1468-1331.2002.00421.x] [PMID: 12099922]

[87] Arendash GW, Lewis J, Leighty RE, et al. Multi-metric behavioral comparison of APPsw and P301L models for Alzheimer's disease: linkage of poorer cognitive performance to tau pathology in forebrain. Brain Res 2004; 1012(1-2): 29-41.
[http://dx.doi.org/10.1016/j.brainres.2004.02.081] [PMID: 15158158]

[88] Ding EL, Hutfless SM, Ding X, Girotra S. Chocolate and prevention of cardiovascular disease: a systematic review. Nutr Metab (Lond) 2006; 3: 2.
[http://dx.doi.org/10.1186/1743-7075-3-2] [PMID: 16390538]

[89] Francis ST, Head K, Morris PG, Macdonald IA. The effect of flavanol-rich cocoa on the fMRI response to a cognitive task in healthy young people. J Cardiovasc Pharmacol 2006; 47 (Suppl. 2): S215-20.
[http://dx.doi.org/10.1097/00005344-200606001-00018] [PMID: 16794461]

[90] Fisher ND, Sorond FA, Hollenberg NK. Cocoa flavanols and brain perfusion. J Cardiovasc Pharmacol 2006; 47 (Suppl. 2): S210-4.
[http://dx.doi.org/10.1097/00005344-200606001-00017] [PMID: 16794460]

[91] Patel AK, Rogers JT, Huang X. Flavanols, mild cognitive impairment, and Alzheimer's dementia. Int J Clin Exp Med 2008; 1(2): 181-91.
[PMID: 19079672]

[92] Luchsinger JA, Mayeux R. Dietary factors and Alzheimer's disease. Lancet Neurol 2004; 3(10): 579-87.
[http://dx.doi.org/10.1016/S1474-4422(04)00878-6] [PMID: 15380154]

[93] Ruitenberg A, den Heijer T, Bakker SL, et al. Cerebral hypoperfusion and clinical onset of dementia: the Rotterdam Study. Ann Neurol 2005; 57(6): 789-94.
[http://dx.doi.org/10.1002/ana.20493] [PMID: 15929050]

[94] Nagahama Y, Nabatame H, Okina T, et al. Cerebral correlates of the progression rate of the cognitive decline in probable Alzheimer's disease. Eur Neurol 2003; 50(1): 1-9.

[http://dx.doi.org/10.1159/000070851] [PMID: 12824705]

[95] Commenges D, Scotet V, Renaud S, Jacqmin-Gadda H, Barberger-Gateau P, Dartigues JF. Intake of flavonoids and risk of dementia. Eur J Epidemiol 2000; 16(4): 357-63.
[http://dx.doi.org/10.1023/A:1007614613771] [PMID: 10959944]

[96] Fernando WM, Martins IJ, Goozee KG, Brennan CS, Jayasena V, Martins RN. The role of dietary coconut for the prevention and treatment of Alzheimer's disease: potential mechanisms of action. Br J Nutr 2015; 114(1): 1-14.
[http://dx.doi.org/10.1017/S0007114515001452] [PMID: 25997382]

[97] Shah KA, Patel MB, Patel RJ, Parmar PK. *Mangifera indica* (mango). Pharmacogn Rev 2010; 4(7): 42-8.
[http://dx.doi.org/10.4103/0973-7847.65325] [PMID: 22228940]

[98] Berardini N, Fezer R, Conrad J, Beifuss U, Carle R, Schieber A. Screening of mango (*Mangifera indica* L.) cultivars for their contents of flavonol O- and xanthone C-glycosides, anthocyanins, and pectin. J Agric Food Chem 2005; 53(5): 1563-70.
[http://dx.doi.org/10.1021/jf0484069] [PMID: 15740041]

[99] Kumar S, Maheshwari KK, Singh V. Effects of *Mangifera indica* fruit extract on cognitive deficits in mice. J Environ Biol 2009; 30(4): 563-6.
[PMID: 20120497]

[100] Wattanathorn J, Muchimapura S, Thukham-Mee W, Ingkaninan K, Wittaya-Areekul S. *Mangifera indica* fruit extract improves memory impairment, cholinergic dysfunction, and oxidative stress damage in animal model of mild cognitive impairment. Oxid Med Cell Longev 2014; 2014: 132097.

[101] Prediger RD. Effects of caffeine in Parkinson's disease: from neuroprotection to the management of motor and non-motor symptoms. J Alzheimers Dis 2010; 20 (Suppl. 1): S205-20.
[PMID: 20182024]

[102] Ross GW, Abbott RD, Petrovitch H, *et al.* Association of coffee and caffeine intake with the risk of Parkinson disease. JAMA 2000; 283(20): 2674-9.
[http://dx.doi.org/10.1001/jama.283.20.2674] [PMID: 10819950]

[103] Schwarzschild MA, Xu K, Oztas E, *et al.* Neuroprotection by caffeine and more specific A2A receptor antagonists in animal models of Parkinson's disease. Neurology 2003; 61(11) (Suppl. 6): S55-61.
[http://dx.doi.org/10.1212/01.WNL.0000095214.53646.72] [PMID: 14663012]

[104] Dall'Igna OP, Porciúncula LO, Souza DO, Cunha RA, Lara DR. Neuroprotection by caffeine and adenosine A2A receptor blockade of beta-amyloid neurotoxicity. Br J Pharmacol 2003; 138(7): 1207-9.
[http://dx.doi.org/10.1038/sj.bjp.0705185] [PMID: 12711619]

[105] Jankovic J. Are adenosine antagonists, such as istradefylline, caffeine, and chocolate, useful in the treatment of Parkinson's disease? Ann Neurol 2008; 63(3): 267-9.
[http://dx.doi.org/10.1002/ana.21348] [PMID: 18383071]

[106] Wolz M, Kaminsky A, Löhle M, Koch R, Storch A, Reichmann H. Chocolate consumption is increased in Parkinson's disease. Results from a self-questionnaire study. J Neurol 2009; 256(3): 488-92.

[http://dx.doi.org/10.1007/s00415-009-0118-9] [PMID: 19277767]

[107] Wolz M, Schleiffer C, Klingelhöfer L, *et al.* Comparison of chocolate to cacao-free white chocolate in Parkinson's disease: a single-dose, investigator-blinded, placebo-controlled, crossover trial. J Neurol 2012; 259(11): 2447-51.
[http://dx.doi.org/10.1007/s00415-012-6527-1] [PMID: 22584952]

[108] Yaman M, Eser O, Cosar M, *et al.* Oral administration of avocado soybean unsaponifiables (ASU) reduces ischemic damage in the rat hippocampus. Arch Med Res 2007; 38(5): 489-94.
[http://dx.doi.org/10.1016/j.arcmed.2007.01.008] [PMID: 17560453]

[109] Buitrago-Lopez A, Sanderson J, Johnson L, *et al.* Chocolate consumption and cardiometabolic disorders: systematic review and meta-analysis. BMJ 2011; 343: d4488.
[http://dx.doi.org/10.1136/bmj.d4488] [PMID: 21875885]

[110] Buijsse B, Weikert C, Drogan D, Bergmann M, Boeing H. Chocolate consumption in relation to blood pressure and risk of cardiovascular disease in German adults. Eur Heart J 2010; 31(13): 1616-23.
[http://dx.doi.org/10.1093/eurheartj/ehq068] [PMID: 20354055]

[111] Rautiainen S, Larsson S, Virtamo J, Wolk A. Total antioxidant capacity of diet and risk of stroke: a population-based prospective cohort of women. Stroke 2012; 43(2): 335-40.
[http://dx.doi.org/10.1161/STROKEAHA.111.635557] [PMID: 22135074]

[112] Shah ZA, Li RC, Ahmad AS, *et al.* The flavanol (-)-epicatechin prevents stroke damage through the Nrf2/HO1 pathway. J Cereb Blood Flow Metab 2010; 30(12): 1951-61.
[http://dx.doi.org/10.1038/jcbfm.2010.53] [PMID: 20442725]

[113] ADA (American Diabetes Association). Diagnosis and classification of diabetes mellitus. Diabetes Care 2013; 36 (Suppl. 1): S67-74.
[http://dx.doi.org/10.2337/dc13-S067] [PMID: 23264425]

[114] Donath MY, Shoelson SE. Type 2 diabetes as an inflammatory disease. Nat Rev Immunol 2011; 11(2): 98-107.
[http://dx.doi.org/10.1038/nri2925] [PMID: 21233852]

[115] Xie W, Du L. Diabetes is an inflammatory disease: evidence from traditional Chinese medicines. Diabetes Obes Metab 2011; 13(4): 289-301.
[http://dx.doi.org/10.1111/j.1463-1326.2010.01336.x] [PMID: 21205111]

[116] Crook M. Type 2 diabetes mellitus: a disease of the innate immune system? An update. Diabet Med 2004; 21(3): 203-7.
[http://dx.doi.org/10.1046/j.1464-5491.2003.01030.x] [PMID: 15008827]

[117] Donath MY. Targeting inflammation in the treatment of type 2 diabetes. Diabetes Obes Metab 2013; 15 (Suppl. 3): 193-6.
[http://dx.doi.org/10.1111/dom.12172] [PMID: 24003937]

[118] Mahmoud F, Al-Ozairi E. Inflammatory cytokines and the risk of cardiovascular complications in type 2 diabetes. Dis Markers 2013; 35(4): 235-41.
[http://dx.doi.org/10.1155/2013/931915] [PMID: 24167372]

[119] Patel PS, Buras ED, Balasubramanyam A. The role of the immune system in obesity and insulin resistance. J Obes 2013; 2013: 616193.

[120] Gratas-Delamarche A, Derbré F, Vincent S, Cillard J. Physical inactivity, insulin resistance, and the oxidative-inflammatory loop. Free Radic Res 2014; 48(1): 93-108.
[http://dx.doi.org/10.3109/10715762.2013.847528] [PMID: 24060092]

[121] Prior RL, E Wilkes S, R Rogers T, Khanal RC, Wu X, Howard LR. Purified blueberry anthocyanins and blueberry juice alter development of obesity in mice fed an obesogenic high-fat diet. J Agric Food Chem 2010; 58(7): 3970-6.
[http://dx.doi.org/10.1021/jf902852d] [PMID: 20148514]

[122] Tsuda T, Horio F, Uchida K, Aoki H, Osawa T. Dietary cyanidin 3-O-beta-D-glucoside-rich purple corn color prevents obesity and ameliorates hyperglycemia in mice. J Nutr 2003; 133(7): 2125-30.
[PMID: 12840166]

[123] Yin J, Zhang H, Ye J. Traditional chinese medicine in treatment of metabolic syndrome. Endocr Metab Immune Disord Drug Targets 2008; 8(2): 99-111.
[http://dx.doi.org/10.2174/187153008784534330] [PMID: 18537696]

[124] Hung HY, Qian K, Morris-Natschke SL, Hsu CS, Lee KH. Recent discovery of plant-derived anti-diabetic natural products. Nat Prod Rep 2012; 29(5): 580-606.
[http://dx.doi.org/10.1039/c2np00074a] [PMID: 22491825]

[125] Schippmann U, Leaman DJ, Cunningham AB. Impact of cultivation and gathering of medicinal plants on biodiversity: global trends and issues. Biodiversity and the Ecosystem Approach in Agriculture, Forestry and Fisheries. Rome: FAO 2002; pp. 1-21.

[126] Muthu C, Ayyanar M, Raja N, Ignacimuthu S. Medicinal plants used by traditional healers in Kancheepuram district of Tamil Nadu, India. J Ethnobiol Ethnomed 2006; 2: 43-53.
[http://dx.doi.org/10.1186/1746-4269-2-43] [PMID: 17026769]

[127] ProFound Advisers in Development. Market Brief in the European Union for selected natural ingredients derived from native species. Crescentia cujete 2005.

[128] Grant WC. Influence of avocados on serum cholesterol. Proc Soc Exp Biol Med 1960; 104: 45-7.
[http://dx.doi.org/10.3181/00379727-104-25722] [PMID: 13828982]

[129] Colquhoun DM, Moores D, Somerset SM, Humphries JA. Comparison of the effects on lipoproteins and apolipoproteins of a diet high in monounsaturated fatty acids, enriched with avocado, and a high-carbohydrate diet. Am J Clin Nutr 1992; 56(4): 671-7.
[PMID: 1414966]

[130] Alvizouri-Muñoz M, Carranza-Madrigal J, Herrera-Abarca JE, Chávez-Carbajal F, Amezcua-Gastelum JL. Effects of avocado as a source of monounsaturated fatty acids on plasma lipid levels. Arch Med Res 1992; 23(4): 163-7.
[PMID: 1308699]

[131] Lerman-Garber I, Ichazo-Cerro S, Zamora-González J, Cardoso-Saldaña G, Posadas-Romero C. Effect of a high-monounsaturated fat diet enriched with avocado in NIDDM patients. Diabetes Care 1994; 17(4): 311-5.
[http://dx.doi.org/10.2337/diacare.17.4.311] [PMID: 8026287]

[132] Carranza J, Alvizouri M, Alvarado MR, Chávez F, Gómez M, Herrera JE. [Effects of avocado on the level of blood lipids in patients with phenotype II and IV dyslipidemias]. Arch Inst Cardiol Mex 1995;

65(4): 342-8. [Effects of avocado on the level of blood lipids in patients with phenotype II and IV dyslipidemias].
[PMID: 8561655]

[133] López Ledesma R, Frati Munari AC, Hernández Domínguez BC, *et al*. Monounsaturated fatty acid (avocado) rich diet for mild hypercholesterolemia. Arch Med Res 1996; 27(4): 519-23.
[PMID: 8987188]

[134] Brindis F, González-Trujano ME, González-Andrade M, Aguirre-Trujano E, Villalobos-Molina R. Aqueous extract of *Annona macroprophyllata:* a potential α-glucosidase inhibitor. Biomed Res Int 2013; 2013: 591313.
[http://dx.doi.org/10.1155/2013/591313] [PMID: 24298552]

[135] Budinsky A, Wolfram R, Oguogho A, Efthimiou Y, Stamatopoulos Y, Sinzinger H. Regular ingestion of *opuntia robusta* lowers oxidation injury. Prostaglandins Leukot Essent Fatty Acids 2001; 65(1): 45-50.
[http://dx.doi.org/10.1054/plef.2001.0287] [PMID: 11487308]

[136] Díaz-Medina EM, Martín-Herrera D, Rodríguez-Rodríguez EM, Díaz-Romero C. Chromium (III) in cactus pad and its possible role in the antihyperglycemic activity. J Funct Foods 2012; 4: 311-4.
[http://dx.doi.org/10.1016/j.jff.2011.12.009]

[137] Wu SY, Leske MC. Antioxidants and cataract formation: a summary review. Int Ophthalmol Clin 2000; 40(4): 71-81.
[http://dx.doi.org/10.1097/00004397-200010000-00006] [PMID: 11064858]

[138] Christen WG. Antioxidant vitamins and age-related eye disease. Proc Assoc Am Physicians 1999; 111(1): 16-21.
[http://dx.doi.org/10.1046/j.1525-1381.1999.09231.x] [PMID: 9893153]

[139] Bhuyan KC, Bhuyan DK. Molecular mechanism of cataractogenesis: III. Toxic metabolites of oxygen as initiators of lipid peroxidation and cataract. Curr Eye Res 1984; 3(1): 67-81.
[http://dx.doi.org/10.3109/02713688408997188] [PMID: 6317286]

[140] Intensive blood-glucose control with sulphonylureas or insulin compared with conventional treatment and risk of complications in patients with type 2 diabetes (UKPDS 33). Lancet 1998; 352(9131): 837-53.
[http://dx.doi.org/10.1016/S0140-6736(98)07019-6] [PMID: 9742976]

[141] Maschio G, Alberti D, Janin G, *et al*. Effect of the angiotensin-converting-enzyme inhibitor benazepril on the progression of chronic renal insufficiency. N Engl J Med 1996; 334(15): 939-45.
[http://dx.doi.org/10.1056/NEJM199604113341502] [PMID: 8596594]

[142] Bidani AK, Griffin KA, Bakris G, Picken MM. Lack of evidence of blood pressure-independent protection by renin-angiotensin system blockade after renal ablation. Kidney Int 2000; 57(4): 1651-61.
[http://dx.doi.org/10.1046/j.1523-1755.2000.00009.x] [PMID: 10760100]

[143] Martínez-Maldonado M. Hypertension in end-stage renal disease. Kidney Int Suppl 1998; 68: S67-72.
[http://dx.doi.org/10.1046/j.1523-1755.1998.06816.x] [PMID: 9839287]

[144] Mogensen CE. ACE inhibitors and antihypertensive treatment in diabetes: focus on microalbuminuria and macrovascular disease. J Renin Angiotensin Aldosterone Syst 2000; 1(3): 234-9.

[http://dx.doi.org/10.3317/jraas.2000.035] [PMID: 11881030]

[145] Estacio RO, Jeffers BW, Gifford N, Schrier RW. Effect of blood pressure control on diabetic microvascular complications in patients with hypertension and type 2 diabetes. Diabetes Care 2000; 23 (Suppl. 2): B54-64.
[PMID: 10860192]

[146] Wang C, Zhao X, Mao S, Wang Y, Cui X, Pu Y. Management of SAH with traditional Chinese medicine in China. Neurol Res 2006; 28(4): 436-44.
[http://dx.doi.org/10.1179/016164106X115044] [PMID: 16759447]

[147] Stengel B, Billon S, Van Dijk PC, *et al.* Trends in the incidence of renal replacement therapy for end-stage renal disease in Europe, 1990-1999. Nephrol Dial Transplant 2003; 18(9): 1824-33.
[http://dx.doi.org/10.1093/ndt/gfg233] [PMID: 12937231]

CHAPTER 8

Bioactive Compounds from Amazonian Fruits and their Antioxidant Properties

Renan C. Chisté[a,b,*], Eduarda Fernandes[a]

[a] *UCIBIO, REQUIMTE, Department of Chemical Sciences, Faculty of Pharmacy, University of Porto, Porto, Portugal*

[b] *Faculty of Food Engineering (FEA), Institute of Technology (ITEC), Federal University of Pará (UFPA), Belém, Pará, Brazil*

Abstract: An adequate intake of fruits has long been correlated to a lower occurrence of chronic degenerative diseases triggered by oxidative stress. These health benefits are extensively claimed in the literature to be owing to the scavenging capacity of some bioactive compounds, such as ascorbic acid, tocopherols, carotenoids and phenolic compounds, against the oxidizing effect of reactive oxygen and reactive nitrogen species with physiological relevance. The Amazon is the largest reserve of biodiversity in the world and is also the largest Brazilian biome, occupying almost half of Brazil (49%). In this way, Amazon hosts numerous fruit species, which are consumed by the local population and distributed through the local economy at trade markets, but most of them are unknown to the wider population. Analysis of bioactive compounds has an essential role in the study of biodiversity, and in the evaluation of food safety and nutritional properties. To contribute for a better knowledge of Amazonia biome, this chapter gathers scientific information concerning the phytochemical composition of some Amazonian fruits with potential biological properties.

Keywords: Amazonian biome, Ascorbic acid, *Astrocaryum aculeatum*, *Bactris gasipaes*, *Byrsonima crassifolia*, Carotenoids, *Caryocar villosum*, *Endopleura uchi*, *Eugenia stipitata*, *Euterpe oleracea*, *Mauritia flexuosa*, *Myrciaria dubia*, *Oenocarpus bacaba*, Oxidative stress, Phenolic compounds, Reactive nitrogen

---

* **Corresponding author Renan C. Chisté**: Faculty of Food Engineering (FEA), Institute of Technology (ITEC), Federal University of Pará (UFPA), Belém, Pará, Brazil; Email: rcchiste@ufpa.br.

species, Reactive oxygen species, *Solanum sessiliflorum, Theobroma grandiflorum.*

INTRODUCTION

The Amazon, characterised by the Amazon River basin, is an area predominantly covered by dense and moist tropical forest, with inclusions of several other vegetation types, such as floodplain forests, savannas, swamps, grasslands, palm forests and bamboos.

Fig. (1). Amazon biome and Amazon basin shared by Brazil, Peru, Bolivia, French Guiana, Suriname, Guyana, Venezuela, Colombia and Ecuador (modified from [1]).

The biome covers an area of 6.7 million km² and comprises nine countries, namely Brazil (which houses ≈ 60% of total area), Peru (≈ 13%), Bolivia, French Guiana, Suriname, Guyana, Venezuela, Colombia and Ecuador (Fig. **1**) [1]. The Amazon biome drains 20% of the world's fresh water and is the largest reserve of biodiversity in the world and is also the largest Brazilian biome (49%). It is dominated by a warm, humid climate, with an average temperature of 25 °C and torrential rains that are well distributed throughout the year [2]. Considering all these particular characteristics, the Amazon hosts a huge number of fruit species that are consumed by the local population and distributed through the local economy at trade fairs, but mostly unknown to the wider population [2].

The usual diet of fruits and vegetables provides, in addition to macro and micronutrients, some chemical compounds that may have potent biological activities. These compounds are called bioactive compounds (or sometimes phytochemicals) and may play many important roles in human health. The bioactive compounds from plants may be grouped into primary and secondary metabolites. Briefly, primary metabolites are necessary for the basic metabolism and growth in all plants, such as carbohydrates, lipids, and amino acids, whilst secondary metabolites (1-5% of the dry weight), such as phenolic compounds, carotenoids, tocopherols and ascorbic acid are not essential, but they are pivotal compounds for the wellbeing of plants by interacting with the ecosystems. In human health, dietary ingestion of bioactive compounds from fruits and vegetables has been linked to the immune system improvement and decrease of the risk of development of chronic degenerative diseases, such as macular degeneration, cataracts, cardiovascular diseases and also certain types of cancer [3]. In general, these observed health effects are attributed to the capacity of some bioactive compounds in inhibiting the oxidizing effects of reactive oxygen species (ROS) and reactive nitrogen species (RNS). Due to the high importance of their biological activities, secondary metabolites of plants have been applied for centuries in traditional medicine practices. Currently, they find applications in fine chemicals, cosmetics and more recently in functional food or nutraceutics [3].

Dietary antioxidants have a broad scope and may be defined as substances that significantly decrease the deleterious effects of ROS/RNS on the human normal physiological functions [4]. These ROS/RNS are generated as a consequence of naturally occurring processes (exercise and mitochondrial electron transport, as examples), environmental pollutants, changed atmospheric conditions (hypoxia), environmental stimuli (sun ionizing radiation) and lifestyle stressors (excessive alcohol consumption and cigarette smoking).

The consumption of fruits and vegetables is globally inadequate and should be encouraged, in association with a good scientific knowledge of its chemical composition and biological properties. In this sense, the analysis of bioactive compounds has an essential role in the evaluation of food safety and nutritional properties. Additionally, it corresponds to an essential ground in the study of biodiversity. To contribute for a better knowledge of Amazonia biome, this chapter gathers scientific information concerning the phytochemical composition of selected Amazonian fruits and its scavenging capacities against physiologically relevant ROS and RNS to provide an overview about the research of bioactive compounds with potential biological properties.

Amazonian Fruits

Açai

Açai fruits (*Euterpe oleracea* Mart.), from Arecaceae family, have received much attention in the recent years as ''superfruits'' and have been promoted as a dietary food supplement alleging to exhibit remarkable benefits to human health, such as the induction of weight loss, fighting cardiovascular diseases, improving digestion and delaying the ageing process [5, 6]. The *Euterpe* genus has about 28 species distributed throughout the Amazon basin, but among them, *E.oleracea*, *E. precatoria* and *E. edulis* occur most frequently. However, only *E. oleracea* and *E. precatoria* are commercially used for their fruits [6]. *E. oleracea* is a monoecious palm with a tall, multi-stemmed, slender stem that grows up to 25 m in height. The fruits form green bunches during unripe stages and become deep purple when fully ripe. Each fruit has one large seed (7–10 mm). The purple epicarp is a very thin hard layer, the mesocarp has 1–2 mm in thickness and the major part of the fruit is represented by the seeds (80–95%) [5, 7].

The health promoting properties of açai are claimed to be due to the phytochemical composition of its fruits (and its resulting products), and its widely reported antioxidant properties. The major bioactive compounds in açai are polyphenols, mostly anthocyanins and flavonoids. The content of anthocyanins in açai pulp is very high (282–303 mg/100 g of total anthocyanins) with cyanidin 3-rutinoside and cyanidin 3-glucoside accounting for 87% and 13%, respectively [8]. Additionally, (-)-epicatechin, (+)-catechin, rutin, orientin, homoorientin, isovitexin, scoparin, taxifolin deoxyhexose, apigenin, chrysoeriol, dihydro-kaempferol, velutin, luteolin diglycoside and procyanidin dimers were already reported in the literature to be present in açai pulp [5, 6]. The presence of ferulic acid, *p*-hydroxy benzoic acid, (-)-epicatechin, protocatechuic acid, gallic acid, ellagic acid, vanillic acid, (+)-catechin and *p*-coumaric acid at concentrations ranging from 17 to 212 mg/L was also reported [9]. Low levels of carotenoids (β-carotene, α-carotene and lutein) (0.5 mg/100 g) were detected in açai pulp [10]. However, acai pulp was considered a good source of tocopherols (α-, β-, γ-tocopherol), mainly α-tocopherol (394 μg/g pulp) [11].

Regarding the antioxidant capacity, in a study with 18 frozen fruit pulps commonly traded in Brazil, the frozen pulp of açai, which exhibited the highest total phenolic compounds contents (251 mg gallic acid equivalent/100 g) and total flavonoids (134 mg catechin equivalent/100 g), was considered the most efficient ROS scavenger, since it presented the highest capacity against peroxyl radicals (ROO$^{\bullet}$) (7498 μmol trolox equivalent/100 g) (ORAC assay) and also the highest efficiency against the oxidizing effects of hydroxyl radical (HO$^{\bullet}$) (IC$_{50}$ of 3 μg/mL), as generated by the Fenton reaction [12]. The frozen pulp of açai was also able to inhibit the *in vitro* deleterious effects of hydrogen peroxide (H$_2$O$_2$) (IC$_{50}$ of 259 μg/mL) with antioxidant capacity about 6 times superior than orange and pineapple fruit pulps, but almost 2 times less efficient than lemon [12].

An extract with very high contents of anthocyanin, which was derived from açai powder was investigated concerning the proliferation of glioma cells in brain of C6 rats and the breast cancer in human cell line (MDA-468) using the MTT-based assay [13]. These authors concluded that the anthocyanin-rich açai extract suppressed the proliferation of brain glioma cells of the C6 rats (IC$_{50}$ of 121 μg/mL), in a dose-dependent manner, but no effect was observed on the growth of breast cancer cells (MDA-468). In another study, the protective effects of açai pulp in swiss albino mice were evaluated by two different treatments, acute and subacute, when administered prior to doxorubicin, an anti-tumour agent which the efficiency of treatment decreases due to side effects limited by the dose (mainly cardiotoxicity) [10]. These authors reported a great efficiency of açai pulp in protecting liver and kidney cells against the DNA damage induced by doxorubicin for the subacute treatment, and that the phytochemicals present in açai pulp can explain these protective effects.

Camu-Camu

Camu-camu [*Myrciaria dubia* (Kunth) McVaugh], from Myrtaceae family, is a small shrub (1 to 3 m height) found in flooded or swampy areas of Amazonia biome. The globular fruits (1.0–3.2 cm diameter) have a shiny and thin skin, an extremely acidic pink juicy pulp surrounding 1 to 4 seeds. In general, the fruits are not consumed in its natural state due to its very sour taste; preferably, it is normally consumed as purees, juices and the pulp is usually directed to support the production of beverages and also powder as food additives [7, 14]. Camu-camu can be seen as a versatile berry, since its pulp, skin and seeds all they were reported to exhibit antioxidant potential [14].

Camu-camu fruits are known as the richest natural source of ascorbic acid in Brazil and its contents depends on the maturity stages: camu-camu fruits in the fully ripe stage (red stage) were shown to contain 2010 mg ascorbic acid/100 g pulp, whereas at full unripe stage it contained 2280 mg/100 g [15]. The content of ascorbic acid of camu-camu was reported to be 1.5 and 11 times superior to that values described for acerola (*Malpighia emarginata*) and cashew (*Anacardium occidentale*), respectively, two other Brazilian fruits considered as good sources of ascorbic acid [16].

In relation to other bioactive compounds, all-*trans*-lutein is the major carotenoid in camu-camu fruits, with values varying from 45% to 55% of the total carotenoid contents (160 to 602 µg/100 g), followed by all-*trans*-β-carotene, all-*trans*-violaxanthin and all-*trans*-luteoxanthin. Although camu-camu cannot be seen as a good source of provitamin A carotenoids, its lutein contents are comparable to other good lutein sources, such as leafy vegetables [17]. Regarding anthocyanins, the fresh peel of camu-camu was reported to be 30-54 mg anthocyanins/100 g, with cyanidin 3-glucoside as the major compound (89% of total anthocyanins), followed by delphinidin 3-glucoside (4-5%) [18]. Furthermore, other different kinds of flavonoids, such as flavanones, flavonols and flavanols were already identified in camu-camu fruits, namely (+)-catechin, quercetin, rutin, myricetin, kaempferol, naringenin, eriodictyol, as well as ellagic and galic acid derivatives, ellagitanins and proanthocyanidins and the total sum of these compounds ranged from 8 mg/100 g fresh weight (pulp) to 336 mg/100 g (seeds) [19, 20].

Concerning the antioxidant potential of camu-camu fruits, the powder obtained from the pulp exhibited higher scavenging capacity against ROO• (ORAC) (337 μmol trolox equivalent/g, dry matter) than strawberry (153 μmol trolox/g) and blackberry (204 μmol trolox equivalent/g) [20]. Inoue *et al.*, [21] suggested that the juice of camu-camu fruits may possess great antioxidant and anti-inflammatory potential, compared to tablets of ascorbic acid with equivalent vitamin C contents and these effects might be due to the presence of other bioactive compounds, besides ascorbic acid, such as anthocyanins and β-carotene.

Cupuaçu

Cupuaçu [*Theobroma grandiflorum* (Willd. ex Spreng.) K. Schum], from Sterculiaceae family, is a very popular fruit in city markets of Amazon region. The *Theobroma* genus is remarkable because it includes "cacao" (*Theobroma cacao* L.), an important commodity with high economic importance. Although *T. grandiflorum* and *T. cacao* are morphologically distinct, both the species can be hybridized to achieve plants with high productivity in terms of fruits. Thus, this is the reason for which *T. grandiflorum* is considered a very close relative of *T. cacao* [22]. The fruits of cupuaçu are oblong (12–25 cm long and 10–12 cm of diameter) and they weigh from 1 to 2 kg with 20-50 seeds surrounded by a mucilaginous pulp (45% of the fruit). The shell is hard, covered by a brown dust and 1–3 cm thick [7, 23].

Cupuaçu fruits are highly appreciated due to their strongly acidic and aromatic pulp, which is not typically consumed alone, but rather, in the preparation of juice, ice cream, liquors and jellies. Furthermore, *T. grandiflorum* seeds have received much attention due to their potential to be used as a substitute for chocolate production [22, 23].

Regarding the bioactive compounds of cupuaçu, the fresh pulps were reported to contain a high ascorbic acid content (17 mg ascorbic acid/100 g), but this content seems to decrease during pulp processing (11 mg/100 g in commercial frozen pulps) [24]. In agreement, the ascorbic acid content in frozen pulp of cupuaçu, commercialized in São Paulo State (Brazil) for juice preparation, was about 1 mg/100 g [12]. Concerning the phenolic compounds profile, some flavonoids were already identified in cupuaçu seeds, namely quercetin, quercetin 3-O-β-D-glucuronide 6″-methyl ester, quercetin 3-O-β-D-glucuronide, kaempferol, isoscutellarein 8-O-β-D-glucuronide 6″-methyl ester, isoscutellarein 8-O-β-D -glucuronide, hypolaetin 8-O-β-glucuronide, (−)-epicatechin and (+)-catechin, along with theograndins I and II, two sulphated flavonoid glycosides [22, 24]. No data related to carotenoid composition and tocopherol contents were found for cupuaçu fruits.

The fresh seeds of cupuaçu have 4-5 times more phenolic and 6- 8 times proanthocyanidins contents than that found in fresh pulps, and, therefore, fresh seeds were shown to present higher scavenging capacity against ROO•(ORAC) (568 μmol trolox equivalent/g, dry weight) than fresh pulp (217 μmol trolox equivalent/g) and commercial frozen pulps (109 μmol trolox equivalent/g) [24]. Among 18 commercial frozen fruit pulps from Brazil, cupuaçu frozen pulp was more efficient in inhibiting the oxidizing effects of H_2O_2 (IC_{50} of 700 μg/mL) and HO• (IC_{50} of 37 μg/mL) than commercial frozen pulps of cacao, mango, orange, pineapple and watermelon [12].

Piquiá

Piquiá [*Caryocar villosum* (Aubl.) Pers.] is a large tree with 40−50 m high that belongs to Caryocaraceae family and its pulp is usually cooked in regional dishes with rice. The consumption of piquiá is very similar to Pequi (*Caryocar brasiliense* Camb.), which is another fruit commonly found in the Northeast and Midwest regions of Brazil and that belongs to the same Caryocaraceae family [25]. The endocarp normally involves one seed, but sometimes up to four seeds may be found. Some uses to its edible oil were mentioned as substitute for butter and also for cosmetic applications [25, 26].

Recent studies have investigated the bioactive composition of piquiá [25] and its ability to scavenge ROS and RNS [27]. Gallic and ellagic acids were the major phenolic compounds detected in piquiá pulp, while all-*trans*-antheraxanthin and all-*trans*-zeaxanthin were the main carotenoids [25]. Moreover, some new galloyl and ellagic acid derivatives were also identified in *C. villosum* and *C. glabrum* stem barks [28]. Concerning other phytochemicals, triterpenoid saponins were extracted and isolated from the stem bark, pulp and peel of *C. villosum* (methanol extracts) and their cytotoxic, lipolytic and antimicrobial activities, as well as their tyrosinase inhibitory potential were reported [29, 30]. In relation to the tocopherol derivatives, the only compound which was detected in the pulp of piquiá was α-tocopherol (1.2 mg/100 g, dry basis) [31].

The antioxidant capacity of piquiá pulp was demonstrated by Chisté and Mercadante [25] through the scavenging capacity against $ROO^{\bullet}$ (ORAC) (3.7 mmol trolox equivalent/100 g), which was higher than the average value (2.7 mmol trolox equivalent/100 g) reported for 41 fruits consumed in the United States, but lower than açai pulp (*Euterpe oleracea* Mart) (99 mmol trolox equivalent/100 g) [32]. In another study, five freeze-dried extracts of piquiá were obtained after solid-liquid extraction with distinct solvents and the *in vitro* scavenging capacities against H_2O_2, superoxide anion radical ($O_2^{\bullet-}$), singlet oxygen ($^1O_2$), hypochlorous acid (HOCl), peroxynitrite ($ONOO^-$) and nitric oxide ($^{\bullet}NO$) were determined [27]. According to these authors, the extracts obtained with water and ethanol/water showed the highest phenolic compounds yields and exhibited the highest ROS/RNS scavenging activity than the less polar extracts, highlighting that the antioxidant capacity of the extracts of *C. villosum* was highly associated to the phenolic compounds contents [27].

The ingestion of piquiá pulp significantly decreased the doxorubicin-induced DNA damage in kidney, bone narrow, liver and heart cells of rats [31]. These same authors concluded that the administration for 14 days (by gavage) with pulp doses of 75 mg/kg (body weight) suppressed the damage in DNA by 56% in kidney, 81% in liver and 43% in heart cells.

Murici

Murici (or muruci) [*Byrsonima crassifolia* (L.) Kunth], belongs to the Malpighiaceae family, and may be characterized as a fruit with a cheese-like aroma, which is normally consumed *in natura* or as juice, ice cream, liquor and jelly [33]. The ripe murici fruits (1–2 cm diameter) have yellow colour, soft pulp and the local habitants use the fruits, seeds, leaves and barks in the treatment of snake bites and also for gynaecological, gastrointestinal and skin inflammations [34].

In a study carried out with eighteen tropical fruits, murici fruit (without the peel) exhibited appreciable contents of ascorbic acid (0.4 mg/100 g), total carotenoids (1.2 mg/100 g), total flavonoids (319 mg catechin equivalents/100 g) and total phenolic compounds (385 mg gallic acid equivalents/100 g) [35]. Regarding the phenolic compounds profile, several compounds were identified in murici fruits, such as quercetin derivatives, proanthocyanidins, quinic acid gallates, gallotannins [36], kaempferol, taxifolin, rutin, (−)-epicatechin and (+)-catechin [33]. Murici fruit was characterized as a good carotenoid source, being zeaxanthin and lutein the predominant compounds, representing 94% of the total carotenoids [34].

The antioxidant potential of leaf, bark and fruit extracts of murici were evaluated and the leaf extract, that presented the highest yield of total flavonoids (11.7 mg/g, fresh weight) and total phenolic compounds (45 mg/g) presented higher efficiency against the oxidizing effects of ROO$^{\bullet}$ (ORAC) (779 μmol trolox equivalent/g) than bark (591 μmol trolox equivalent/g) and fruit extracts (12 μmol trolox equivalent/g) [37]. Later, Mariutti *et al.* [33], reported that the high efficiency of hydrophilic extracts of murici in scavenging HO$^{\bullet}$, HOCl, ROO$^{\bullet}$, ONOO^{-} and H$_2$O$_2$ may be attributed to the presence of phenolic compounds [33].

The hypoglycemic effects of seed and fruit extracts of murici were investigated by oral administration to normoglycemic and streptozotocin-induced severe diabetic rats and the researchers concluded that after 4 h of a single oral dose, the murici extracts showed significant anti-hyperglycemic effects. Moreover, a relevant inhibitory activity against the formation of advanced glycation end products was observed for both extracts (IC$_{50}$ from 94 to 138 μg/mL) [38].

Tucumã

There are two species popularly known as tucumã, whose fruits are edible and marketed in northern Brazil: *Astrocaryum vulgare* Mart., which is mainly found and traded in Pará State, and *Astrocaryum aculeatum* G. Mey., also known as "tucumã-açu", mainly found in central Amazonia and traded in Amazonas State [39]. Both species belong to the Arecaceae family and their fruits are characterized as smooth drupe with an epicarp and mesocarp that may vary greatly in colour (yellowish to dark orange and red, or whitish), shape (oval, spherical), mesocarp thickness (2.8 to 10 mm), endocarp, and seed type. The consistency of the fruit is mucilaginous, oily, and sweet to the taste with a characteristic flavour. Its weight varies from 20 to 100 g, of which 22% is edible [40]. The raw pulp of tucumã presents high provitamin A content [41], and is highly appreciated as sandwich filling in Amazonas State culinary [39].

Since the mesocarps of tucumã fruits possess high oil contents (12%, dry basis), with high contents of monounsaturated fatty acids (65% of C18:1) [42], similar to that found in olive oil, most researches are directed to provide nutritional use to tucumã oil. For this reason, the carotenoid composition of tucumã pulp was completely elucidated and all-*trans*-β-carotene, accounting for 75% of total carotenoids, was reported as the major compound, followed by all-*trans*-α-cryptoxanthin, all-*trans*-β-cryptoxanthin, all-*trans*-α-carotene and 13-*cis*-β-carotene, each accounting for 2-3% of the total carotenoid content [41]. The alkaloids, total tannins, total flavonoids and total phenolic compounds were determined in extracts obtained from tucumã peel and pulp by Sagrillo *et al.* [43]. According to these authors, the peel extract presented higher content of bioactive compounds than pulp extract and the major phenolic compound identified in both extracts was rutin (14-25 mg/100 g, fresh weight), followed by quercetin (5-10 mg/100 g) and gallic acid (3-11 mg/100 g) [43]. In addition, (+)-catechin, caffeic acid and chlorogenic acid were also detected in extracts of tucumã fruits [43, 44].

Based on the biological potential of the bioactive compounds found in tucumã fruits, the cytoprotective effect of tucumã extracts (peel and pulp) was evaluated in lymphocyte cultures exposed to H_2O_2, and both extracts (300 to 900 µg/mL) increased cells viability and decrease the activity of caspases 1, 3, and 8 [43].

Furthermore, another study with tucumã suggests that the pulp oil may prevent an *in vivo* chronic and acute inflammatory response in an efficient way, and this action may be probably mediated by some minor compounds and also by fatty acids [45].

Buriti

Buriti (*Mauritia flexuosa* L.f.) is a palm from Arecaceae family, which is widely found in the Amazon region of Brazil and its fruits contain a pulp with a peculiar flavour and aroma, which is usually consumed in the regions where it grows, in the form of sweets, jams, ice creams, compotes and wines [46]. Buriti fruits can be highlighted among other fruits since it is an excellent source of vegetable oil with oleic acid accounting for 73-76% of total fatty acid composition [47, 48]. Additionally, the pulp is a very promising bioactive compounds source, due to its high levels of carotenoids in fresh weight (23-53 mg/100 g of pulp), tocopherols (117 mg/100 g, dry weight), ascorbic acid (57 mg/100 g, fresh weight) and phenolic compounds (23800 mg/100 g pulp, dry weight) [46 - 49].

Regarding the bioactive compounds profile, quinic acid (cyclic polyol) was reported to be much more abundant (231 mg/g, dry weight) than protocatechuic (2.1 mg/g) and chlorogenic acids (1.1 mg/g), which are the other main phenolic compounds found in buriti pulp [49]. Moreover, quercetin, kaempferol, myricetin, luteolin, apigenin, (-)-epicatechin, (+)-catechin, caffeic acid, ferulic acid, *p*-coumaric acid [49], as well as naringenin, vitexin, scoparin, rutin, cyanidin 3-rutinoside and cyanidin 3-glucoside [50] were also detected in extracts of buriti pulp. Concerning the carotenoid composition, the major compound identified in buriti pulp was all-*trans*-β-carotene (72% of total carotenoids), followed by 9-*cis*-β-carotene and 13-*cis*-β-carotene, which accounted for 3.6% and 11% of total carotenoids, respectively [41]. Due to the very high content of carotenoids with vitamin A activity in buriti pulp (> 400 μg/g pulp and 3640 μg retinol activity equivalent (RAE)/100 g pulp) [41], buriti fruit can be certainly considered, among all other fruits, the richest source of provitamin A until today.

The biological potentials of buriti fruits are typically attributed to the tocopherols and carotenoids, which are the main bioactive compounds in its pulp oil.

Noteworthy, a high positive correlation ($r \geq 0.95$) was also observed between the phenolic contents of buriti pulp and its antioxidant capacity through different antioxidant assays (ABTS, DPPH, FRAP and ORAC) [46]. A recent study carried out with the oil extracted from peel of buriti fruits demonstrated the inhibition of the platelet secretion and aggregation, in a concentration-dependent manner (0.1 to 1 mg/mL), when induced by collagen, ADP and TRAP-6 (thrombin receptor activator peptide 6), without the participation of the adenylyl cyclase pathway, and diminished the rolling and firm adhesion of the platelet under flow conditions [51]. Moreover, these same authors reported that the oil extract induced a marked increase in the rolling speed of leukocytes retained on the surface of the platelet, resulting in the reduction of adhesion and rolling and that the oil extract also significantly decreased platelet release of sP-selectin, an atherosclerotic-related inflammatory mediator, along with thrombus growth inhibition at the same concentration as that of aspirin, a classical reference drug [51]. Furthermore, the antimicrobial activity of leaf, fruit and trunk extracts of buriti against some pathogenic bacteria (*Micrococcus luteus, Pseudomonas aeruginosa, Staphylococcus aureus*) demonstrated its high ability to inhibit the growth of pathogens with MIC values (minimal inhibitory concentrations) varying from 50-200 µg/mL, where the leaf extract exhibited the best results against *P. aeruginosa* (MIC = 50 µg/mL) [50].

Pupunha

Pupunha or peach palm (*Bactris gasipaes* Kunth) is a palm tree that belongs to the Arecaceae family and only preliminary studies concerning the bioactive compounds and antioxidant potential of its fruits are available. Pupunha fruits are small (30g, 3.4 cm diameter and 3.1 cm length) and the pulp accounts for 72% of the fruit weight, followed by seeds (21%) and peel (6%) [52]. The use of pupunha fruits in Amazonia region is essentially limited to direct consumption after cooking in salty water. The mesocarp of pupunha fruits is fleshy, with a yellow to orange colour and these fruits have high nutritional value due to its high energy value, essential aminoacids, oils and content of fibers [53, 54].

Concerning the bioactive compounds profile of pupunha fruits, only the carotenoid composition was reported, with the predominance of all-*trans*-β-carotene (28%), and also other compounds such as *cis*-γ-carotene (14%), all-*trans*-γ-carotene (18%) and all-*trans*-δ-carotene (23%), [41]. Its vitamin A activity (745.5 μg RAE/100 g) was higher than that reported to piquiá (6 μg RAE/100 g) [25] and tucumã fruits (425 μg RAE/100 g), but much lower than buriti (3640 μg RAE/100 g) [41].

Mana-cubiu

Mana-cubiu (*Solanum sessiliflorum* Dun.) is a fruit that belongs to the same family of potato and tomato (Solanaceae family). The fruits may have a great variety of weights (30-400 g), with a hairy and leathery skin colour that goes from green (unripe) to orange (ripe) during ripening. The pulp of mana-cubiu is light yellow due to the presence of carotenoids [55] and may be used *in natura* in juices and salads, but it may be also cooked and then used as a substitute for apples, in cakes or in compotes. Nowadays, this fruit is not familiar to most Amazonian people and is scarcely cultivated in small gardens, mainly in the western part of the Amazon basin [56].

The bioactive compounds of mana-cubiu, namely the profile of phenolic compounds and carotenoids, was recently reported for the first time and 5-caffeoylquinic acid (chlorogenic acid) was the major phenolic compound, accounting for more than 78% of the phenolic compounds, followed by two dihydrocaffeoyl spermidines, while all-*trans*-β-carotene and all-*trans*-lutein were the major carotenoids [55]. Additionally, these authors reported that the carotenoid extract of mana-cubiu exhibited high scavenging capacity against ROO˙, while the hydrophilic extract was efficient against the *in vitro* oxidizing effects of H_2O_2 and HOCl [55]. The *in vivo* genotoxicity of mana-cubiu fruits was assayed to figure out whether maná-cubiu is safe to consumption and the results showed that the pulp fruit (up to 500 mg/kg, body weight) did not present any genotoxic or cytotoxic effect in Wistar rats [57]. These authors stated that the antigenotoxic effects of mana-cubiu fruit are probably attributed to the antioxidant capacity of the pulp [57].

Other Amazonian fruits

As mentioned before, Amazon biome hosts numerous fruit species, which are consumed by the local people and distributed through the local economy at trade markets, but most of them are unknown to the wider population and also to the scientific community. In this sense, there are a huge number of other Amazonia fruits with promising antioxidant properties to human health, but only few preliminary investigations concerning the bioactive compounds and its biological potential are available to other less known fruits.

Among these less studied fruits, araçá-boi (*Eugenia stipitata* McVaugh), from Myrtaceae family, which is a fruit with an intense yellow colour and unique sensory characteristics used to produce jellies, jams, nectars and juices, had the carotenoid composition only recently elucidated (lutein as the major carotenoid in both peel and pulp fruits) (Garzón *et al.*, 2012). In addition, another research reported that the ethanolic extract of araçá-boi showed higher antimutagenic and antigenotoxic properties (in male Swiss albino mice) at the highest tested concentration (300 mg/kg of body weight), and this effect was associated to the yield of phenolic compounds in the extract (184 mg gallic acid equivalent/100 g) [58].

Bacaba (*Oenocarpus bacaba* Mart) is a palm tree that belongs to the Arecaceae family and produces purple-coloured fruits. These fruits are commonly consumed after processing into ice cream, jelly, drinks and also *in natura* [59]. Bacaba fruits had been considered a potential source of bioactive compounds due to its similarity with açai fruits; however, the phenolic compounds profile (including anthocyanins) was only recently characterized and the following compounds were detected: diosmetin rutinoside, isorhamnetin acetylhexoside, quercetin hexoside, rutin, isorhamnetin dihexoside (or rhamnetin), vitexin, orientin (or isoorientin), vicenin-2, cyanidin 3-rutinoside and cyanidin 3-glucoside [59].

Uxi [*Endopleura uchi* (Huber) Cuatrecasas], belongs to the Humiriaceae family and some studies suggest that methanolic extracts of its bark exhibited anti-inflammatory properties associated to the presence of bergenin, a *C*-glucoside derived from 4-*O*-methyl gallic acid [60], as well as antioxidant [61] and antimicrobial activities [62]. Regarding the presence of bioactive compounds in uxi fruits, α-tocopherol was found in high amounts in the pulp (201 μg/g) and among other Brazilian fruits studied by [63], its α-tocopherol content was only lower than that found for buriti fruits (347 μg/g).

CONCLUDING REMARKS

Unquestionably, the scientific interest on the nutritional and biological potentialities of fruits from Amazonia biome has increased over the years. These fruits, among other Amazonian fruits not mentioned in this chapter, can be seen as promising and reliable sources of bioactive compounds, whose individual and summated biological activities, suggest high benefits to human health. Notwithstanding, further research is clearly required to understand the molecular mechanisms of its bioactive compounds.

CONFLICT OF INTEREST

The authors confirm that they have no conflict of interest to declare for this publication.

ACKNOWLEDGEMENTS

We would like to thank FCT (*Fundação para a Ciência e Tecnologia,* Portugal) for the financial support to UCIBIO-REQUIMTE through projects UID/Multi/04378/2013 and PTDC/QEQQAN/1742/2014.

REFERENCES

[1]   WWF.panda.org [homepage on the Internet]. About the Amazon 2015 [cited 2015 04/30/2015]. Available from: http://wwf.panda.org/what_we_do/where_we_work/amazon/about_the_amazon/ 2015.

[2]   De Rosso VV. Bioactivities of Brazilian fruits and the antioxidant potential of tropical biomes. Food Public Health 2013; 3(1): 37-51.

[3]   Lavecchia T, Rea G, Antonacci A, Giardi MT. Healthy and adverse effects of plant-derived functional metabolites: the need of revealing their content and bioactivity in a complex food matrix. Crit Rev Food Sci Nutr 2013; 53(2): 198-213.
[http://dx.doi.org/10.1080/10408398.2010.520829] [PMID: 23072533]

[4]   Huang D, Ou B, Prior RL. The chemistry behind antioxidant capacity assays. J Agric Food Chem 2005; 53(6): 1841-56.
[http://dx.doi.org/10.1021/jf030723c] [PMID: 15769103]

[5]   Heinrich M, Dhanji T, Casselman I. Açai (*Euterpe oleracea* Mart.) - A phytochemical and pharmacological assessment of the species' health claims. Phytochem Lett 2011; 4: 10-21.
[http://dx.doi.org/10.1016/j.phytol.2010.11.005]

[6]   Yamaguchi KK, Pereira LF, Lamarão CV, Lima ES, da Veiga-Junior VF. Amazon acai: chemistry and biological activities: a review. Food Chem 2015; 179(0): 137-51.
[http://dx.doi.org/10.1016/j.foodchem.2015.01.055] [PMID: 25722148]

[7]   FAO. Fao R, Ed. Food and fruit-bearing forest species. 3: Examples from Latin America. Forestry Paper No 44/31986.

[8]   De Rosso VV, Hillebrand S, Montilla EC, *et al.* Determination of anthocyanins from acerola (*Malpighia emarginata* DC.) and acai (*Euterpe oleracea* Mart.) by HPLC-PDA-MS/MS. J Food Compos Anal 2008; 21(4): 291-9.
[http://dx.doi.org/10.1016/j.jfca.2008.01.001]

[9]   Del Pozo-Insfran D, Brenes CH, Talcott ST. Phytochemical composition and pigment stability of Açai (*Euterpe oleracea* Mart.). J Agric Food Chem 2004; 52(6): 1539-45.
[http://dx.doi.org/10.1021/jf035189n] [PMID: 15030208]

[10]   Ribeiro JC, Antunes LM, Aissa AF, *et al.* Evaluation of the genotoxic and antigenotoxic effects after acute and subacute treatments with açai pulp (*Euterpe oleracea* Mart.) on mice using the erythrocytes micronucleus test and the comet assay. Mutat Res 2010; 695(1-2): 22-8.
[http://dx.doi.org/10.1016/j.mrgentox.2009.10.009] [PMID: 19892033]

[11]   Darnet S, Serra JL, Rodrigues AM, Silva LH. A high-performance liquid chromatography method to measure tocopherols in assai pulp (*Euterpe oleracea*). Food Res Int 2011; 44(7): 2107-11.
[http://dx.doi.org/10.1016/j.foodres.2010.12.039]

[12]   Vissotto LC, Rodrigues E, Chisté RC, Benassi MD, Mercadante AZ. Correlation, by multivariate statistical analysis, between the scavenging capacity against reactive oxygen species and the bioactive compounds from frozen fruit pulps. Food Sci Technol (Campinas) 2013; 33: 57-65.
[http://dx.doi.org/10.1590/S0101-20612013000500010]

[13]   Hogan S, Chung H, Zhang L, *et al.* Antiproliferative and antioxidant properties of anthocyanin-rich extract from açai. Food Chem 2010; 118(2): 208-14.
[http://dx.doi.org/10.1016/j.foodchem.2009.04.099]

[14]   Langley PC, Pergolizzi JV Jr, Taylor R Jr, Ridgway C. Antioxidant and associated capacities of Camu camu (*Myrciaria dubia*): a systematic review. J Altern Complement Med 2015; 21(1): 8-14.
[http://dx.doi.org/10.1089/acm.2014.0130] [PMID: 25275221]

[15]   Chirinos R, Galarza J, Betalleluz-Pallardel I, Pedreschi R, Campos D. Antioxidant compounds and antioxidant capacity of Peruvian camu camu (*Myrciaria dubia* (HBK) McVaugh) fruit at different maturity stages. Food Chem 2010; 120(4): 1019-24.
[http://dx.doi.org/10.1016/j.foodchem.2009.11.041]

[16]   Rufino MD, Alves RE, de Brito ES, *et al.* Bioactive compounds and antioxidant capacities of 18 non-traditional tropical fruits from Brazil. Food Chem 2010; 121(4): 996-1002.
[http://dx.doi.org/10.1016/j.foodchem.2010.01.037]

[17]   Zanatta CF, Mercadante AZ. Carotenoid composition from the Brazilian tropical fruit camu-camu (*Myrciaria dubia*). Food Chem 2007; 101(4): 1526-32.
[http://dx.doi.org/10.1016/j.foodchem.2006.04.004]

[18]   Zanatta CF, Cuevas E, Bobbio FO, Winterhalter P, Mercadante AZ. Determination of anthocyanins

from camu-camu (*Myrciaria dubia*) by HPLC-PDA, HPLC-MS, and NMR. J Agric Food Chem 2005; 53(24): 9531-5.
[http://dx.doi.org/10.1021/jf051357v] [PMID: 16302773]

[19]   Akter MS, Oh S, Eun JB, Ahmed M. Nutritional compositions and health promoting phytochemicals of camu-camu (*Myrciaria dubia*) fruit: A review. Food Res Int 2011; 44(7): 1728-32.
[http://dx.doi.org/10.1016/j.foodres.2011.03.045]

[20]   Fracassetti D, Costa C, Moulay L, Tomás-Barberán FA. Ellagic acid derivatives, ellagitannins, proanthocyanidins and other phenolics, vitamin C and antioxidant capacity of two powder products from camu-camu fruit (*Myrciaria dubia*). Food Chem 2013; 139(1-4): 578-88.
[http://dx.doi.org/10.1016/j.foodchem.2013.01.121] [PMID: 23561148]

[21]   Inoue T, Komoda H, Uchida T, Node K. Tropical fruit camu-camu (*Myrciaria dubia*) has anti-oxidative and anti-inflammatory properties. J Cardiol 2008; 52(2): 127-32.
[http://dx.doi.org/10.1016/j.jjcc.2008.06.004] [PMID: 18922386]

[22]   Yang H, Protiva P, Cui B, *et al.* New bioactive polyphenols from *Theobroma grandiflorum* ("cupuaçu"). J Nat Prod 2003; 66(11): 1501-4.
[http://dx.doi.org/10.1021/np034002j] [PMID: 14640528]

[23]   Rogez H, Buxant R, Mignolet E, *et al.* Chemical composition of the pulp of three typical Amazonian fruits: araca-boi (*Eugenia stipitata*), bacuri (*Platonia insignis*) and cupuacu (*Theobroma grandiflorum*). Eur Food Res Technol 2004; 218(4): 380-4.
[http://dx.doi.org/10.1007/s00217-003-0853-6]

[24]   Pugliese AG, Tomas-Barberan FA, Truchado P, Genovese MI. Flavonoids, proanthocyanidins, vitamin C, and antioxidant activity of *Theobroma grandiflorum* (Cupuassu) pulp and seeds. J Agric Food Chem 2013; 61(11): 2720-8.
[http://dx.doi.org/10.1021/jf304349u] [PMID: 23431956]

[25]   Chisté RC, Mercadante AZ. Identification and quantification, by HPLC-DAD-MS/MS, of carotenoids and phenolic compounds from the Amazonian fruit *Caryocar villosum*. J Agric Food Chem 2012; 60(23): 5884-92.
[http://dx.doi.org/10.1021/jf301904f] [PMID: 22612541]

[26]   Chisté RC, Benassi MD, Mercadante AZ. Efficiency of different solvents on the extraction of bioactive compounds from the Amazonian fruit *Caryocar villosum* and the effect on its antioxidant and colour properties. Phytochem Anal 2014; 25(4): 364-72.
[http://dx.doi.org/10.1002/pca.2489]

[27]   Chisté RC, Freitas M, Mercadante AZ, Fernandes E. The potential of extracts of *Caryocar villosum* pulp to scavenge reactive oxygen and nitrogen species. Food Chem 2012; 135(3): 1740-9.
[http://dx.doi.org/10.1016/j.foodchem.2012.06.027] [PMID: 22953916]

[28]   Magid AA, Voutquenne-Nazabadioko L, Harakat D, Moretti C, Lavaud C. Phenolic glycosides from the stem bark of *Caryocar villosum* and *C. glabrum*. J Nat Prod 2008; 71(5): 914-7.
[http://dx.doi.org/10.1021/np800015p] [PMID: 18412393]

[29]   Alabdul Magid A, Voutquenne L, Harakat D, *et al.* Triterpenoid saponins from the fruits of *Caryocar villosum*. J Nat Prod 2006; 69(6): 919-26.
[http://dx.doi.org/10.1021/np060097o] [PMID: 16792411]

[30] Alabdul Magid A, Voutquenne-Nazabadioko L, Renimel I, Harakat D, Moretti C, Lavaud C. Triterpenoid saponins from the stem bark of *Caryocar villosum*. Phytochemistry 2006; 67(19): 2096-102.
[http://dx.doi.org/10.1016/j.phytochem.2006.07.009] [PMID: 16930644]

[31] Almeida MR, Darin JD, Hernandes LC, *et al.* Antigenotoxic effects of piquiá (*Caryocar villosum*) in multiple rat organs. Plant Foods Hum Nutr 2012; 67(2): 171-7.
[http://dx.doi.org/10.1007/s11130-012-0291-3] [PMID: 22562095]

[32] Schauss AG, Wu X, Prior RL, *et al.* Antioxidant capacity and other bioactivities of the freeze-dried Amazonian palm berry, *Euterpe oleraceae* mart. (acai). J Agric Food Chem 2006; 54(22): 8604-10.
[http://dx.doi.org/10.1021/jf0609779] [PMID: 17061840]

[33] Mariutti LR, Rodrigues E, Chisté RC, Fernandes E, Mercadante AZ. The Amazonian fruit *Byrsonima crassifolia* effectively scavenges reactive oxygen and nitrogen species and protects human erythrocytes against oxidative damage. Food Res Int 2014; 64: 618-25.
[http://dx.doi.org/10.1016/j.foodres.2014.07.032]

[34] Mariutti LR, Rodrigues E, Mercadante AZ. Carotenoids from *Byrsonima crassifolia*: identification, quantification and *in vitro* scavenging capacity against peroxyl radicals. J Food Compos Anal 2013; 31(1): 155-60.
[http://dx.doi.org/10.1016/j.jfca.2013.05.005]

[35] Barreto GP, Benassi MT, Mercadante AZ. Bioactive compounds from several tropical fruits and correlation by multivariate analysis to free radical scavenger activity. J Braz Chem Soc 2009; 20(10): 1856-61.
[http://dx.doi.org/10.1590/S0103-50532009001000013]

[36] Gordon A, Jungfer E, da Silva BA, Maia JG, Marx F. Phenolic constituents and antioxidant capacity of four underutilized fruits from the Amazon region. J Agric Food Chem 2011; 59(14): 7688-99.
[http://dx.doi.org/10.1021/jf201039r] [PMID: 21662239]

[37] Silva EM, Souza JN, Rogez H, Rees JF, Larondelle Y. Antioxidant activities and polyphenolic contents of fifteen selected plant species from the Amazonian region. Food Chem 2007; 101(3): 1012-8.
[http://dx.doi.org/10.1016/j.foodchem.2006.02.055]

[38] Perez-Gutierrez RM, Muñiz-Ramirez A, Gomez YG, Ramírez EB. Antihyperglycemic, antihyperlipidemic and antiglycation effects of *Byrsonima crassifolia* fruit and seed in normal and streptozotocin-induced diabetic rats. Plant Foods Hum Nutr 2010; 65(4): 350-7.
[http://dx.doi.org/10.1007/s11130-010-0181-5] [PMID: 20734144]

[39] Didonet AA, Ferraz ID. Fruit trade of tucuma (*Astrocaryum aculeatum* G. Mey - Arecaceae) at local market places in Manaus (Amazonas, Brazil). Rev Bras Frutic 2014; 36(2): 353-62.
[http://dx.doi.org/10.1590/0100-2945-108/13]

[40] Maia GCHM, Campos MS, Barros-Monteiro J, *et al.* Effects of *Astrocaryum aculeatum* Meyer (tucumã) on diet-induced dyslipidemic rats. J Nutr Metab 2014; 2014 Article ID 202367 (9 pages)

[41] De Rosso VV, Mercadante AZ. Identification and quantification of carotenoids, by HPLC-PD--MS/MS, from Amazonian fruits. J Agric Food Chem 2007; 55(13): 5062-72.

[http://dx.doi.org/10.1021/jf0705421] [PMID: 17530774]

[42]   Santos MF, Marmesat S, Brito ES, Alves RE, Dobarganes MC. Major components in oils obtained from Amazonian palm fruits. Grasas Aceites 2013; 64(3): 328-34.
[http://dx.doi.org/10.3989/gya.023513]

[43]   Sagrillo MR, Garcia LF, de Souza Filho OC, *et al.* Tucumã fruit extracts (*Astrocaryum aculeatum* Meyer) decrease cytotoxic effects of hydrogen peroxide on human lymphocytes. Food Chem 2015; 173: 741-8.
[http://dx.doi.org/10.1016/j.foodchem.2014.10.067] [PMID: 25466084]

[44]   De Souza Schmidt Gonçalves AE, Lajolo FM, Genovese MI. Chemical composition and antioxidant/antidiabetic potential of Brazilian native fruits and commercial frozen pulps. J Agric Food Chem 2010; 58(8): 4666-74.
[http://dx.doi.org/10.1021/jf903875u] [PMID: 20337450]

[45]   Bony E, Boudard F, Brat P, *et al.* Awara (*Astrocaryum vulgare* M.) pulp oil: chemical characterization, and anti-inflammatory properties in a mice model of endotoxic shock and a rat model of pulmonary inflammation. Fitoterapia 2012; 83(1): 33-43.
[http://dx.doi.org/10.1016/j.fitote.2011.09.007] [PMID: 21958966]

[46]   Cândido TL, Silva MR, Agostini-Costa TS. Bioactive compounds and antioxidant capacity of buriti (*Mauritia flexuosa* L.f.) from the Cerrado and Amazon biomes. Food Chem 2015; 177: 313-9.
[http://dx.doi.org/10.1016/j.foodchem.2015.01.041] [PMID: 25660891]

[47]   Darnet SH, da Silva LH, Rodrigues AM, Lins RT. Nutritional composition, fatty acid and tocopherol contents of buriti (*Mauritia flexuosa*) and patawa (*Oenocarpus bataua*) fruit pulp from the Amazon region. Ciencia Tecnol Aliment 2011; 31(2): 488-91.
[http://dx.doi.org/10.1590/S0101-20612011000200032]

[48]   Manhaes LR, Sabaa-Srur AU. Centesimal composition and bioactive compounds in fruits of buriti collected in Para. Ciencia Tecnol Aliment 2011; 31(4): 856-63.
[http://dx.doi.org/10.1590/S0101-20612011000400005]

[49]   Bataglion GA, da Silva FM, Eberlin MN, Koolen HH. Simultaneous quantification of phenolic compounds in buriti fruit (*Mauritia flexuosa* L.f.) by ultra-high performance liquid chromatography coupled to tandem mass spectrometry. Food Res Int 2014; 66: 396-400.
[http://dx.doi.org/10.1016/j.foodres.2014.09.035]

[50]   Koolen HH, da Silva FM, Gozzo FC, de Souza AQ, de Souza AD. Antioxidant, antimicrobial activities and characterization of phenolic compounds from buriti (*Mauritia flexuosa* L. f.) by UPLC-ES--MS/MS. Food Res Int 2013; 51(2): 467-73.
[http://dx.doi.org/10.1016/j.foodres.2013.01.039]

[51]   Fuentes E, Rodriguez-Perez W, Guzman L, *et al.* Mauritia flexuosa presents in vitro and in vivo antiplatelet and antithrombotic activities. Evid-based Complement Altern Med. 2013; 18.

[52]   Ferreira CD, Pena RS. Hygroscopic behavior of the pupunha flour. Ciencia Tecnol Aliment 2003; 23: 251-5.
[http://dx.doi.org/10.1590/S0101-20612003000200025]

[53]   Yuyama LK, Aguiar JP, Yuyama K, *et al.* Chemical composition of the fruit mesocarp of three peach

palm (*Bactris gasipaes*) populations grown in central Amazonia, Brazil. Int J Food Sci Nutr 2003; 54(1): 49-56.
[PMID: 12701237]

[54]   Leterme P, Garcia MF, Londono AM, *et al.* Chemical composition and nutritive value of peach palm (*Bactris gasipaes* Kunth) in rats. J Sci Food Agric 2005; 85(9): 1505-12.
[http://dx.doi.org/10.1002/jsfa.2146]

[55]   Rodrigues E, Mariutti LR, Mercadante AZ. Carotenoids and phenolic compounds from *Solanum sessiliflorum*, an unexploited Amazonian fruit, and their scavenging capacities against reactive oxygen and nitrogen species. J Agric Food Chem 2013; 61(12): 3022-9.
[http://dx.doi.org/10.1021/jf3054214] [PMID: 23432472]

[56]   Marx F, Andrade EH, Maia JG. Chemical composition of the fruit of *Solanum sessiliflorum*. Z Lebensm Unters F A 1998; 206(5): 364-6.
[http://dx.doi.org/10.1007/s002170050274]

[57]   Hernandes LC, Aissa AF, de Almeida MR, *et al. In vivo* assessment of the cytotoxic, genotoxic and antigenotoxic potential of mana-cubiu (*Solanum sessiliflorum* Dunal) fruit. Food Res Int 2014; 62: 121-7.
[http://dx.doi.org/10.1016/j.foodres.2014.02.036]

[58]   Neri-Numa IA, Carvalho-Silva LB, Morales JP, *et al.* Evaluation of the antioxidant, antiproliferative and antimutagenic potential of araca-boi fruit (*Eugenia stipitata* Mc Vaugh - Myrtaceae) of the Brazilian Amazon forest. Food Res Int 2013; 50(1): 70-6.
[http://dx.doi.org/10.1016/j.foodres.2012.09.032]

[59]   Abadio Finco FD, Kammerer DR, Carle R, Tseng WH, Böser S, Graeve L. Antioxidant activity and characterization of phenolic compounds from bacaba (*Oenocarpus bacaba* Mart.) fruit by HPLC-DAD-MS(n). J Agric Food Chem 2012; 60(31): 7665-73.
[http://dx.doi.org/10.1021/jf3007689] [PMID: 22788720]

[60]   Nunomura RC, Oliveira VG, Da Silva SL, Nunomura SM. Characterization of bergenin in *Endopleura uchi* bark and its anti-inflammatory activity. J Braz Chem Soc 2009; 20(6): 1060-4.
[http://dx.doi.org/10.1590/S0103-50532009000600009]

[61]   Politi FA, de Mello JC, Migliato KF, Nepomuceno AL, Moreira RR, Pietro RC. Antimicrobial, cytotoxic and antioxidant activities and determination of the total tannin content of bark extracts *Endopleura uchi*. Int J Mol Sci 2011; 12(4): 2757-68.
[http://dx.doi.org/10.3390/ijms12042757] [PMID: 21731469]

[62]   Silva SL, Oliveira VG, Yano T, Nunomura RC. Antimicrobial activity of bergenin from *Endopleura uchi* (Huber). Cuatrec Acta Amaz 2009; 39(1): 187-92.
[http://dx.doi.org/10.1590/S0044-59672009000100019]

[63]   Da Costa PA, Ballus CA, Teixeira J, Godoy HT. Phytosterols and tocopherols content of pulps and nuts of Brazilian fruits. Food Res Int 2010; 43(6): 1603-6.
[http://dx.doi.org/10.1016/j.foodres.2010.04.025]

CHAPTER 9

Bioactive Compounds of Banana as Health Promoters

Aline Pereira[*], **Rodolfo Moresco, Marcelo Maraschin**

Federal University of Santa Catarina, Natural Products Core, Plant Morphogenesis and Biochemistry Laboratory, Florianopolis, Brazil

Abstract: Bananas (*Musa* spp) are sources of income and food for people who have been cultivating them all over the world. Banana is assumed to be a fruit with nutraceutical properties, mainly the pulp which is worldwide consumed. In the traditional medicine of some countries the treatment of burn wounds and the prevention of depression, for example, have been reported by compounds from banana peel. Banana peel is an under-explored waste from the food industry and could be considered a source of bioactive compounds with potential application in health care, for instance. This chapter provides updated information about *Musa* spp fruit and peel as chemically complex matrices sources of high-value secondary metabolites with claimed health promoters properties. Bananas are source of antioxidant compounds such as phenolic acids, biogenic amines (*i.e.*, L-dopa and dopamine), and pro-vitamin A carotenoids, which are important for the treatment of wounds, Parkinson's disease, and as dietary supplement, respectively. Therefore, the pulp and peel of banana fruits are biomasses riches in human health promoters compounds with biotechnological importance.

Keywords: Amine compounds, Bioactive compounds, Carotenoids, Flavonoids, *Musa* spp, Parkinson's disease, Phenolic compounds, Pro-vitamin A, Wound healing.

INTRODUCTION

The term *banana* describes hybrid plants of the genus *Musa* known as dessert bananas and plantains [1, 2].

[*] **Corresponding author Aline Pereira:** Federal University of Santa Catarina, Natural Products Core, Plant Morphogenesis and Biochemistry Laboratory, Florianópolis, Brazil; Email: allinep@gmail.com

Bananas have sweet flavor, exuberant appearance, and nutraceutical properties which makes them a fruit worldwide appreciated.

The differences among the genotypes of bananas reveal important particularities in their nutritional potential (Table **1**). Bananas have been one of the most commonly eaten fruit in Europe, North America, and in the tropics where are usually grown. In addition, bananas are forwarded to the food, pharmaceutical, and cosmetic industries as a high added-value raw material [3].

Table 1. Nutritional composition (100g of pulp) of banana cultivars typically grown in Brazil [4].

Composition	Maçã cultivar (AAB*)	Nanica cultivar (AAA*)	Prata cultivar (AAB*)
Calories (kcal)	97.70	99.00	100.00
Protein (g)	1.44	2.56	2.30
Fat (g)	0.20	0.29	0.20
Carbohydrates (g)	26.4	20.80	29.60
Calcium (mg)	0.30	0.02	0.01
Iron (mg)	60.00	1.00	0.60
Potassium (g)	0.03	0.03	0.03
Vitamin A (UI)	127.00	127.00	127.00
Vitamin B1 (mg)	0.40	0.37	0.79
Vitamin B2 (mg)	0.30	0.78	0.90
Vitamin C (mg)	12.70	4.10	17.30

Source [4]. * Genomic group

Banana has a high nutritive status as source of carbohydrates, mostly starch, vitamin A, iron, calcium, and neuroamines, *e.g.* serotonin and dopamine [5, 6].

Plants have been used as sources of bioactive compounds since a long time ago for the development of medicines to treat human beings diseases [7]. Banana peel could be considered a biomass with health promoter properties, despite being an under-explored residue (*i.e.* pomace) from the food industry, due to its chemical composition rich in bioactive compounds.

In the Brazilian folk medicine, the banana peel's inner face when applied topically on a wound, mainly caused by burns, acts as a healing promoter [8], reducing the pain and swelling [9].

This chapter provides updated information about *Musa* sp. fruit pulp and peel as matrices complex in their chemical compositions and sources of high-value secondary metabolites of interest for human health. Emphasis is given on the banana's peel and pulp chemical profiles as to their phenolic and carotenoid compounds, as well as biogenic amines. Besides, the biological activities of those secondary metabolites as health promoters, *e.g.* wound healing and for the treatment of depression are also addressed.

BANANAS AS SOURCE OF BIOACTIVE COMPOUNDS

Phenolic Compounds and Flavonoids

Phenolic compounds are chemically characterized by having in their structures one or more benzenic rings and one or more hydroxyl groups linked to the ring as substituents, as well as methyl, methoxyl, and amino groups (Fig. **1**). Most of them occur naturally as glycosides, forming conjugates with monosaccharides and polysaccharides. Phenolic compounds may be esters and methyl esters which are functional derivatives [10]. They can be found in several plant species with a typical electron transfer function related to their ability of capturing hydroxyl and superoxide radicals [11].

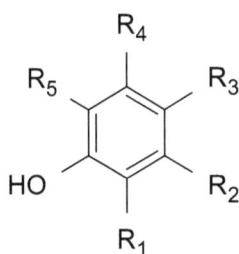

Fig. (1). General structure of phenolic compounds. R1, R2, R3, R4, and R5: benzenic ring, hydroxyl, methyl, methoxyl, amino groups, and/or mono-(poly)saccharides [10].

Since bananas are sources of bioactive phenolics relevant to the food industry, a series of studies has addressed the analysis of *Musa* spp genotypes cultivated in

Malaysia [13, 14] and Brazil [15], for example, regarding their (poly)phenolic profiles. The phenolic scavenging activity detected in extracts of banana tissues is due to the redox potential of those molecules, which is related to donate hydrogens, acting as reducing agents and singlet oxygen quenchers. The number of hydroxyl radicals in the aromatic structure is directly related to the antioxidant activity of those secondary metabolites [12]. The reports in literature reveal a wide results reveal a wide range of concentrations of phenolic compounds among the genotypes, indicating the clear possibility of exploring the potential of *Musa* germplasm for biotechnological applications.

The flavonoids' chemical structure consists of pyran, benzene rings, and diphenylpropanes (C6-C3-C6). Diphenylpropanes consist of an aromatic B-ring linked to a carbon-carbon connected through an aromatic A-ring and a heterocyclic C-ring (Fig. **2**). The classification of flavonoids comprises the conjugation between rings A and B, a glucosyl, a hydroxyl or a methoxyl group as the substituent side-chain. In foods, flavonoids exist as 3-*O*-glycosides and polymers. They are proanthocyanins, anthocyanidins, catechins, glycosides, flavones, and flavonols characterized by a flavone nucleus present in the chemical structure [17].

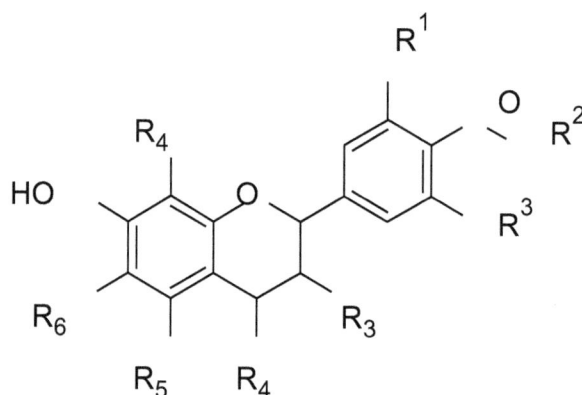

Fig. (2). Details of the basic chemical structure of flavonoids and radical substituents. R^2: H, R_5: OH, and R_6: H. Some flavonoids are characterized by functional groups at R_3, R^1, and R^3 [17].

Flavonoids are a class of polyphenols and can be found in flowers, leaves, seeds, and fruit peels and pulp. There are four more important classes: 4-oxoflavonoids (flavones and flavonols; flavones differ from flavonols because they do not have a

OH-group at C3-atom of the center ring), anthocyanins, isoflavones, and flavan-*3*-ol derivatives (catechin and tannins) [14, 17]. These compounds protect against herbivores, pathogens , and ultraviolet radiation. Some properties like chelating of metals [18], antimutagenic and antitumoral activities [11, 19], and antioxidant potential [27] are the most beneficial effects of flavonoids on human health. Flavonoids inhibit a variety of enzyme systems, *i.e.* phosphatases, oxido-reductases, oxygenase, hydrolases, DNA synthetases, RNA polymerases, protein phosphokinases, and amino acid oxidases. According to biochemical investigations about mechanisms of action of these compounds, they also play roles as hormones, immune as hormones, immune modulators, neurotransmitters, electron transfer catalyst, and as scavenger of free radicals [20, 21].

Antioxidant and Wound Healing Potential

The potential of banana peel as source of bioactive compounds was described by the authors [22] which detected high amounts of antioxidant compounds in extracts of banana peel flour (*Musa* AAA, Cavendish type), *i.e.*, total phenolic content (29 mg/g, as gallic acid equivalent - GAE), flavonoids (proan-thocyanidins: 3952 mg/kg), flavonol glycosides (129 mg/kg, as quercetin 3-rutinoside), and B-type procyanidin dimers and monomeric flavan-*3*-ols (126 mg/kg, as (+)-catechin). The antioxidant activity expressed as Trolox equivalents measured by FRAP (14 μM/g pulp), ABTS (242 μM/g pulp), and ORAC (436 μM/g pulp.) assays in the banana peel flour corroborates the set of compounds found [22]. Indeed, in the past years several studies [23, 24, 25, 26, 27] on the chemical constituents of banana peel point to that biomass as a raw material rich in bioactive molecules that act as health promoters through antioxidant mechanism of action as free radicals scavengers [26, 27].

According to the authors [27]: "The role of antioxidants in the removal of inflammation products is already known and these compounds are also beneficial in wound healing for several reasons [26]. Antioxidants work against the excess of proteases and reactive oxygen species (ROS), protecting protease inhibitors from oxidative damage [26]. In addition, antioxidants can prevent destruction of fibroblasts and other cells caused by ROS generation, playing an important role in the successful treatment of lesions [26, 27, 28]. As described by the authors [29,

30], extracts of *Musa* sp. and *Musa sapientum* peels showed a positive effect in rats wound healing process. Similarly, the wound healing of mice treated with aqueous extract of the Prata Anã (*Musa* sp.) cultivar has been reported [27, 28]. The Brazilian traditional knowledge described by the author [8] corroborated the role of antioxidant compounds found in banana peel [26, 27].

Biogenic Amines

Catecholamines, *e.g.* dopamine, norepinephrine (noradrenaline), and epinephrine (adrenaline), are biogenic amines which contain a 3, 4-dihydroxy-substituted phenyl ring (Fig. **3**).

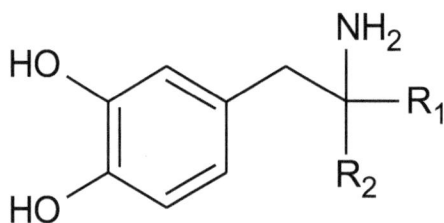

Fig. (3). Chemical structures of catecholamines. L-dopa: R_1 = COOH and R_2 = H; dopamine: R_1 = R_2 = H; norepinephrine: R_1 = H and R_2 = OH; and epinephrine: R_1 = H, R_2 = OH, and CH_3 replacing one of the amine group's hydrogen [17].

The amino acids tyramine, L-phenylalanine, and L-tyrosine are the precursors of biogenic amines in the biosynthetic pathway (Fig. **4**), which is similar both in plants and mammals [31]. In the later, important metabolic pathways, such as the glycogen metabolism, are regulated by neurotransmitters, *i.e.* the biogenic amines [32]. On the other hand, in plants, chemodiversity translates the possibility of increase the catecholamine contents by a large amount of biosynthetic pathways susceptible to stimulation [33].

Parkinson's Disease

In India, the traditional use of bananas is associated to prevent or treat diseases as depression [35], for example, as further supported by [36, 37]. Catecholamines have been found in several plant species, conferring to them an important medicinal potential according to the authors [34, 38, 39] eventually associated with the treatment of Parkinson's disease (PD).

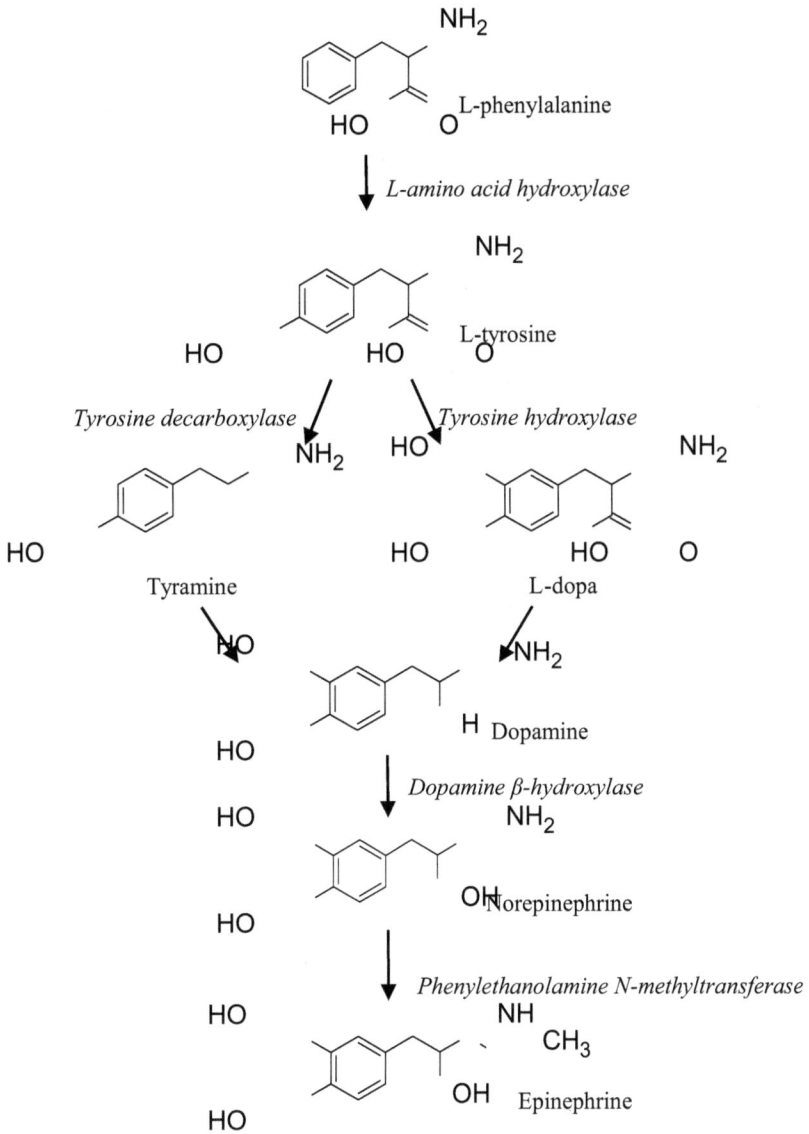

Fig. (4). Catecholamine biosynthesis pathway in plants. Source [34].

Described by James Parkinson in 1817, "PD is a neurodegenerative disease characterized as a movement disorder, and its diagnosis is based on the presence of two or more cardinal motor signs: resting tremor, decrease of voluntary movements, bradykinesia, muscle rigidity, stooped posture, postural instability, and handwriting changes. The symptoms become progressively worse as the neurodegeneration continues until patients are virtually unable to move" [40].

Recently, it was designed the D-phenylglycine-L-dopa (D-PhG-L-dopa) by the authors [41], dopamine dipeptide mimetic pro-drug 31-fold higher oral bioavailability than L-dopa in rats. D-PhG-L-dopa and L-dopa reach the brain quickly (maximum 1 min) after intravenous administration. D-PhG-L-dopa showed higher systemic bioavailability (1.97-fold) in comparison to L-dopa, which could be due to its larger volume of distribution and lower clearance rate. Terminal half-life of brain dopamine was 2.5 times longer when D-PhG-L-dopa was administrated than that upon L-dopa. Pharmaceutical formulations to treat PD might be eventually considered having bananas as raw material for its development. As source of biogenic amines, the concentration of dopamine in banana's peel is higher than the pulp [42]. The dopamine pro-drug D-PhG-L-dopa might works as dopamine-releasing system usefulness for treating PD, considering the prevention or prolongation of the L-dopa decarboxylation process [41].

Carotenoids

Carotenoids are yellow-orange colored chemical compounds, occurring in a large number of biological systems, and playing a role as accessory pigments in photosynthesis, for instance. The carotenoid's chemical structure has a set of conjugated double-bonds which are light-absorbing chromophores. Cartenoid are plant secondary metabolites and belong to the class of terpenes, which are classified by the number of C5 units and also known as tetraterpenes (with 40 carbons, *i.e.*, 8 units of C5). The carotenoids are more stable in their *trans* isomeric form and could be divided in carotenes and xanthophylls. The carotenes are hydrocarbon carotenoids, such as α- and β-carotene and lycopene (Fig. **5**). The xanthophylls are oxygenated derivatives of carotenes and include, among other, lutein, β-cryptoxanthin, and zeaxanthin (Fig. **6**) [43].

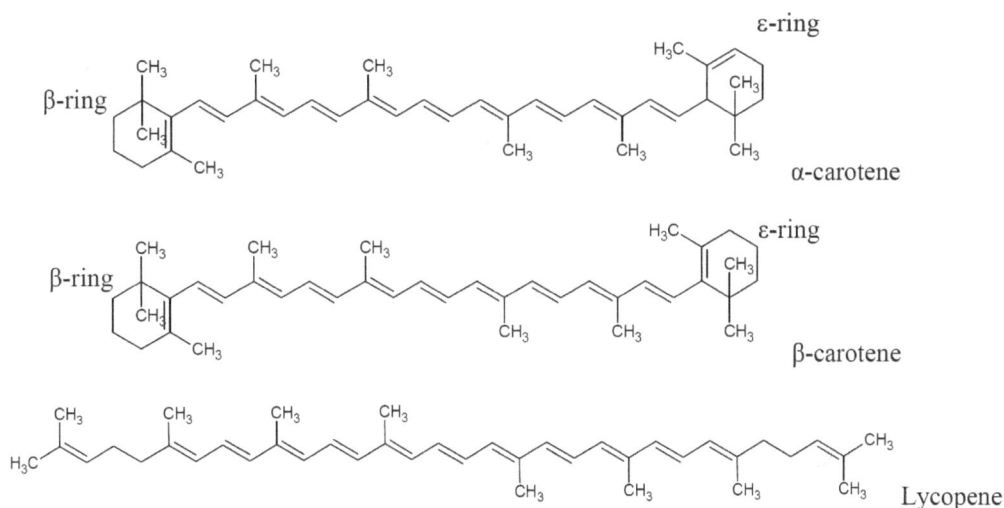

Fig. (5). Chemical structures of biologically important carotenes [43].

Carotenoid biosynthesis begins with a C5 primer. The addition of C5 units elongates the hydrocarbon chain, yielding C10 (monoterpenes), C15 (sesquiterpenes), C20 (diterpenes), C30 (triterpenes), C40 (tetraterpenes), and [C5]n (polyterpenes) compounds. Phytoene is a result of the C20 dimerization and produces the C40 carotenoid. The compound lycopene has 11 conjugated double bonds (Fig. **5**) and is synthesized from phytoene. Among the biosynthetic pathway, the cyclization of both ends of the molecule produces α-carotene and β-carotene (Fig. **5**). β-cryptoxanthin results from the addition the addition of a hydroxyl group in β-carotene chemical structure. The introduction of two hydroxyl groups results in the zeaxanthin (Fig. **5**). Other chemical modifications in the α-carotene produce lutein (Fig. **6**) [43]. Some carotenoids are also precursors to vitamin A [44]. The enzymatically conversion of pro-vitamin A carotenoids into vitamin A occurs mainly in the intestinal mucosa during the absorption of those precursors, protecting against vitamin A deficiency. The authors [45, 46, 47] suggest that carotenoids are health-promoting compounds by estimulating the immune system, to reduce the risk of developing cardiovascular and degenerative diseases, cataract, macular degeneration, and cancer. According to the authors [48], the carotenoids with pro-vitamin A activity are α-carotene, β-carotene, and β-cryptoxanthin. Lycopene do not have pro-vitamin A activity. The colors yellow, orange or red are associated with the pro-vitamin A activity in foods [43].

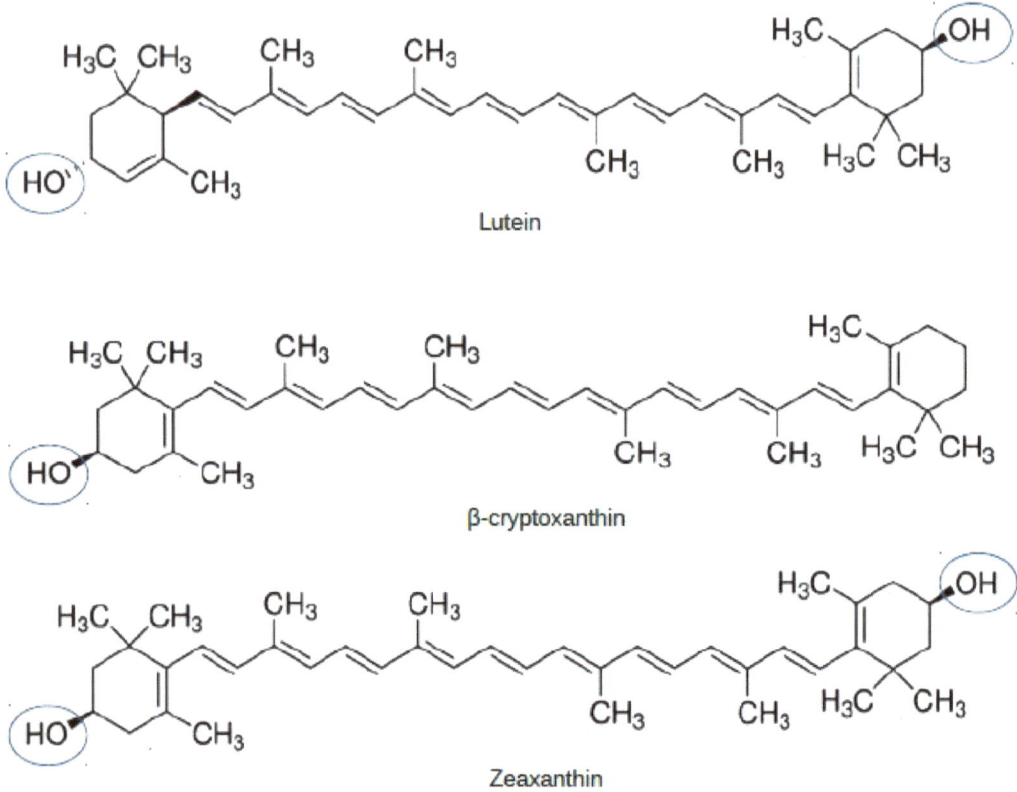

Lutein

β-cryptoxanthin

Zeaxanthin

Fig. (6). Details of the chemical structures of bioactive xanthophylls. Contrarily to carotenes which are polyunsaturated pure hydrocarbons, xanthophylls contain oxygen in their structures forming hydroxyl groups (circles) that make them more polar than carotenes [43].

Banana Fruit and Pro-Vitamin A Supplementation

Animals need a dietary supply of vitamin A because they are unable to synthesize retinoids. The pro-vitamin A carotenoids from plants and the retinyl esters from animal food sources are the mainly ways that vitamin A occurs in Nature. β-carotene is the most abundant carotenoid in plants and through its conversion into retinol in human body, it works as a precursor of vitamin A, which describes its pro-vitamin A activity (Fig. **7**) [49]. In fact, β-carotene is the main pro-vitamin A carotenoid contributing for most of the vitamin A status in human beings. β-carotene conversion involves a central cleavage reaction affording two molecules of retinal [50], followed by the reduction of retinal to retinol [51]. Retinol is further esterified esterified mainly with long chain fatty acids and absorbed *via* the lymphatic route in association with lymph chylomicrons [50].

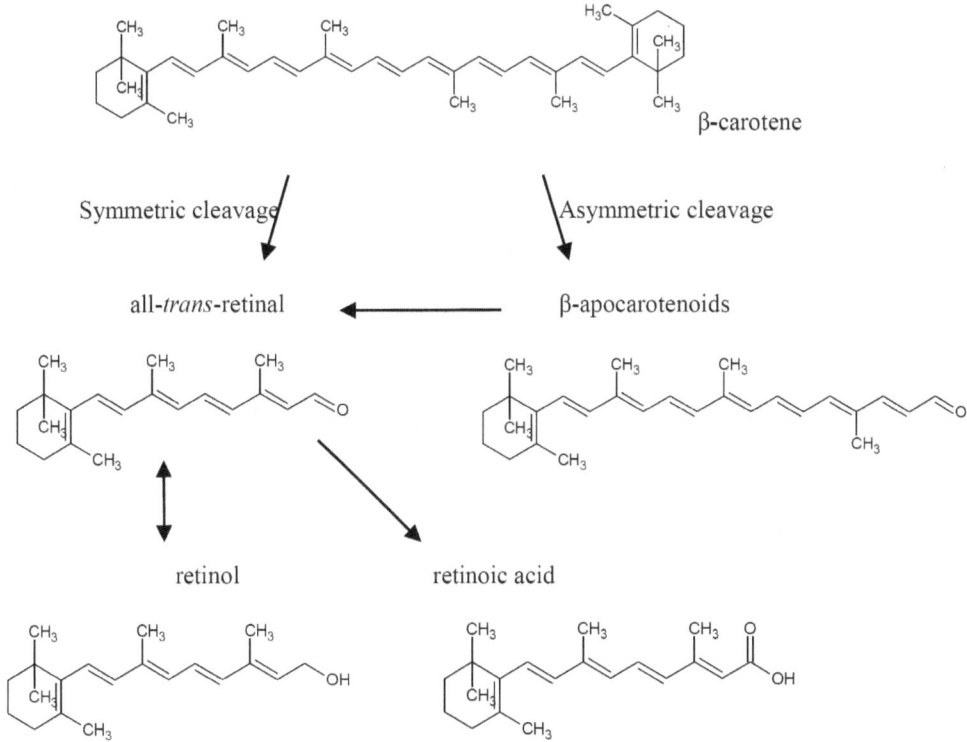

Fig. (7). Symmetrical oxidative cleavage of β-carotene (pro-vitamin A carotenoid) by β, β-carotene 15, 15′-monooxygenase, an intestinal enzyme, resulting into two all-*trans*-retinal molecules [49].

The differences of chemical structures in pro-vitamin A carotenoids contribute to their range of activities. This is translated in the retinol activity equivalent (RAE) concept. RAE considers that 12 µg of *trans*-β-carotene or 24 µg of other pro-vitamin A carotenoids correspond to 1 µg of retinol. The amount of 400 RAE is the daily vitamin A dietary recommended for children under 6 years old. For women in reproductive reproductive age the amount recommended is 700 RAE [52].

Diseases related to vitamin A deficiency are among the major nutritional problems in developing countries. It is estimated that 190 million children in preschool age have low retinol activity in plasma (<0.70 μmol.L-1), subclinical symptom of this deficiency [53]. Three strategies have traditionally been used in the prevention of vitamin A deficiency: dietary diversification, food fortification, and supplementation. These strategies have been ineffective and have failed to complete eradication of the problem in different populations and ethnic groups in the world. Over the past years, a series of institutions worldwide have started nutritional programs to implement a fourth strategy, *i.e.*, biofortification [54]. The biofortification links scientific and technical knowledge of agronomy and health studies. In this approach, the first step is the selection of carotenoid-rich crop varieties that already have positive agronomic and consumption characteristics to allow access to supplements, healthy diets, and commercially biofortified foods for people. The carotenoid-rich varieties have been maintained in germplasm banks in some countries due to their higher levels of pigments, in order to be exploited in genetic breeding programs [55].

Banana is one of the most important food crops with a wide range of varieties - cooking and dessert types, providing a staple and nutritious inexpensive food which has gained increase importance in programs for the selection of superior genotypes for breeding, aiming at generating *Musa* spp biofortified cultivars to mitigate vitamin A deficiency in developing countries [56 - 59]. According to the authors [16], banana fruit carotenoid content comprises up to ~86% in pro-vitamin A compounds, *e.g., trans*-α-carotene, *trans*-β-carotene, and *cis*-β-carotene. Importantly, the proportion of carotenoids identified in banana pulps is very different from that found in crops considered rich in those pigments such as maize, in which the major compounds are lutein and zeaxanthin and only 10–20% are pro-vitamin A carotenoids [60].

Several factors such as the matrix, fat, and the fiber contents, as well as the processing/cooking method affect the efficiency of carotenoid micellarization in the human digestive tract [61]. Depending on the banana cultivar, the bioaccessibility of pro-vitamin A carotenoids can be different [62]. The bioaccessibility of pro-vitamin A carotenoids from prepared Musa-based dishes (10% to 32%) is higher than other starchy foods, such as orange-fleshed (0.6%) and sweet potato (3%) [63]. According to the authors [56], 100 g of fresh weight pulp of banana cultivars with the highest pro-vitamin A carotenoids contents (*i.e.*, Bantol red, Pusit, and Henderneyargh) are sufficient to provide 95% of the RAE for children and 47% of RAE for adults. This finding highlights the importance of that plant species for programs aiming at to increase the consumption of vitamin A by populations with great need of that nutrient.

OTHER HEALTH PROMOTERS EFFECTS ATTRIBUTED TO THE PRIMARY METABOLITES - STARCH

The association of high resistant starch content and the consumption of unripe bananas is reported by several studies to offer beneficial effects for human health. Resistant starch is the portion of starch molecule and its products that are not metabolized along the gastrointestinal tract. The most resistant starch-rich non-processed food is the unripe banana [64]. Resistant starch protects against many diseases such as colorectal cancer, type II diabetes, and other diet-related chronic diseases [65, 66].

Resistant starch has physiological effects similar to dietary fiber and prebiotics which are directly related to its nutritional implications. Resistant starch can lower the gut's lumen pH by stimulation of the growth of beneficial bacteria and increases the production of short-chain fatty acids, creating a less susceptible environment for the development of tumors as observed in the colorectal cancer [67].

Reduction in the number of episodes of hypoglycemia and postprandial glucose peaks, lower concentration of glycosylated hemoglobin and fructosamine, and greater insulin sensitivity are epidemiological study suggestions to diabetes management [68]. It has been recommended the slowly digestible starch intake for the prevention and management of diabetes due to the beneficial metabolic response observed [69, 70]. The authors [71] reported that in a group of obese type 2 diabetics, the intake of banana starch (24 g/day) during a period of four weeks resulted in the increase of insulin sensitivity and body weight loss. Thus, banana starch supplementation could be a cheap alternative to be included in a dietary plan of obese type 2 diabetic patients [71].

CONCLUDING REMARKS

Banana is a fruit consumed worldwide, because of its particularities regarding its potential as nutraceutical food and mainly its low cost. Banana has great value as an interesting source of secondary metabolites with bioactive potential. Bananas are fruits indicated for the development of phytomedicines considering their contents of biogenic amines, phenolic compounds, and carotenoids, as well as the pharmacological evidences associated with the ethnopharmacological data available about pulp and peel extracts usage. Indeed, bananas could be useful to treat PD and depression, in the wound healing process, and to help mitigate hypovitaminosis A as an important source of pro-vitamin A carotenoids. Besides the health benefits derived from the banana's secondary metabolites presented in this chapter, other benefits can be attributed to primary metabolites as noted regarding its resistant starch contents, emphasizing the importance of this plant species as a functional food with real benefits to human health.

CONFLICT OF INTEREST

The author confirms that author has no conflict of interest to declare for this publication.

ACKNOWLEDGEMENTS

The authors gratefully acknowledge the financial support of the Coordenação de Aperfeiçoamento de Pessoal de Nível Superior (CAPES) and the Fundação de Amparo à Pesquisa e Inovação do Estado de Santa Catarina (FAPESC) (CHAMADA PÚBLICA FAPESC Nº 04/2012 UNIVERSAL) for financial support for part of this work. The researcher grant from CNPq on behalf of the last author is acknowledged.

REFERENCES

[1]   Robinson JC. Distribution and importance; taxonomic classification, cultivars and breeding. In: Robinson JC, Ed. Banana and plantains Crop production science in horticulture 5. Oxon: CAB Intl. Wallingford 1996; pp. 1-33.

[2]   Stover RH, Simmonds NW. Classification of banana cultivars. In: Stover RH, Simmonds NW, Eds. Bananas. London: Longmans 1987; pp. 12-7.

[3]   Serviço Brasileiro de Apoio às Micro e Pequenas Empresas. Série perfil de projetos: industrialização

de banana 1999; 28.

[4]   Serviço Brasileiro de Apoio às Micro e Pequenas Empresas. Estudos de mercado SEBRAE/ESPM: Banana relatório completo 2008; 88.

[5]   Chauhan KS, Pundir JP, Singh S. Studies on the mineral composition of certain fruits. Haryana J of Hort Sci 1991; 20: 210-3.

[6]   Waalkes TP, Sjoerdsma A, Creveling CR, Weissbach H, Udenfriend S. Serotonin, norepinephrine, and related compounds in bananas. Science 1958; 127(3299): 648-50.
        [http://dx.doi.org/10.1126/science.127.3299.648] [PMID: 17808884]

[7]   Phillipson JD. Phytochemistry and medicinal plants. Phytochemistry 2001; 56(3): 237-43.
        [http://dx.doi.org/10.1016/S0031-9422(00)00456-8] [PMID: 11243450]

[8]   Balbach A. As frutas na medicina doméstica. São Paulo: Editora M. V. P. 1945; pp. 32-4.

[9]   International Network For The Improvement Of Banana And Plantain. Celebrating 20 years of networking banana and plantain Annual Report. France 2006; p. 35.

[10]   Harborne JB. Phytochemical dictionary: Handbook of bioactive compounds from plants. London: Taylor and Francis 1999.

[11]   Rice-Evans CA, Miller NJ, Paganga G. Structure-antioxidant activity relationships of flavonoids and phenolic acids. Free Radic Biol Med 1996; 20(7): 933-56.
        [http://dx.doi.org/10.1016/0891-5849(95)02227-9] [PMID: 8743980]

[12]   Cook NC, Sammon S. Flavanoids chemistry, metabolism, cardioprotective effects, and dietary sources. Nutr Biochem 1996; 7: 66-76.
        [http://dx.doi.org/10.1016/0955-2863(95)00168-9]

[13]   Sulaiman SF, Yusoff NA, Elden IM, Seow EM, Supriatno AA, Ooi KL. Correlation between total phenolic and mineral contents with antioxidant activity of eight Malaysian bananas (*Musa* sp.). J Food Compos Anal 2011; 4: 1-10.
        [http://dx.doi.org/10.1016/j.jfca.2010.04.005]

[14]   Alothman M, Bhat R, Karim AA. Antioxidant capacity and phenolic content of selected tropical fruits from Malaysia, extracted with different solvents. Food Chem 2009; 115: 785-8.
        [http://dx.doi.org/10.1016/j.foodchem.2008.12.005]

[15]   Mèlo EA, Lima VL, Maciel MI. Polyphenol, ascorbic acid, and total carotenoid contents in common fruits and vegetables. Braz J Food Technol 2006; 9: 89-94.

[16]   Borges CV, Amorim VB, Ramlov F, *et al.* Characterisation of metabolic profile of banana genotypes, aiming at biofortified *Musa spp.* cultivars. Food Chem 2014; 145: 496-504.
        [http://dx.doi.org/10.1016/j.foodchem.2013.08.041] [PMID: 24128506]

[17]   Rhodes MJ, Price KR. Analytical problems in the study of flavonoid compounds in onions. Food Chem 1996; 57: 113-7.
        [http://dx.doi.org/10.1016/0308-8146(96)00147-1]

[18]   Heim KE, Tagliaferro AR, Bobilya DJ. Flavonoid antioxidants: chemistry, metabolism and structure-activity relationships. J Nutr Biochem 2002; 13(10): 572-84.
        [http://dx.doi.org/10.1016/S0955-2863(02)00208-5] [PMID: 12550068]

[19] Middleton E, Harborner JB. Plant flavonoids biology and medicine. New York: AlanLiss 1986.

[20] Korkina LG, Afanas'ev IB. Antioxidant and chelating properties of flavonoids. Adv Pharmacol 1997; 38: 151-63.
[http://dx.doi.org/10.1016/S1054-3589(08)60983-7] [PMID: 8895808]

[21] Havsteen BH. The biochemistry and medical significance of the flavonoids. Pharmacol Ther 2002; 96(2-3): 67-202.
[http://dx.doi.org/10.1016/S0163-7258(02)00298-X] [PMID: 12453566]

[22] Rebello LP, Ramos AM, Pertuzatti PB, Barcia MT, Castillo-Muñoz N, Hermosín-Gutiérrez I. Flour of banana (Musa AAA) peel as a source of antioxidant phenolic compounds. Food Res Int 2014; 55: 397-403.
[http://dx.doi.org/10.1016/j.foodres.2013.11.039]

[23] Babbar N, Oberoi HS, Uppal DS, Patil RT. Total phenolic content and antioxidant capacity of extracts obtained from six important fruit residues. Food Res Int 2011; 44: 391-6.
[http://dx.doi.org/10.1016/j.foodres.2010.10.001]

[24] Someya S, Yoshiki Y, Okubo K. Antioxidant compounds from bananas (Musa cavendish). Food Chem 2002; 79: 351-4.
[http://dx.doi.org/10.1016/S0308-8146(02)00186-3]

[25] Novak FR, Almeida JA, Silva RS. Casca da banana: uma possível fonte de infecção no tratamento de fissuras mamilares. J Pediatr 2003; 79: 221-6.
[http://dx.doi.org/10.2223/JPED.1023]

[26] Pereira A. Avaliação das atividades cicatrizante e antitumoral de extratos provenientes da casca de banana cultivar Prata Anã (Musa spp) Dissertação de mestrado. Florianópolis: Universidade Federal de Santa Catarina 2010.

[27] Pereira A, Maraschin M. Banana (*Musa* spp) from peel to pulp: ethnopharmacology, source of bioactive compounds and its relevance for human health. J Ethnopharmacol 2015; 160: 149-63.
[http://dx.doi.org/10.1016/j.jep.2014.11.008] [PMID: 25449450]

[28] Houghton PJ, Hylands PJ, Mensah AY, Hensel A, Deters AM. *In vitro* tests and ethnopharmacological investigations: wound healing as an example. J Ethnopharmacol 2005; 100(1-2): 100-7.
[http://dx.doi.org/10.1016/j.jep.2005.07.001] [PMID: 16040217]

[29] Agarwal PK, Goel RK. Wound healing activity of extracts of plantain banana (Musa sapientum var. Paradisiaca) in rats. Indian J Pharmacol 2008; 40: S66-91.
[PMID: 21593982]

[30] Atzingen DA, Gragnani A, Veiga DF, *et al.* Unripe Musa sapientum peel in the healing of surgical wounds in rats. Acta Cir Bras 2013; 28(1): 33-8.
[http://dx.doi.org/10.1590/S0102-86502013000100006] [PMID: 23338111]

[31] Steiner U, Schliemann W, Strack D. Assay for tyrosine hydroxylation activity of tyrosinase from betalain-forming plants and cell cultures. Anal Biochem 1996; 238(1): 72-5.
[http://dx.doi.org/10.1006/abio.1996.0253] [PMID: 8660589]

[32] Kimura M. Fluorescence histochemical study on serotonin and catecholamine in some plants. Jpn J Pharmacol 1968; 18(2): 162-8.

[http://dx.doi.org/10.1254/jjp.18.162] [PMID: 5303514]

[33]   Smith T. Secondary plant products. In: Bell BV Charlwood (Eds), Encyclopedia of Plant Physiology New Series. Berlin: Berlin: Springer-Verlag 1980.

[34]   Kulma A, Szopa J. Catecholamines are active compounds in plants. Plant Sci 2007; 172: 433-40.
[http://dx.doi.org/10.1016/j.plantsci.2006.10.013]

[35]   Kumar KP, Bhowmik D, Duraivel S, Umadevi M. Traditional and medicinal uses of banana. J Pharmacog Phytochem 2012; 1: 51-63.

[36]   Chemuturi NV, Donovan MD. Metabolism of dopamine by the nasal mucosa. J Pharm Sci 2006; 95(11): 2507-15.
[http://dx.doi.org/10.1002/jps.20724] [PMID: 16917843]

[37]   Wang C, Fan Y, Lee S, Lian J, Liou J, Wang H. Systemic and brain bioavailabilities of D-phenylglycine-L-Dopa, a sustained dopamine-releasing prodrug. J Food Drug Anal 2013; 21: 136-41.
[http://dx.doi.org/10.1016/j.jfda.2013.05.001]

[38]   Roepenack-Lahaye E, Newman MA, Schornack S, *et al.* p-Coumaroylnoradrenaline, a novel plant metabolite implicated in tomato defense against pathogens. J Chem Biol 2003; 278: 43373-83.
[http://dx.doi.org/10.1074/jbc.M305084200]

[39]   Swiedrych A, Lorenc-Kukuła K, Skirycz A, Szopa J. The catecholamine biosynthesis route in potato is affected by stress. Plant Physiol Biochem 2004; 42(7-8): 593-600.
[http://dx.doi.org/10.1016/j.plaphy.2004.07.002] [PMID: 15331087]

[40]   Barrio JR, Huang SC, Phelps ME. Biological imaging and the molecular basis of dopaminergic diseases. Biochem Pharmacol 1997; 54(3): 341-8.
[http://dx.doi.org/10.1016/S0006-2952(97)00031-2] [PMID: 9278092]

[41]   Wang C, Fan Y, Lee S, Lian J, Liou J, Wang H. Systemic and brain bioavailabilities of D-phenylglycine-L-dopa, a sustained dopamine-releasing prodrug. J Food Drug Anal 2013; 21: 136-41.
[http://dx.doi.org/10.1016/j.jfda.2013.05.001]

[42]   Kanazawa K, Sakakibara H. High content of dopamine, a strong antioxidant, in Cavendish banana. J Agric Food Chem 2000; 48(3): 844-8.
[http://dx.doi.org/10.1021/jf9909860] [PMID: 10725161]

[43]   Rodriguez-Amaya DB. A guide to carotenoid analysis in foods. Washington, D.C.: OMNI Research 2001.

[44]   Kurilich AC, Juvik JA. Quantification of carotenoid and tocopherol antioxidants in Zea mays. J Agric Food Chem 1999; 47(5): 1948-55.
[http://dx.doi.org/10.1021/jf981029d] [PMID: 10552476]

[45]   Krinsky NI, Johnson EJ. Carotenoid actions and their relation to health and disease. Mol Aspects Med 2005; 26(6): 459-516.
[http://dx.doi.org/10.1016/j.mam.2005.10.001] [PMID: 16309738]

[46]   Tapiero H, Townsend DM, Tew KD. The role of carotenoids in the prevention of human pathologies. Biomed Pharmacother 2004; 58(2): 100-10.
[http://dx.doi.org/10.1016/j.biopha.2003.12.006] [PMID: 14992791]

[47] Voutilainen S, Nurmi T, Mursu J, Rissanen TH. Carotenoids and cardiovascular health. Am J Clin Nutr 2006; 83(6): 1265-71.
[PMID: 16762935]

[48] Erdman JW Jr, Bierer TL, Gugger ET. Absorption and transport of carotenoids. Ann N Y Acad Sci 1993; 691: 76-85.
[http://dx.doi.org/10.1111/j.1749-6632.1993.tb26159.x] [PMID: 8129321]

[49] Chichili GR, Nohr D, Schäffer M, von Lintig J, Biesalski HK. β-Carotene conversion into vitamin A in human retinal pigment epithelial cells. Invest Ophthalmol Vis Sci 2005; 46(10): 3562-9.
[http://dx.doi.org/10.1167/iovs.05-0089] [PMID: 16186334]

[50] Goodman DS, Huang HS. Biosynthesis of vitamin A with rat intestinal enzymes. Science 1965; 149(3686): 879-80.
[http://dx.doi.org/10.1126/science.149.3686.879] [PMID: 14332853]

[51] Fidge NH, Smith FR, Goodman DS. Vitamin A and carotenoids. The enzymic conversion of beta-carotene into retinal in hog intestinal mucosa. Biochem J 1969; 114(4): 689-94.
[http://dx.doi.org/10.1042/bj1140689] [PMID: 4981032]

[52] Food and Agriculture Organization/World Health Organization. Requirements of Vitamin A, iron, folate and vitamin B12. Report of a Joint FAO/WHO Expert Consultation Report of a Joint FAO/WHO Expert Consultation Food and Nutrition Series 1988; 23

[53] World Health Organization. Global Prevalence of vitamin A deficiency in populations at risk 1995–2005. Geneva: WHO Global Database on Vitamin A deficiency 2012.

[54] Ceballos H, Luna J, Escobar AF, et al. Spatial distribution of dry matter in yellow fleshed cassava roots and its influence on carotenoid retention upon boiling. Food Res Int 2012; 45: 52-9.
[http://dx.doi.org/10.1016/j.foodres.2011.10.001]

[55] Saltzman A, Birol E, Bouis HE, et al. Biofortification: Progress toward a more nourishing future. Glob Food Secur 2013; 2: 9-17.
[http://dx.doi.org/10.1016/j.gfs.2012.12.003]

[56] Davey MW, Bergh IV, Markham R, Swennen R, Keulemans J. Genetic variability in Musa fruit pro-vitamin A carotenoids, lutein, and mineral micronutrient contents. Food Chem 2009; 115: 806-13.
[http://dx.doi.org/10.1016/j.foodchem.2008.12.088]

[57] Englberger L, Darnton-Hill I, Coyne T, Fitzgerald MH, Marks GC. Carotenoid-rich bananas: a potential food source for alleviating vitamin A deficiency. Food Nutr Bull 2003; 24(4): 303-18.
[http://dx.doi.org/10.1177/156482650302400401] [PMID: 14870618]

[58] Englberger L, Wills RB, Blades B, Dufficy L, Daniells JW, Coyne T. Carotenoid content and flesh color of selected banana cultivars growing in Australia. Food Nutr Bull 2006; 27(4): 281-91.
[http://dx.doi.org/10.1177/156482650602700401] [PMID: 17209469]

[59] Englberger L, Grahan L, Foley W, Daniells J, Aalbersberg UD, Watoto C. Carotenoid and riboflavin content of banana cultivars from Makira, Solomon Islands. J Food Compos Anal 2010; 23: 624-32.
[http://dx.doi.org/10.1016/j.jfca.2010.03.002]

[60] Kuhnen S, Ogliari JB, Dias PF, et al. ATR-FTIR spectroscopy and chemometric analysis applied to discrimination of landrace maize flours produced in southern Brazil. Food Sci Technol (Campinas)

2010; 45: 1673-8.
[http://dx.doi.org/10.1111/j.1365-2621.2010.02313.x]

[61]   O'Connell OF, Ryan L, O'Brien NM. Xanthophyll carotenoids are more bioaccessible from fruits than dark green vegetables. Nutr Res 2007; 27: 258-64.
[http://dx.doi.org/10.1016/j.nutres.2007.04.002]

[62]   Ekesa B, Poulaert M, Davey MW, *et al.* Bioaccessibility of provitamin A carotenoids in bananas *(Musa spp.)* and derived dishes in African countries. Food Chem 2012; 133: 1471-7.
[http://dx.doi.org/10.1016/j.foodchem.2012.02.036]

[63]   Failla ML, Thakkar SK, Kim JY. *In vitro* bioaccessibility of beta-carotene in orange fleshed sweet potato (Ipomoea batatas, Lam.). J Agric Food Chem 2009; 57(22): 10922-7.
[http://dx.doi.org/10.1021/jf900415g] [PMID: 19919124]

[64]   Rodriguez-Ambriz SL, Islas-Hernández JJ, Agama-Acevedo E, Tovar J, Bello-Perez LA. Characterization of a fiber-rich powder prepared by liquefaction of unripe banana flour. Food Chem 2008; 107: 1515-21.
[http://dx.doi.org/10.1016/j.foodchem.2007.10.007]

[65]   Niba LL. Resistant starch: a potential functional food ingredient. Nutr Food Sci 2002; 32: 62-7.
[http://dx.doi.org/10.1108/00346650210416985]

[66]   Topping DL, Clifton PM. Short-chain fatty acids and human colonic function: roles of resistant starch and nonstarch polysaccharides. Physiol Rev 2001; 81(3): 1031-64.
[PMID: 11427691]

[67]   Yao N, Paez AV, White PJ. Structure and function of starch and resistant starch from corn with different doses of mutant amylose-extender and floury-1 alleles. J Agric Food Chem 2009; 57(5): 2040-8.
[http://dx.doi.org/10.1021/jf8033682] [PMID: 19256560]

[68]   Wolever TM. Carbohydrate and the regulation of blood glucose and metabolism. Nutr Rev 2003; 61(5 Pt 2): S40-8.
[http://dx.doi.org/10.1301/nr.2003.may.S40-S48] [PMID: 12828191]

[69]   Axelsen M, Arvidsson Lenner R, Lönnroth P, Smith U. Breakfast glycaemic response in patients with type 2 diabetes: effects of bedtime dietary carbohydrates. Eur J Clin Nutr 1999; 53(9): 706-10.
[http://dx.doi.org/10.1038/sj.ejcn.1600837] [PMID: 10509766]

[70]   Ells LJ, Seal CJ, Kettlitz B, Bal W, Mathers JC. Postprandial glycaemic, lipaemic and haemostatic responses to ingestion of rapidly and slowly digested starches in healthy young women. Br J Nutr 2005; 94(6): 948-55.
[http://dx.doi.org/10.1079/BJN20051554] [PMID: 16351772]

[71]   Ble-Castillo JL, Aparicio-Trápala MA, Francisco-Luria MU, *et al.* Effects of native banana starch supplementation on body weight and insulin sensitivity in obese type 2 diabetics. Int J Environ Res Public Health 2010; 7(5): 1953-62.
[http://dx.doi.org/10.3390/ijerph7051953] [PMID: 20623003]

SUBJECT INDEX

A

Absorption 10, 33, 54, 76, 82, 104, 117, 119, 136, 153, 156, 164, 174, 180, 204, 205, 212, 273, 282

Almond 110, 111, 128, 131, 132, 236

Amazonian biome 244

Amine compounds 265

Anticancer 18, 19, 21, 23, 27, 40, 50, 53, 55, 60, 63, 64, 67, 69, 86, 92, 96, 98, 110, 137, 138, 141, 147, 149, 161, 169, 170, 174, 180, 182, 214, 216, 217, 219, 220, 234, 235

Antidiabetic 12, 17, 27, 110, 134, 172, 175, 229, 263

Antiinflammatory 110, 122

Antioxidant iii, 4, 6, 8, 10, 16, 17, 26, 29, 39, 40, 55, 64, 79, 80, 82, 85, 86, 88, 90, 113, 118, 119, 138, 180, 182, 188, 204, 210, 212, 215, 216, 227, 230, 231, 236, 240, 242, 244, 252, 253, 279-281

Antioxidants 5, 16, 17, 25, 26, 39, 53, 74, 79, 93, 107, 113, 122, 124, 130, 139, 142, 144, 145, 148, 153, 165, 167, 169, 207, 208, 217, 220, 222, 224, 227, 231, 232, 234, 242, 246, 269, 279, 281

Apricot 107, 110, 111, 129, 131, 133, 136-138

Ascorbic acid 29, 50, 51, 85, 93, 100, 103, 107, 150, 192, 203, 222, 229, 244, 246, 253, 255, 279

Astrocaryum aculeatum 244, 254, 262, 263

B

Bactris gasipaes 244, 256, 264

Berries 8, 97, 126, 208, 211

Bioactive-compounds 99

Bioactive compounds i, iii, 3, 4, 17, 24, 79, 87, 89, 102, 108, 110, 112, 122, 129, 132, 147, 170, 176, 180, 182, 189, 200, 218, 221, 244, 269, 279, 280

Bioavailability 75, 86, 95, 100, 109, 144, 145, 147, 148, 170, 201, 203, 204, 231, 237, 272

Biological activity 24, 91, 97, 163, 180, 188, 198, 201, 214, 218

Byrsonima crassifolia 244, 253, 262

C

Cancer i, iii, 4, 12, 18, 19, 21, 23, 33, 39, 40, 67, 69, 75, 76, 78, 86, 92, 93, 102, 107, 133, 158, 168, 171, 172, 198, 201, 204, 205, 208, 230, 246, 248, 273, 277

Cardiovascular diseases i, iii, 4, 52, 86, 99, 119, 127, 129, 136, 144, 145, 147, 148, 151, 178, 179, 187, 199, 200, 216, 246, 247

Cardiovascular protection 110, 112, 178, 230

Carotenoids 29, 31, 33, 57, 59, 94, 95, 97, 100, 107, 112, 113, 129, 130, 180, 182, 183, 188, 209, 222, 229, 233, 244, 246, 248, 249, 257, 261, 262, 264, 265, 281-283

Caryocar villosum 244, 251, 261, 262

Cherry 110, 111, 120, 129, 131, 137, 176

Chronic-diseases 99

Clementine 29, 57, 58, 67, 68, 71, 73, 87, 94

Clinical 21, 98, 99, 125, 128, 138, 160, 161, 169, 178, 199, 200, 213, 221, 238

Coumarin 29, 68, 204